# Introduction to Real Estate Development and Finance

This book provides readers with a basic understanding of the principles that underlie real estate development. A brief historical overview and an introduction to basic principles are followed by examples from practice. Case studies focus on how cities change and respond to the economic, technological, social, and political forces that shape urban development in North America. It is important to have a framework for understanding the risks and rewards in real estate investing. In measuring return, consideration must be given to both investment appreciation and the cash flow generated over the life of a project. In addition, metrics are presented that can be useful in assessing the financial feasibility of a real estate development proposal. This book also provides an overview of the forces of supply and demand that gauge the potential market for a new project. In determining the size of "residual demand", estimates for population growth, family formation, and new development are important.

All development projects fall under the auspices of one or several jurisdictions. Though every jurisdiction has different rules and procedures, basic knowledge of the planning process is critical to the success of all development projects regardless of location. Furthermore, all projects have a legal component. Basic issues of land ownership, property rights, property transfer, and land registration are reviewed, all of which need to be considered when a property is sold or purchased. This book also provides a primary on the design and construction process. In constructing a building, a team of experts is first required to design the architectural, structural, and heating, ventilation, and air conditioning (HVAC) systems for a building. An overview is provided of each building system: wood, concrete, and steel. Critical to a successful real estate development, project management principles for the processes of design, bidding, and construction are explored, with close attention given to budgeting, scheduling, and resource management. Essential reading for anyone involved in the development of our built environment, this is a must-read introduction for students and professionals in architecture, urban planning, engineering, or real estate seeking an approachable and broad view of real estate development and finance.

**Richard M. Levy,** M.Arch, PhD, was formerly Professor of Planning at the University of Calgary, Canada, where he taught courses in urban planning and real estate development for 26 years. He currently teaches in the College of Business at the University of Rhode Island.

# Introduction to Real Estate Development and Finance

Richard M. Levy

LONDON AND NEW YORK

First published 2020

by Routledge
2 Park Square, Milton Park, Abingdon, Oxon OX14 4RN

and by Routledge
52 Vanderbilt Avenue, New York, NY 10017

*Routledge is an imprint of the Taylor & Francis Group, an informa business*



*British Library Cataloguing-in-Publication Data*
A catalogue record for this book is available from the British Library

*Library of Congress Cataloging-in-Publication Data*
A catalog record has been requested for this book

ISBN: 978-1-138-60244-1 (hbk)
ISBN: 978-1-138-60245-8 (pbk)
ISBN: 978-0-429-46956-5 (ebk)

Typeset in Times
by Swales & Willis, Exeter, Devon, UK

**When I was a boy aged 8, I wanted to be either a bridge designer or an architect. Perhaps, in retrospect, this is not surprising. Having grown up in a suburb of New York, my parents would often take us to New York City to shop or to visit one of the great architectural monuments of the age. For me, the George Washington Bridge was the great portal to a city filled with wondrous places. I still remember vividly our visits to the Guggenheim Museum, the Metropolitan Museum of Fine Arts, the New York Public Library, the Rockefeller Center, and Grand Central Station. Taking the elevator to the observation deck of the Empire State Building was an event I never forgot. Surveying the vista, the obvious question of "How did all this get built?" provided the underlying motivation for my pursuing degrees in engineering and architecture. To my parents, who first introduced me to the great architectural monuments of New York, I will always be indebted.**

**To my wife, Elizabeth E. Dickson, without whose love, support, and editorial assistance this book would never have been completed.**

# Contents

# Figures

## Chapter 1



## Chapter 2

## Chapter 3



## Chapter 4



## Chapter 5



## Chapter 6

## Chapter 7



## Chapter 8



## Chapter 9

# Preface

Cities are in a constant state of demolition and rebuilding. Within an urban landscape, we discover buildings and structures from every period in its history. Even in cities that were only established a hundred years ago, we will find a history marked by periods of development followed by periods of decline or characterized by little or no activity. In general, the time before and after World War I had periods of heightened activity marked by the construction of the first skyscrapers in America. This era was followed by the Great Depression and World War II, when construction was on hold for several decades. Today, many cities are finding new uses for buildings that have sat vacant and idle for decades. Though we may come to appreciate the history of an individual building by taking a tour suggested by an architectural guide, we may learn very little about the process responsible for its planning, financing, design, and construction.

An appreciation of the complexity of building in the 21st century is essential to understanding the role of the private sector in our cities. First and foremost, we must acknowledge that real estate development is an investment. Like other investments, real estate development must compete for capital, offering rates of return commensurate with its risk. However, unlike other asset classes, it is fixed in space. It cannot be relocated to a new location if market conditions change or if the demand for space is better in another city. In addition, real estate is a long-term investment. Profitability in some cases can be determined only when the building is sold after many years have passed. Creating a successful plan requires knowledge of accepted financial practices. For a project to be built, investors and lending institutions will first subject projects to numerous financial tests. These financial metrics are used to determine the creditworthiness of any proposed project. Projects must be able to generate sufficient cash flow to pay the expenses, make loan commitments, and have reserves for unexpected expenditures. Underlying these projections are assumptions made about population growth and employment opportunities. In translating these numbers into future demand, consideration must be given to the scale of the development. Competition from existing and proposed developments, access to transportation, and proximity to retail, social, and recreational services should all be considered in the evaluation of any project. Real estate development is designed to meet the need for space in specific markets: residential, commercial, retail, industrial, hospitality, health care, and government services. The economic environment may change over the life of the project. If the plan is to sell the project in 10 or 15 years, it cannot be predetermined whether the market will have appreciated significantly to generate a reasonable rate of return after all the positive and negative cash flows are tallied. For those investing in real estate, the rewards must be commensurate with their risk. Subject to local supply and demand for space, rents can vary, tax rates can increase, borrowing costs can change, and utilities and other costs can increase. In being prudent, it is important to consider both the best and worst case scenarios. If there is a significant probability that we may not achieve our expected rate of return, then the proposal will need to be revised to secure financing.

In preparing development plans, the role of government is an important factor to understand. At the federal level, taxation, banking, and securities practices can have a critical bearing on the financial outlook for a project. At the local and state levels, controls over land use constrain the type and physical form of the development. Municipal zoning restricts not only what can be built, but also how it will be built. Though it is always possible to apply for a variance that allows a building to be taller or to have fewer parking spaces than required, there are no guarantees that an application for a variance will be approved. Legal constraints in the forms of easements and covenants can also shape the direction of development. For example, there may be the need to adhere to architectural guidelines, preserve views or guarantee access through the property for adjacent property owners. In some cities, regulations may require developers to set aside a percentage of residential units in a development for "affordable housing" or to pay an additional fee. Projects located along waterways and coastlines will have to abide by local state and federal regulations. Within all these constraints, architects working closely with engineers and contractors must develop plans that satisfy the basic space and budget requirements. Once plans are approved, it is important to understand how project management principles are used in the construction of the actual physical asset. This book cannot address in detail the design and execution of each type of development: multifamily, commercial, industrial, recreational, hotels, and resorts. However, it can provide the reader with an overview of real estate development from inception to completion.

This book provides a basic understanding of the principles and tools used by professionals in creating, reviewing, and managing real estate development proposals. Concepts are introduced that are needed to determine the risks and rewards associated with any investment. In analyzing the details of any proposal, it is also important to have an understanding of the basic elements of a balance sheet and profit-and-loss statement. In analyzing financial statements, borrowing rates, debt level, vacancy factors, expense ratios, depreciation, and tax rates can ultimately impact the financial feasibility of a project. Return on equity, current ratio, and debt coverage are a few of the metrics used when reviewing a pro forma. When projects are evaluated over their lifetimes, the application of discounted cash flows becomes an important tool. Financial functions available in popular spreadsheet programs makes it convenient to calculate net present value and internal rate of return for a series of project cash flows. An important tool in decision making, we can compare different scenarios for a particular site. In determining the most appropriate design for a particular site, we can apply the concept known to all appraisers of real property as "highest and best use". In applying this concept, we must consider the value of the land separately from that of any building. In addition, we can use statistical and spatial models to help us to understand how proximity to centers of commerce influence value. Having an appreciation for these concepts will also give us insight into how properties are valued by taxing authorities and appraisers.

This book has been designed for anyone interested in real estate development. In teaching a course on real estate development and finance to students with backgrounds in business, planning, architecture, and engineering, it has been challenging to find a single book that is comprehensive and affordable. Though there are monographs that are excellent textbooks for the student pursuing a degree in business, finding a book that can serve as either an introductory text or as a primary for those interested in an overview of real estate development and finance is difficult. Real estate development touches on many subjects: finance, urban modeling, urban planning, architecture, construction, and project management. For students in planning, architecture, and engineering, this book should provide an overview of the subject. It is my hope that the book will also give practitioners and students alike the necessary background to understand the development process as it relates to their profession. For the investor who wants to become better acquainted with the unique aspects of real estate investments, this book should also serve as a useful introduction.

For those who desire a more in-depth knowledge of the subject, the bibliography at the end of each chapter offers suggested readings in finance, mortgage lending, and real estate sales and marketing.

# Organization of this book

Chapter 1 focuses on the forces of supply and demand in the context of a dynamic marketplace. In taking a case study approach, it is possible to understand how cities change and respond to economic, technological, social, and political forces. It is the developer who seizes opportunity who is successful. Development proposals, if they are to become a reality, will demand a plan developed by a team of experts. Ultimately, financial backing must be secured and the required approval by regulatory bodies granted.

Chapter 2 considers the application of risk and reward to our understanding of real estate investment. In measuring return, consideration must be given to both the appreciation and the cash flow generated over its lifetime. In addition, portfolio management demands an understanding of volatility. In real estate, like other financial assets, we can only estimate a range for future returns. In this chapter, there is an emphasis on understanding the basic components of a balance sheet and profit-and-loss statement. It is particularly important to understand how leverage and borrowing rates impact the financial metrics used to assess the creditworthiness of an investment.

Chapter 3 first begins with a discussion of modeling and its value, and then considers how modeling techniques can be used in real estate development projects. Like other financial assets, we can use various modeling techniques to help us to estimate the value of an existing property. In addition, explorations of various spatial modeling techniques can help us to appreciate the importance of location in the appraisal of a property's value. In real estate development, two questions emerge in most modeling discussions. The first question, simply stated, is: what is the most profitable development for an existing location? The second question addresses the issue of location: where would be the best spot for a particular proposed enterprise? The modeling techniques presented in this chapter can help us to answer both these questions.

In Chapter 4, financial modeling is introduced as a tool for testing the feasibility of proposed projects. Capitalization rates, discounted cash flows, net present value, and internal rate of return are important to the financial analysis of a real estate development proposal. If we are acquiring property for an investment fund, we need a basis for choosing between several prospective projects for our portfolio. Often, these projects extend over many years, each project with different costs and revenue streams. Financial models can be useful tools in evaluating the potential success of a real estate development project under a number of potential scenarios. In a case study on a residential development project, best, worst, and expected case scenarios are tested against possible changes to the underlying assumptions for a projection. Though, in the short term, we expect businesses to proceed as usual, sudden changes in the economy can impact the success or failure of a project in the long term.

Chapter 5 provides an overview of an approach to developing market research needed to evaluate a project and provide direction in the details of its design. Gauging the forces of supply and demand will be helpful in determining the market for new projects. In determining the size of this "residual demand", estimates for population growth, family formation, and new development are important. Understanding how population projections are made using cohort analysis can help in defining the market for a particular segment of the residential housing market. In determining this demand, issues of affordability become an important consideration. In this chapter, a case study focusing on a condo apartment development highlights the factors to be considered in a market study of future demand. Though the focus of this chapter is on residential development, a brief discussion is given on developing a marketing study for retail and commercial space.

Chapter 6 considers the planning process. All development projects fall under the auspices of one or several jurisdictions. For a project to be successful, getting to the stage at which development and building permits are granted can be both time-consuming and costly. Though every jurisdiction has different rules, understanding the roles played by zoning, area redevelopment plans, area structure plans, master plans, planned unit developments, direct controls, and architectural and historical guidelines can aid in the planning process. A case study on multifamily development provides the reader with an example of how the theory works in practice.

Chapter 7 presents an overview of the legal aspects of real estate. All aspects of development, from acquisition, through construction, to leasing and management, have a legal component. Issues of land ownership, property rights, property transfer, and land registration will provide a basic understanding of the issues that need to be considered when the property is sold or purchased. Issues of tax, liability, and ownership will need to be considered in selecting the most appropriate form of ownership for a particular business enterprise. Although tax considerations are always critical to this decision, liability exposure must also be weighed in making this judgment. Loans are an integral part of most real estate development projects. In this chapter, the mechanics of mortgages and construction loans are presented in relation to real estate development. This chapter also reviews the basic elements of a lease and considers the issue of tenant and landlord rights. For investment properties, leases are critical to the success of the rental property. They dictate rents and the rights and responsibilities of the tenant and landlord. Finally, a note about construction contracts: even for small projects, a contract will be required for building and repairing a new building.

At some point in the development process, something will be built or renovated. Chapter 8 provides a primary on the design and construction process. In creating a building, a team of experts are assembled to create the architectural, structural, and heating, ventilation, and air conditioning (HVAC) systems for a building. In reviewing the design process, a number of constraints shape the architectural solution for any project. Zoning places constraints on the size and shape of buildings. Geotechnical considerations place constraints on the foundation design and ultimately on the design of the superstructure. Decisions must be based on the most appropriate structural system – wood, concrete or steel – given the intended use and budget. This chapter provides a brief review of each building system, providing the reader with an introduction to each. Unlike the building of the past, today's modern edifices incorporate a myriad of new technologies, including heating, cooling, electrical, plumbing, elevators, and communication. In closing, Chapter 8 discusses the role of design software (building information modeling, or BIM,) as a critical component in managing this design process.

Chapter 9 examines project management techniques, which are critical to a successful real estate development. Keeping a project on track in time and on budget, as well as managing performance, are key when applying project management principles to the process of design, bidding, and construction, as are the tasks of budgeting, scheduling, and resource management. In this chapter, a review of project management terminology serves as a basis for a discussion of network analysis. Network analysis gives the manager a tool for monitoring the progress of any project. Using network analysis to keep projects on track is an important skill for those involved in real estate development. Project management software, with its Gantt charts, budgets, and resourcing tools, offers a solution for managing any real estate development project. The case study of renovating an office building demonstrates the use of Microsoft Project, a popular project management application, and the importance of employing these tools to manage a successful real estate development.

This book would not have been written without sabbatical leave from the University of Calgary. To my wife, Elizabeth E. Dickson, whose editorial efforts were critical to the success of this book, I will always be indebted. A careful and thoughtful reader of the manuscript, her suggestions on all aspects of the work were greatly appreciated. Without her help throughout the process of preparing and completing the manuscript, this book would not have been possible. I greatly appreciate the comments and thoughts of Tang Lee, colleague, and long-time friend at the University of Calgary, on Chapter 8, "Design and Construction". And I offer special thanks to everyone at Routledge who worked on this project.

# 1 Introduction to real estate development and finance

## What is real estate?

Real estate is simply defined as "property consisting of land and the buildings".[1] However, real estate is much more than a community's building stock, having impact on much of our everyday life. We must have places in which to live, work, shop, and spend our recreational time. Only through the collective action of landowners, developers, investors, real estate agents, lawyers, government officials, city planners, design professionals, general contractors, and the trades have we been able to create our homes, towns, villages, and cities. This chapter focuses on the processes responsible for creating the homes we live in, the retail stores we frequent, and the places in which we do our business.

Today, many of us earn our livelihood in ways that are connected to the real estate industry. For developers, investors, financial institutions, insurance companies, and pension funds, real estate is at the core of operations. Through the coordinated actions of financial officers, appraisers, accountants, analysts, and numerous trade and design professionals is created the architecture associated with contemporary life. Architects and engineers are responsible for the design of buildings that must satisfy the needs of society. Planners and government officials who are charged with looking after the public good must attend to the requirements of zoning, use, and public transit. Though often viewed by the developer as an obstacle during the development process, planning seeks to maximize benefits for both the community and the developer by creating efficient land use plans. It would serve no one's interest to have a noxious industrial facility located in a residential neighborhood. In our modern age, utilities, communication, and transportation infrastructures are all part of a critical infrastructure that makes a development project viable. The general contractor orchestrates the many trades that build and maintain our city's infrastructure. Plumbers, electricians, carpenters, roofers, metal workers, heavy equipment operators, and a host of other technically skilled individuals are responsible for the construction and maintenance of residential commercial, governmental, and industrial buildings. Creating a market for these developments requires the efforts of leasing agents and real estate brokers who are working on behalf of owners and their tenants. Lawyers and accountants consider the financial and legal aspects of real estate, sales, and development. But the impact of development goes beyond the mere creation of architectural space. Carpet, furniture, appliances, and lighting fixtures must also be purchased to make an architectural space functional. Once the spaces are completed, those who are employed in our offices and places of work contribute collectively to the gross domestic product (GDP) of a country. In fact, over the lifetime of a commercial building, the actual cost of construction represents only 2% of all cash flow generated based on a lifecycle cost analysis.[2] Over 90% is spent on salaries, while the remaining balance is spent on maintenance and utilities.

From the outset, we must acknowledge that bringing a project to a successful conclusion is far from a simple process. Often, the motivations of stakeholders may appear to be at cross-purposes. For the developer, maximizing profit may be central, while for the financial institutions providing the capital, reducing the risk of failure may be of key concern. For the architect, having an award-winning design celebrated in a leading publication may be as important as seeing the building constructed. For the urban planner, satisfying the requirements of regulations and policy is foremost. For the contractor, finishing on time and on budget may place them at odds with other individuals who have less of a sense of urgency. This could include the planners, elected officials, approving authorities, and financial institutions who work on their own timelines. Ultimately, it is the entrepreneurial activity of the developer that is at the center of this process in the US and most industrialized nations.

Without the developer, the responsibility of building our cities and towns falls on government. In China and countries with a large government involvement in the development process, the role of creating a country's infrastructure is given to bureaucrats and governmental design professionals. In these cases, the government acts as the developer, designer, and contractor. Because the motivations are sometimes political, the construction of unoccupied cities has occurred with little purpose other than keeping the construction industry engaged and their employees busy. In the US and Canada, the private sector is largely responsible for the construction of residential, commercial, and industrial space. The exception is where the government acts as the client. In cities with a significant government presence, real estate development can be critical to the economic development of the region. In more recent years, private–public partnerships (PPPs, 3Ps or P3) have been employed to design and build large-scale public projects. These projects can include public works such as hospitals and prisons, but a PPP can also be used on modest-sized projects such as affordable housing. The goal with these PPPs is to foster innovative solutions to financing and building public works projects.

## What determines value?

Real estate, unlike other investment assets, is fixed in space and time, and it cannot be moved or assigned to a new location. Real estate agents often refer to this obvious feature of real estate in terms everyone will know: "location, location, location…" For these reasons, whether we purchase a home, a rental property, a vacant lot, or a farm on the edge of the metropolitan center, the location is what we are buying. Anyone looking for a home in the city or suburbs, a piece of property on which to build a summer home, or a condo in the inner city appreciates the role the setting plays in determining its value. We may buy a home because of the quality of the school district or its proximity to shopping and employment opportunities. A neighborhood with older trees and less traffic will only increase the value of the property. If we are lucky enough to acquire a vacation property, being adjacent to the beach or a lake will increase its premium over other homes a few miles, or even a few blocks, away. A city apartment or condo with a view of anything other than a brick wall or dumpster will exact a premium. Likewise, homes located along busy noisy streets, in areas with industrial pollution, little shopping, high crime, and poor schools, will have a lower resale price or, in some cases, almost no value at all.

When we finally make the decision to purchase a home, our hope is that property values will continue to increase. Without this hope, you might choose to rent and avoid the hassles of home repairs and maintenance. However, in markets in which rents and property values are moving up in lockstep, buying can also help stabilize housing costs by avoiding paying inflated rents in the future. We buy for a variety of reasons other than minimizing cost. We may want to express our personal design values. We may also desire to renovate and decorate to meet our personal needs and preferences in ways not allowed under our lease. For example, landlords generally do not appreciate any deviation from the standard beige color scheme in their rental units. We may want to live in a community with shared values or send children to a specific school. In some markets, rental properties may not even exist that can satisfy our requirements.

In locales that have been experiencing a rapid increase in prices over a sustained period, the need to get into a market quickly, before being priced out completely, will be a strong incentive to a potential purchaser. The opportunity for gain has not escaped anyone buying into the housing markets of Vancouver, San Francisco, and the East Bay area. Over the last 30 years, the values have averaged over 7% a year, even accounting for the declines in 2001 and 2006[3] (Figures 1.1 and 2.1). For those who are priced out of such markets, their only alternative may be to commute longer distances or to accept housing of a poorer quality in a less desirable neighborhood. A political issue in many cities, mayors and councils have been forced to act to dampen price increases through

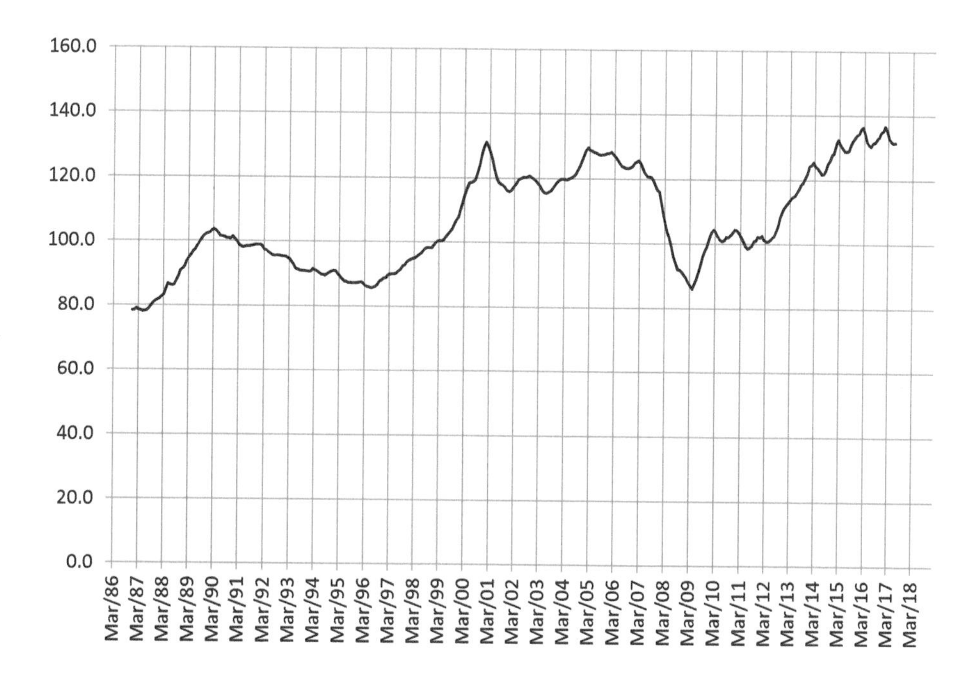


*Figure 1.1* San Francisco S&P/Case-Shiller US national home price index, 1987– 2017 (Jan 2000 = 100).

Data Source: Federal Reserve Economic Data, https://fred.stlouisfed.org; for house prices in 2015, see www.paragon-re.com/trend/san-francisco-home-prices-market-trends-news

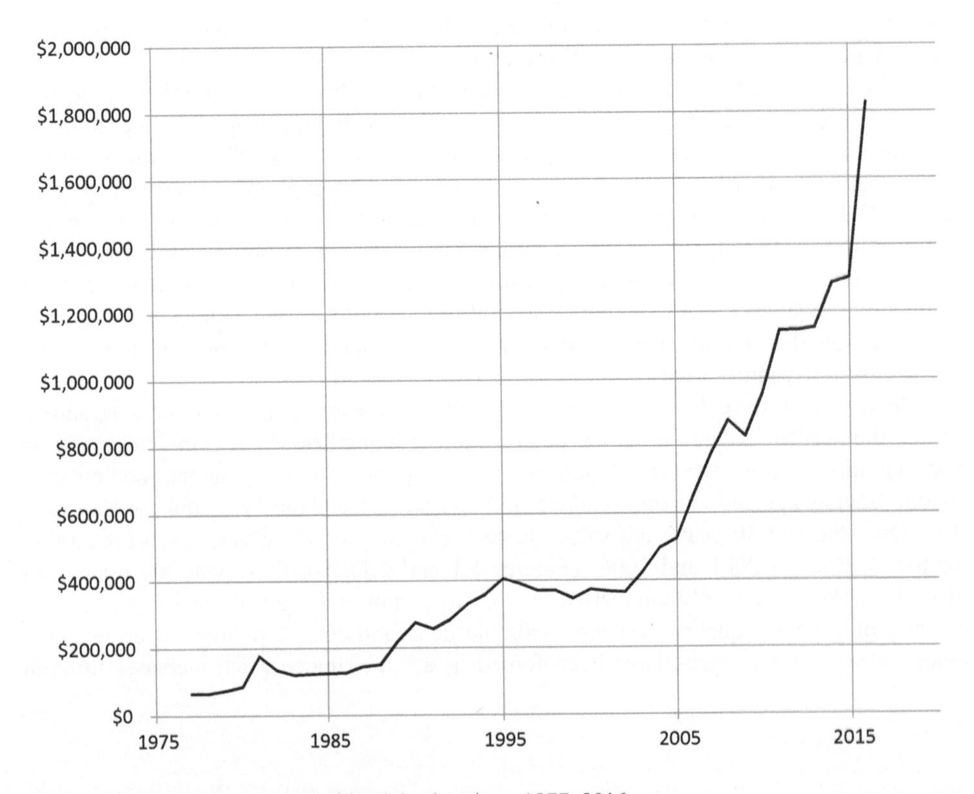


*Figure 2.1* Vancouver average residential sale prices, 1977–2016.

Data Source: CMHC, 2017, www.vancitybuzz.com/2015/11/real-estate-vancouver-october/; www.paragon-re.com/trend/3-recessions-2-bubbles-and-a-baby

a variety of measures including rent control and tax policy. In Vancouver, where prices grew by over 30.1% in 2016,[4] the city council passed measures to increase tax to 15% when a buyer is not Canadian. However, the effort of making Vancouver real estate affordable by discouraging foreign investment may have come too late. With an average price of CAN$1.58 million for a single family home, it is doubtful that this measure will make home properties affordable any time in the near future. However, some were lucky enough to buy their home 20 or 30 years ago, and now have the prospect of selling and receiving a windfall. This encourages others to speculate that this could happen to them. In a sense, we are all speculators with the hope that someday we might benefit from such an auspicious decision.

Though very few of us may be able to benefit from a rapid increase in real estate value, we may still hope that someday we will be able borrow against our equity. Home equity can be used to finance a child's college education or an unexpected medical expense, or, perhaps later in life, to help make retirement more comfortable. Even if the property does not increase in value, paying down the mortgage will allow us to borrow back the money in the form of a home equity loan. If we own a rental property, we can borrow against the property or sell and reinvest our capital gains. However, estimating future value is a tricky exercise. We can model property values using mathematical equations based on past sales data, as many real estate internet sites do. However, projecting forward values based on historical data is always difficult when markets fluctuate. Ultimately, the value of a property is only what someone is willing to pay. The perception of value is critical to this equation. In every real estate market, the collective action of buyers and sellers establishes the ultimate value of real estate. Like a game of musical chairs, when there are not enough chairs for everyone, prices go up; when there are not enough players to fill the seats after each musical selection, nobody is willing to pay a premium for the right to acquire a seat. During an auction in which there are few buyers, the mood is particular depressing for the seller. As a seller, the choice is sometimes to sell at a loss or wait for better times, if they are to come at all. What is interesting about property values is that they can be highly volatile during periods of great speculation. Referred to as animalistic behavior by Keynes and later as "irrational exuberance" by Robert Shiller in his book by the same title, we can be lulled into believing that the direction of the market will continue its upward climb forever, even when history has shown that prices are volatile and are poised for a decline after a significant run-up in prices.[5]

## Learning from history

### *Boom–bust economies*

Communities such as Houston and Calgary illustrate dramatically how boom–bust economies tied to oil and gas have a direct bearing on their local real estate markets. When times are good, we see dramatic increases in property values that are later countered by declines during periods of falling oil prices. This makes sense. After all, when oil prices are up, local companies are hiring employees, and everyone sees a profit from drilling and development activity in new oil and gas fields. Service industries that support this heightened activity also benefit. This includes those who are direct suppliers, such as drilling bit dealers, heavy equipment operators, petroleum engineers, and environmental consultants, in addition to those employed in the service sector, including health, education, and entertainment. During a period of expansion, the construction industry may work overtime to satisfy the demand of new workers for rentals and homes. When the existing housing stock cannot meet demand, prices can rise quickly.

When oil and gas prices fall worldwide, there is little incentive to explore and drill, much less to plan for future development. Employers have no incentive to keep workers around. Those who are fortunate to find work elsewhere leave. Those who stay find jobs at lower wages and will generally have less disposable income. In such an environment, the economy contracts, and there are more layoffs. Restaurants will be some of the first business to suffer from the reduced take-home pay, and with a shrinking population, there is less overall demand for all services including financial, health, and education. Fewer people looking for homes and rentals only depresses prices further. During this recession, construction comes to a halt. Even projects that are only partially completed may be left unfinished for years. Plans for future developments are put on hold. Carpenters, roofers, electricians, plumbers, carpet installers, tilers, house designers, surveyors, engineers, and architects can no longer find work. Housing inventories exceed demand, and sellers mark down prices in the hope of breaking even or just minimizing their losses. Rents are reduced in the hope of meeting some of the fixed costs. The downward spiral stops only when commodity prices return to more profitable levels. Only once there is some assurance that prices are rising will the economy begin to reverse the losses of the previous downturn. Only then will companies begin the process of rehiring and does the economy begin to struggle back to previous levels.

In both Canada and the US, the oil and gas industry has experienced several boom–bust cycles over a 30-year period. Most recently, in 2014, everyone was very optimistic. Oil prices had recovered from previous high of about $30 per barrel just a few years earlier to almost $120 (Figures 3.1, 4.1, and 5.1). The local economies of Houston and Calgary were expanding, and there was a period of significant job creation. In 2014, everyone believed that the trend could only continue upward – even in light of historical data. We are always optimistic about the future, and planning for a decline is something only naysayers do: just ask any chamber of commerce president about the future of the economy when all the signs are positive. Unfortunately, this trend was not to continue forever, and in 2015, with the fall of oil prices, the mood of the market dramatically changed from hope and prosperity to doom and gloom. The bottom had fallen out and, after a few months of denial, all hope had evaporated for a quick recovery. With the continued oil glut and economic slowdown in China and Europe, history repeated itself with a vengeance. Prices for single-family homes fell by double digits in 2015–16 in Calgary and Houston, unemployment increased from 4.8% to over 10% in Calgary[6] and in Houston, from 3.9% to 8.5%,[7] while vacancies for commercial real estate increased into the double-digit range (in Houston, 12%; in Calgary, 20%).

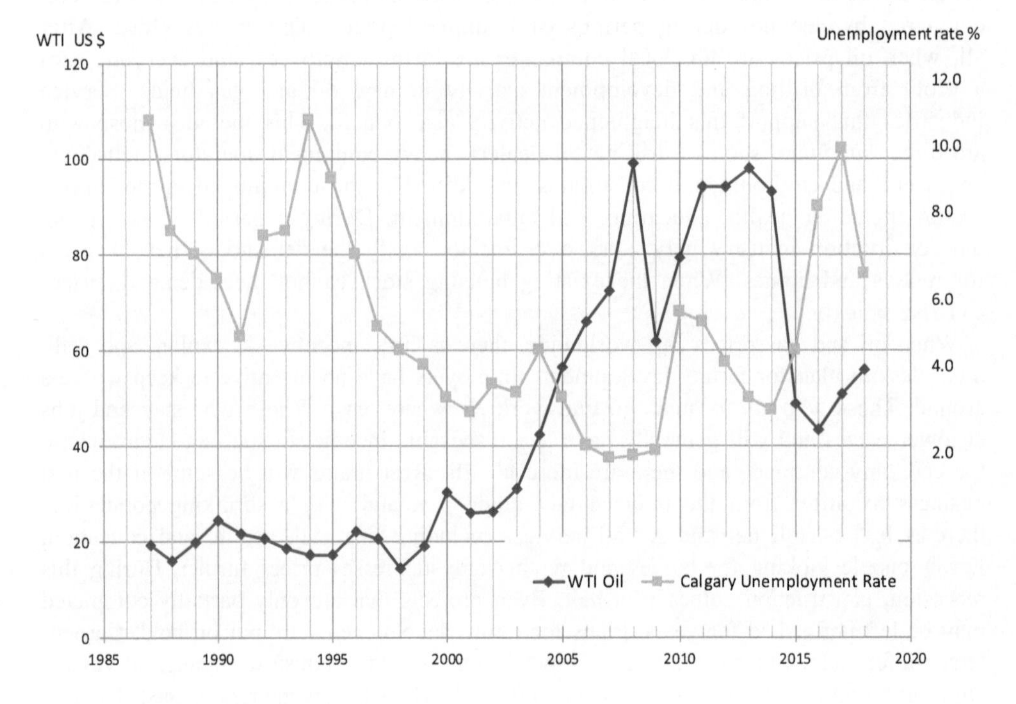


*Figure 3.1* Calgary unemployment, WTI price, 1987–2018.

Data Source: US Energy Information Administration (EIA),[8] Statistics Canada.

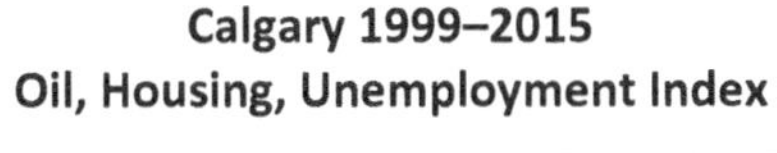


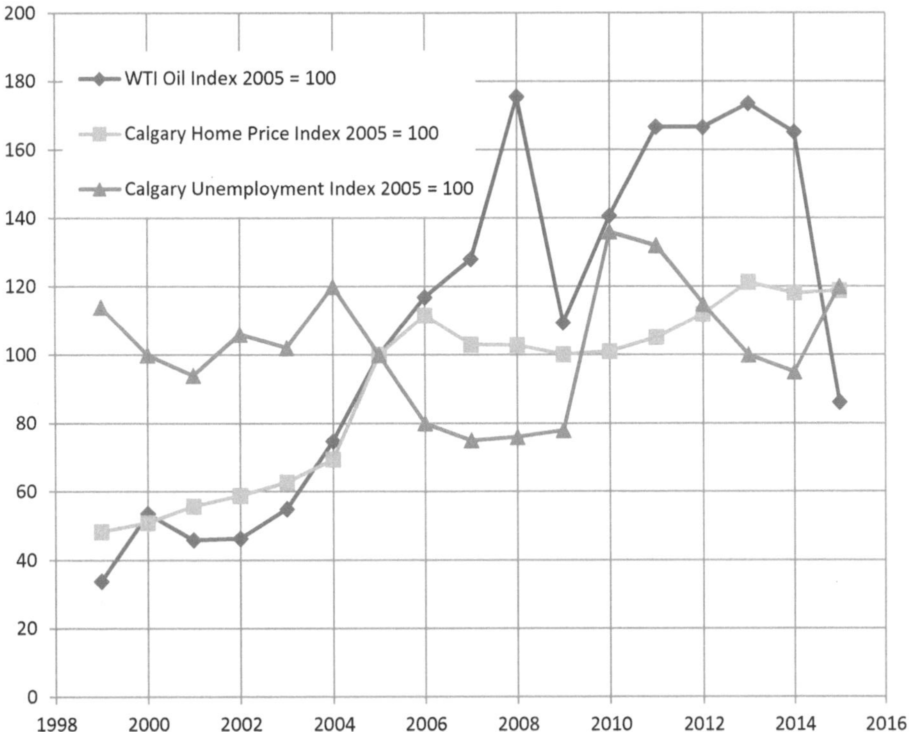


*Figure 4.1* WTI oil prices, home prices, and unemployment, 1999–2015.

Data Source: US Energy Information Administration (EIA),[9] Statistics Canada, CMHC, http://calgaryrealestatereview.com/2015/08/; © 2017 Canada Mortgage and Housing Corporation, www.cbc.ca/news/canada/calgary/calgary-unemployment-rate-still-highest-april-2017-1.4100873

The solution for both cities would be to develop a more diversified economy, one that lessens the dependency on the price of oil and gas, but making this transition to a more balanced economy can be difficult. During periods of economic growth, the competition for labor and rental space makes it difficult for startups and new enterprises to establish a foothold. Labor rates are high, and rental space is scarce and expensive. During prosperous times, the urgency to diversify may also not be the burning issue for the electorate or their officials in government. There is also the downside that the electorate may not favor using taxpayer's dollars to support business ventures that could potentially fail. The criticism that government cannot pick economic winners and losers makes elected officials shy about offering incentives to new businesses in fear of being accused of playing political favorites or making outright payments of graft. Without substantial support by government, making a transition to a more diversified economy may never happen. Things generally do not happen by chance. During good times, there is no need for these programs to diversify the economy, while in bad times, finding the money in the budget for these efforts becomes a challenge, given falling tax revenues.

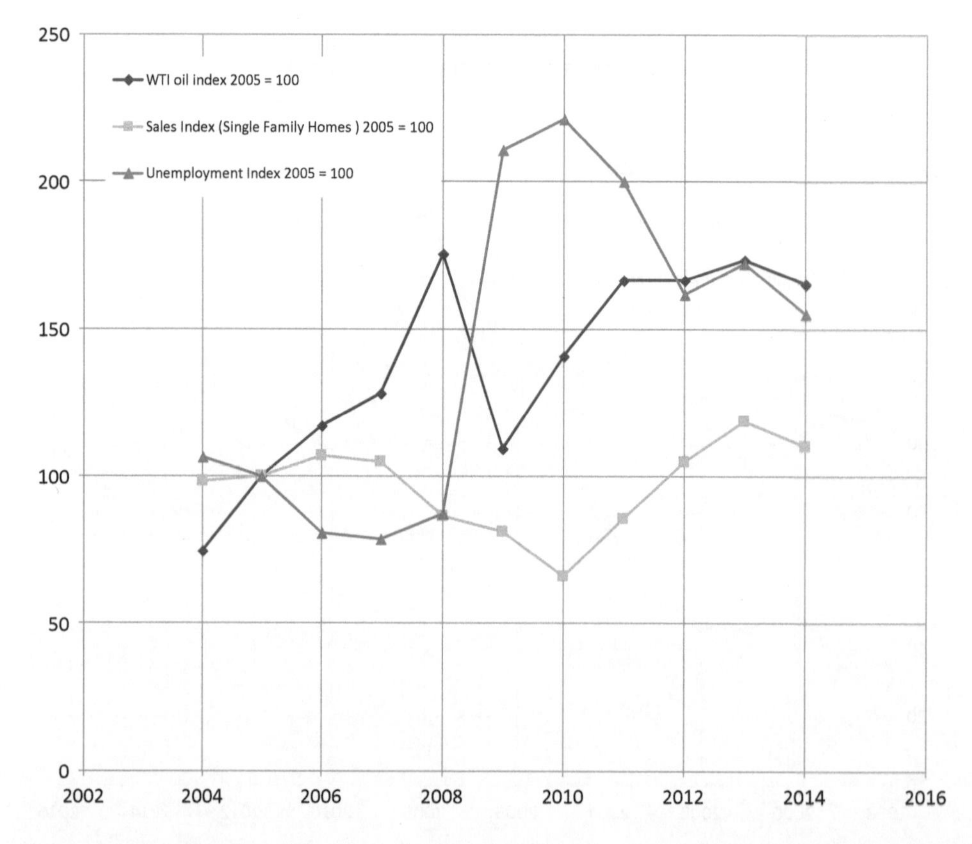


*Figure 5.1* Houston, WTI index, sales of single family homes, unemployment rate index, 2004–14 (2005 = 100)

Data Source: US Energy Information Administration (EIA),[10] US Department of Labor, Bureau of Labor Statistics.

### *When communities are in decline*

Given that the future is never assured, the value we assign to a property is always transitory. What has value today may not be valued as much tomorrow. Opportunities for employment can slip away with changes in the economy. In 1950, Detroit produced over 80% of the world's cars and trucks, and employed more than 150,000 workers in highly paid manufacturing jobs. With a population of more than 1.8 million in 1950, Detroit was rich in comparison with other cities of comparable size, boasting concert halls and art galleries filled with Van Goghs, Rembrandts, and Picassos, upscale residential neighborhoods, and a strong middle class. In the last 2010 census, Detroit's population had slipped to 713,777[11] (Figure 6.1). By 2010, unemployment rates had climbed to 18%, putting many households at risk of losing their homes (Figure 7.1). With less tax revenues and with increasing demands on the city coffers to pay pensions, police, firefighters, and schools, Detroit filed for bankruptcy in 2013 (Figures 8.1 and 9.1).[12] In part, we can point to the balance in world trade as the culprit. Other car manufacturers in the US, Europe, and Asia have placed the big three in Detroit in absolute and relative decline. To be fair, many of these jobs moved to other locations in the US, where labor was

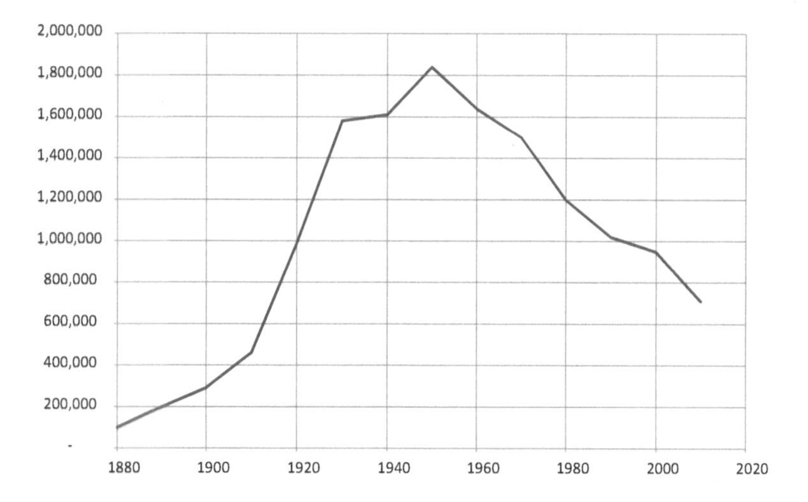


*Figure 6.1* Detroit population, 1880–2010.

Data Source: US Census.

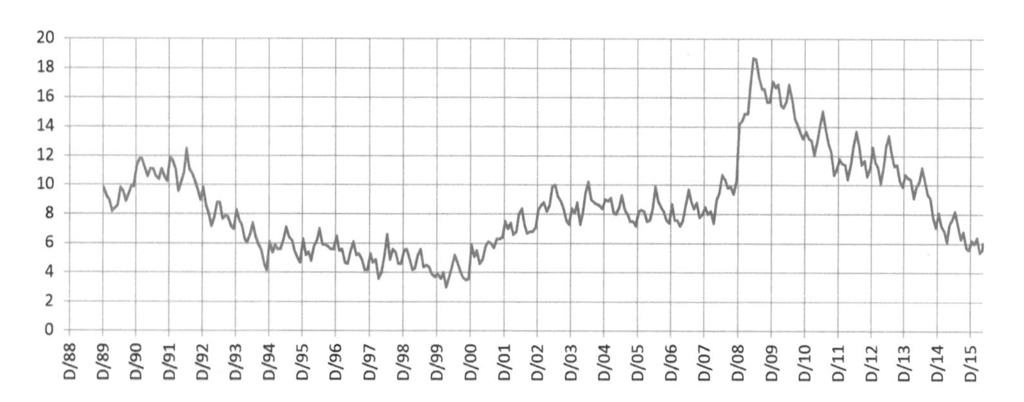


*Figure 7.1* Detroit unemployment rates, 1989–2015.

Data Source: BLS.

cheaper and tax incentives provided by local government were more profitable. In fact, the decline in manufacturing jobs had begun as early as the 1950s. With the end of wartime production, Detroit began to see the closure of manufacturing plants including the Packard Automotive plant and the Hudson Motor Car Company in 1958. In 1950, Detroit employed more than 300,000 in manufacturing, which had fallen by 2010 to about 30,000. During this time period, the proportion of the workforce employed in manufacturing jobs fell from over 75% to less than 15%. What is interesting is that car production in Detroit had declined only a little more than 20% during this time period, from 9 million to 7 million cars.[13] Though car production in the US has actually increased from 7,894,000 in 1950 to 12,105,000 by 2015, the location of much of this new production has been in cities outside of Detroit and Michigan.[14]

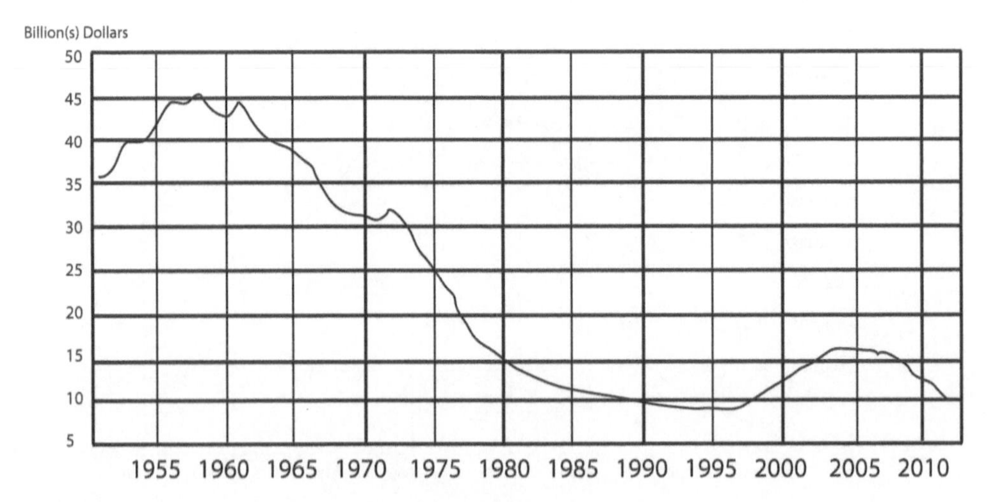


*Figure 8.1* Detroit property values, 1950–2012.

Data Source: Detroit Annual Financial Reports, Moses Harris, Detroit Free Press. Amounts adjusted for inflation, $2013.

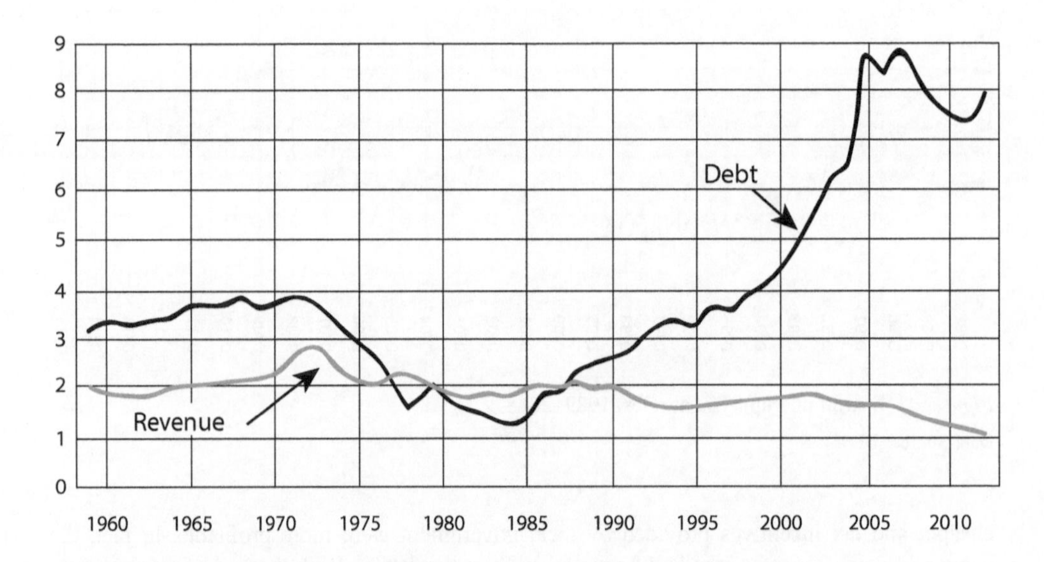


*Figure 9.1* Detroit debt vs revenue, 1959–2014, in billions of dollars.

Data Source: Detroit Annual Financial Reports, Moses Harris/Detroit Free Press.[15]

Changes in technology are partly to blame for the loss of jobs (Figures 10.1 and 11.1). For example, with technological advances, fewer employees are needed to build a car today. Comparing photos of Ford's assembly lines today with those of a century ago, when the first Model Ts were coming off the assembly lines, reveals a stark reduction in the number of workers.

*Figure 10.1* Detroit labor force employment, unemployment, manufacturing sector 1990–2016.

Data Source: Bureau of Labor Statistics (BLS).

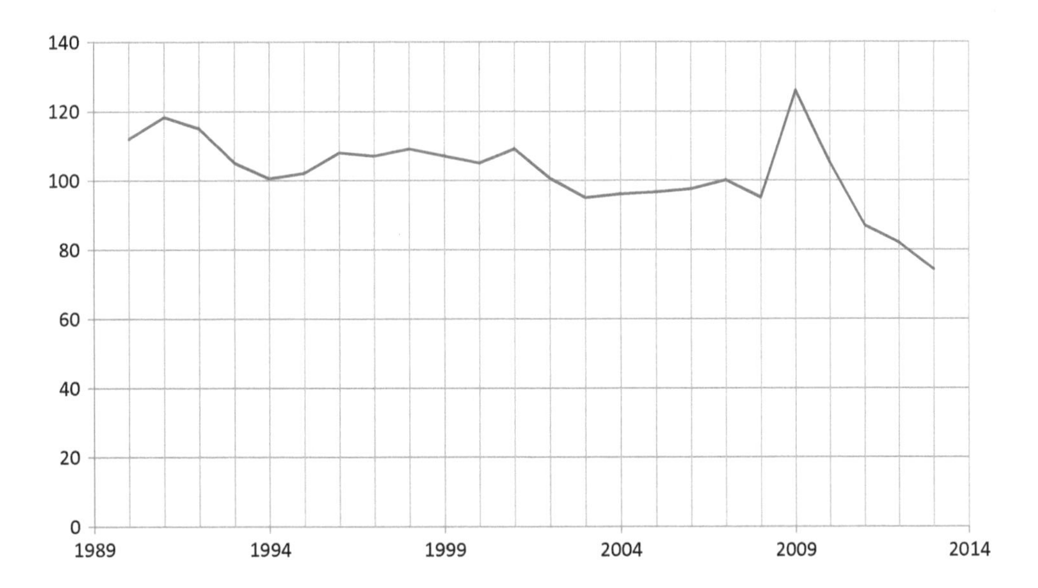


*Figure 11.1* Employees per vehicle produced index, 1990–2013. (2006 = 100)

Data Source: Wards Automotive, Bureau of Labor Statistics, Federal Reserve Board of Chicago, Detroit Branch.

However, it would be unfair to compare the early mass-produced auto with those created today. Today's automobile is more complex, with high-performance engines, and onboard computers and navigation, providing greatly improved personal safety and comfort over its earlier predecessors. In addition, automobile parts are often imported, decreasing the number of employees required locally. The design history of airplanes, telephones, and medical imaging reminds us how rapidly technological change has transformed many of our basic industries. Advanced computing, robots, artificial intelligence (AI), and computer-aided design (CAD) technology have all contributed much to the increased productivity and technological improvements in automobile manufacturing. However, for communities that are experiencing a decline, these benefits may not be self-evident. Robots and computers do not need a place to live, do not send their children to school, or pay real estate taxes to support local community services including fire and police.[16]

If you had acquired a property after World War II, you would have seen real increases in its value up to the early 1960s, only to see its value eroded in the following decades. With a decline in employment opportunities, the citizens of Detroit must look elsewhere for their livelihood. As the number of houses in default rises, tax revenues decline, and basic services cannot be supported (Figures 9.1 and 12.1). The resulting downward spiral takes the value of real estate for a ride. As the saying goes, during a firestorm, you "can't give it away". Owning a property for which you cannot find tenants or a future owner can have even a negative value. Given that you must continue to pay taxes and insurance on it, and maintain the building, the only alternative during a prolonged period of economic decline may be to walk away from the property.

Detroit and cities like it illustrate two basic principles in real estate. First, unless we are retired or independently wealthy, we buy property because of its location relative to employment opportunities. Second, because of its fixed nature, we are optimistic about the future of the community we buy into; otherwise, we would rent. We believe that future prospects will be better than they are today. We assume that the risk associated with a fall in value is modest and temporary, and that the likelihood for improvements in the future is good over the long haul. Nobody buys property in a depressed community unless there is some hope that things will improve tomorrow. In a sense, we are all speculators. We may purchase a home because we need a place to live, but we make this commitment because we have faith in the future of the community we call our home. If we are lucky to live in a community with a solid future, our real estate investment does well, but if we are unlucky in our choice, we can lose what, for many, are our life savings.

### *The road to recovery*

This is not to say that the fate of all cities is to experience a period of rapid growth followed by decline.[17] Pittsburgh was a major center of steel production in the US from the 1830s until the 1960s, when cheaper imports from Asia began to replace American production. Like Detroit, Pittsburgh was a center of great industrial wealth. The home of the Carnegie Steel Company (later US Steel), Pittsburgh was able to support a fine symphony orchestra, a world-class art museum, a leading medical center, and several major research universities.

The demise of US Steel's Homestead Works in 1983 marked the beginning of the end for the supremacy of the steel production in Allegheny County. Even earlier in 1982, layoffs had already begun in every major industrial firm, resulting in an unemployment rate reaching as high as 13.9% in Allegheny County. By 1988, Pittsburgh had lost

56,000 jobs in the steel industry.[18] The loss of these high-earning industrial jobs would ripple through the economy. This trend continued well into the 21st century, with a loss of 10,000 taxpayers in 2005, even when efforts had begun to revitalize Pittsburgh. The PPG glass complex designed by Philip Johnson, one of the leading architects of the Modern period, was an urban redevelopment project that was designed to turn the tide of declining misfortune. Fortunately, Pittsburgh was able to capitalize on its other assets, including its universities, medical centers, and cultural institutions. Carnegie Mellon University and Pittsburgh University are just two of the institutions that supported the growth of high-tech companies in the area. Under the leadership of Mayor Thomas Murphy, Jr., with support from the federal government, it has been possible to take once-toxic industrial sites and transform them into new communities with the amenities of waterfront esplanades and public parks.

Creating a place where high-tech employees would like to live, with an active arts community, high-quality retail, and a strong sense of community is all part of the equation of success. With an emphasis on knowledge-based industries, it is possible for communities to transform themselves, but not without leadership and the support of government.[19] Measuring this type of success is always difficult, but GDP does provide a measure of relative growth. In Pittsburgh, between 2005 and 2015, a steady growth in GDP has occurred – high, when compared with both Pennsylvania as a whole and the US (Figures 13.1, 14.1, and 15.1). Over this period, Pittsburgh has experienced a 20% increase in GDP. This is

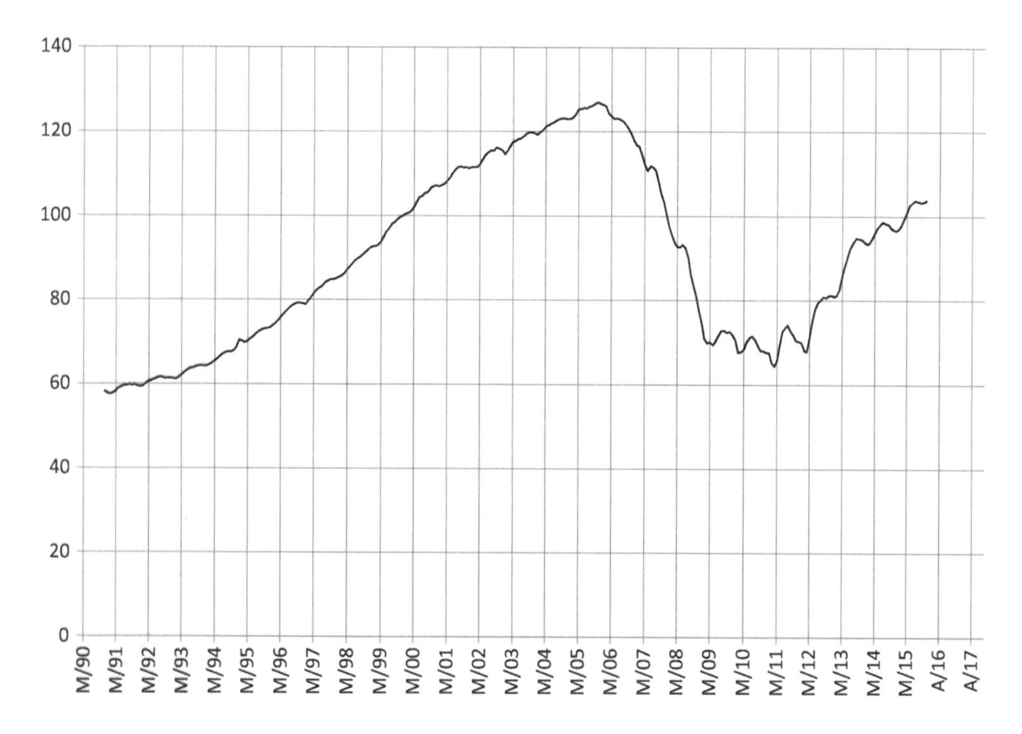


*Figure 12.1* Detroit S&P/Case-Shiller US national home price index, 1990–2016.

Data Source: Datahub, https://datahub.io/core/house-prices-us

*Figure 13.1* Pittsburgh per capita GDP, 2005–14 (2007 = 100).

Data Source: Bureau of Economic Analysis/Haver Analytics.

remarkable, given that, during this same period, the US experienced both a recession followed by a recovery and had GDP growth that was relatively flat for the decade.

One sign of relative prosperity is Pittsburgh's vibrant downtown. Within its Golden Triangle can be found numerous dining and shopping opportunities for its residents Though not without some setbacks, including the closing of Saks Fifth Avenue (2012) and Macy's (2015), with over 11 million visitors and 6,000 hotel rooms in its downtown, Pittsburgh has been able to sustain its growth and appeal as a destination for tourists and residents.[20] Considered one of the ten best housing markets in 2016, and with an ample and affordable housing stock, Pittsburgh should continue to support a positive growth in value and sales in the residential real estate market. In a comparison of Pittsburgh with Detroit over the period 2000–16, the differences are dramatic. Pittsburgh was able to maintain a positive trend even after the housing crisis of 2007, while Detroit experienced a 50% decline in median home prices (Figures 12.1 and 16.1). Demographics also support this positive outlook for Pittsburgh. It is a center of business activity, with information technology (IT) and medical research opportunities for young professionals moving into the area, who will continue to support a residential housing market (Figures 15.1 and 16.1)

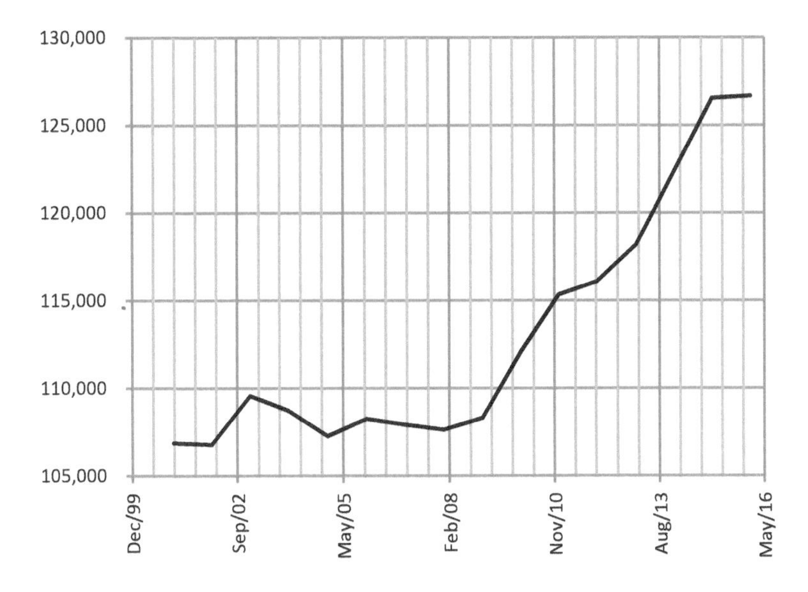


*Figure 14.1* Total real gross domestic product for Pittsburgh, PA (MSA), millions of dollars, annual, not seasonally adjusted.

Data Source: Economic Research Division, Federal Reserve Bank of St. Louis. https://fred.stlouisfed.org

## *Too much of a good thing*

Not all North American cities have been in a state of decline over the last half-century. The San Francisco metropolitan area has seen dramatic growth since World War II. This leading center for high-tech companies in the US includes the area from San Francisco in the north to San Jose in the South, to Berkeley on the east side of the bay. No other metropolitan area in North America, or perhaps in the world, has a greater association with the computer technology industry. As a center of innovation and computer technology, "Silicon Valley" has been unsurpassed in its generation of wealth and employment opportunities since the 1960s. Located in Santa Clara County, Silicon Valley is the home of 39 Fortune 100 companies. It has also been the beneficiary of almost half of all venture capital in the US. A few of the companies on this list have a net worth of over $10 billion. This list of companies includes Apple, Google, Facebook, Microsoft, HP, Cisco, Intel, Adobe, and Intuit, to name a few. Many of these companies have young histories that are measured in years, while others that were founded in the 70s have some of the oldest histories found in this high-tech sector. The area is also the home to Stanford University and the University of California, Berkeley, with major research laboratories including Stanford Research Institute. Silicon Valley researchers have been responsible for the development of the transistor, Internet, semiconductor chip, personal computer, and graphical user interface. The film and entertainment industry was an early direct beneficiary of this revolution: two pioneers in computer graphics, Pixar and Industrial Light and Magic, would not have been able to make the film classics *Star Wars*, *Toy Story*, *Harry Potter*, and *Jurassic Park* without this computer revolution.

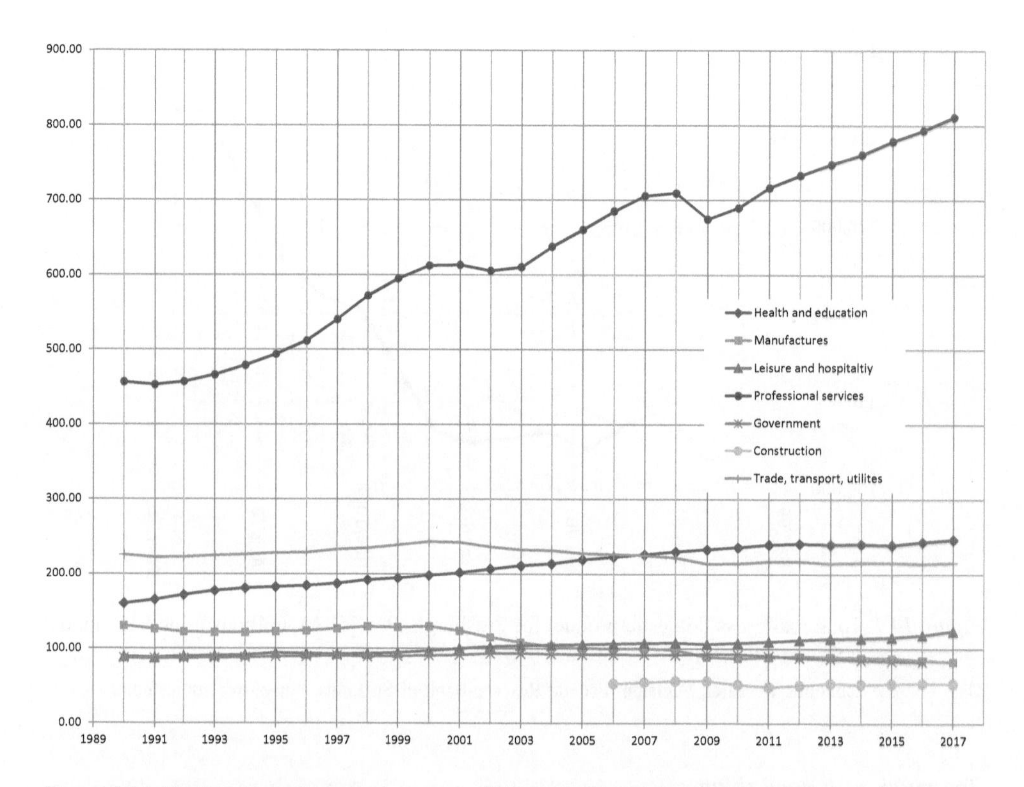


*Figure 15.1* Pittsburgh employment by sector, 1990–2017 (1,000s).

Data Source: Bureau of Labor Statistics, 2017.

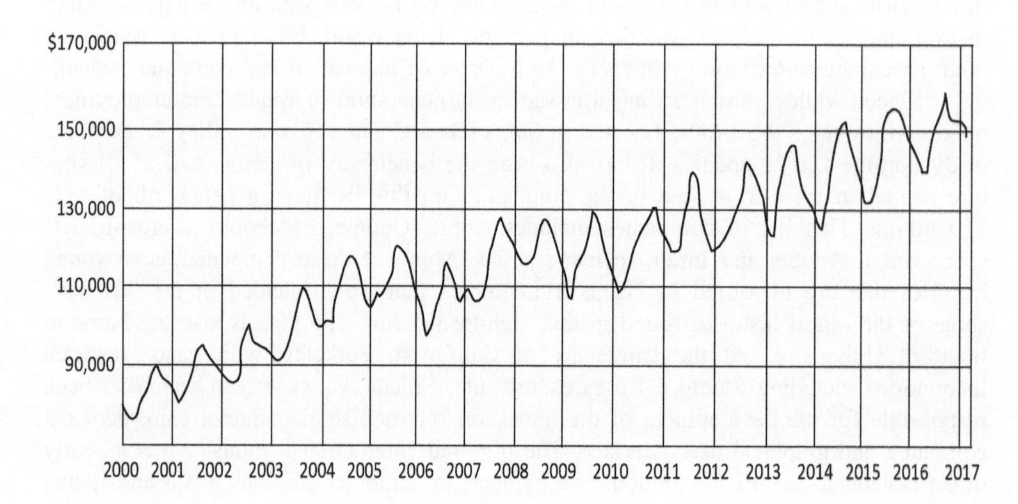


*Figure 16.1* Pittsburgh median home sales price, 2000–17.

Data Source: Trulia.com

The success of San Francisco and the East Bay as a center of technology has made it a magnet for talent and innovation. The home of numerous startup companies, the number of individuals employed in the industry is roughly 30% of the workforce. Average annual salaries of $150,000 are the highest in the US.[21] The region supports a diverse community, rich in culture, entertainment, and recreational opportunities. Temperate weather, beaches, mountains, vineyards, and numerous cultural opportunities make this part of California synonymous with the "good life". However, the good life comes at a cost. The economic success has also produced rapid increases in property values since the 1960s. With only modest increases in housing stock over this time period, the economic growth and development has produced a dramatic housing shortage, especially for those with modest incomes. Geography, in part, limits the growth of new construction. San Francisco is located on a peninsula. Other areas in the East Bay are surrounded by hills, forcing workers to commute from bedroom communities to the east and south. San Francisco and the East Bay, when compared with other metropolitan areas in the US, have the highest rents and house prices in the nation. In the city of San Francisco, the 2016 median home price rose to over $1.3 million and condo prices now average over $1 million. Even with a decline in values after the financial crisis of 2007–8, home prices and rents have escalated significantly over the last decade (Figures 17.1 and 18.1).[22]

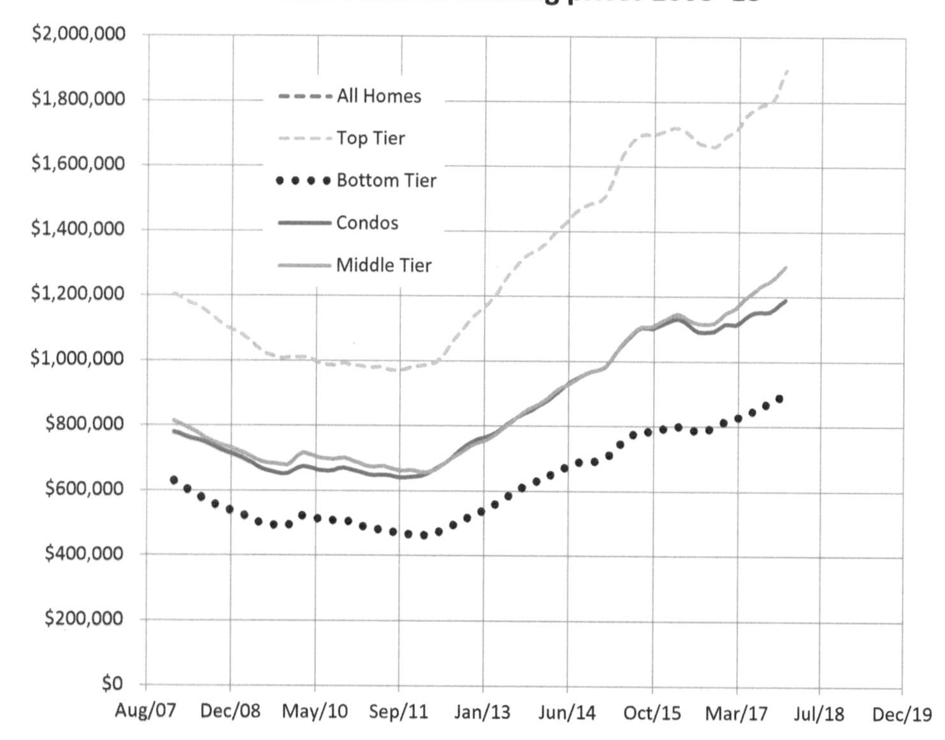


*Figure 17.1* San Francisco median sales prices by year, 2008–18.

Data Source: Zillow, 2018[23]

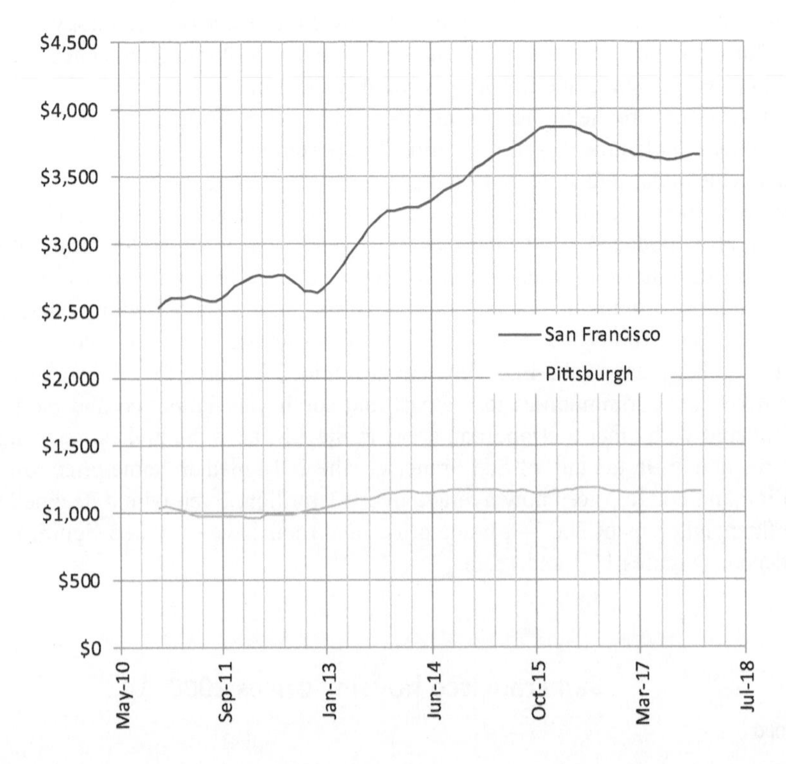


*Figure 18.1* San Francisco and Pittsburgh average rent for a one-bedroom apartment, 2010–17.

Data Source: Zillow, 2017, www.zillow.com/pittsburgh-pa/home-values/

With the increase in demand, some positive adjustment in housing stock would normally be expected. However, the number of sites available for redevelopment within a reasonable commute of work has constrained the growth of the housing market. Highway capacity has been stretched beyond capacity, and the Bay Area Rapid Transit (BART) and bus systems can reach only so far into new areas. In addition to higher construction costs, creating additional residential units has been constrained by zoning and land use restrictions. Such regulations have reduced the number of buildable sites suitable for higher-density affordable housing in many communities. In San Francisco, setback and height limitations have reduced the amount of buildable space that can be placed on lot.[24] The result is a housing market unable to satisfy demand, pushing up prices to stratospheric levels. Most notable are the shortages for households with modest incomes, making it difficult to find an affordable living situation for many families as rents rise faster than incomes. Over one third of renters pay more than 40% of their gross income on rent, while another quarter pay 50%.

The success of the region has made it difficult for many technology companies to be viable in this environment. To attract talented employees, companies need to pay higher salaries to offset the high cost of living. High-tech companies looking to reduce their costs are locating in cities in states that offer lower housing costs. Austin, Seattle, Denver, Boston, Washington, DC, and New York City have all seen growing numbers of tech workers. Cities such as Austin offer high-tech companies, especially startups,

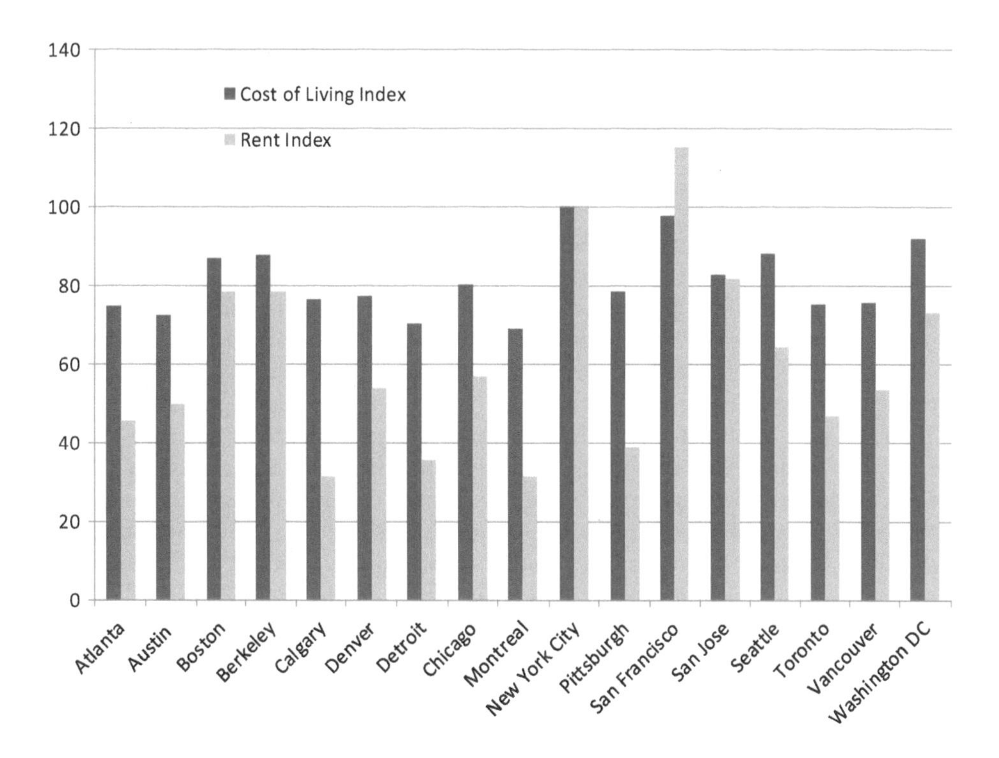


*Figure 19.1* Cost of living in selected American and Canadian cities.

Data Source: Numbeo, www.numbeo.com

several advantages. The home of the University of Texas, a major research university, Austin offers high-tech companies access to a highly educated workforce in a city in which the median sales price of homes is a fraction of the cost of comparable homes in the San Francisco metropolitan area. Austin, over the last two decades, has been transformed by this influx of new companies. Considered one of the most livable cities for those aged 35 and under, it ranks high in jobs, affordability, and cultural opportunities. It is important to note that not all of the cities that are competing for high-tech employees are low-cost environments. New York and Washington, DC are only slightly less expensive than San Francisco and the East Bay (Figure 19.1). The location of potential clients, major research labs, and the federal government are all considerations important in locating a high-tech enterprise. Being where the action is may sometimes be more important than being in a more affordable location.

## The two big questions in real estate development

For real estate developers, there are two big questions: what is the best use for any single parcel; and, given a specific building function, where is the best location? In the first case, we might consider the example of having inherited a property that we are now looking to improve or the current use of which we might like to change. In our attempt to determine the "highest and best use", we may consider a range of possible uses: single-family residence, multifamily residence, commercial, or retail. For each scenario, we can look at simple ratios, including return on capital (ROC), to help us determine which will result in the most profitable option. Of course, the solution to the problem is never simple. Lot shape and size may accommodate only specific uses. Then, there is zoning. Zoning designations will restrict specific uses. Other requirements under the by-laws will specify setbacks, maximum heights and land coverage, and other planning considerations. Subject to these constraints, profitability may vary greatly.

In the case of inheriting a property with a building, we have to consider the costs of renovation against those of demolition. Ultimately, if the cost of renovation or development outweighs the potential increase in value, we may decide simply to continue with the current use. In cities in which properties are no longer generating a positive financial return, it may be difficult to find another use for the property. In this case, we may even consider abandonment. In many older communities in the northeastern US, old warehouses, mills, and tenements cannot be renovated easily *or* demolished without assuming great cost. Idle for decades, these properties sit empty, waiting for changes in the economic climate to generate a positive value. In the interim, the property continues to deteriorate, waiting for its final fate: the wrecking ball. Cities with a large portion of properties in this category endure the weight of these underperforming assets that generate little or no taxes when they are abandoned by their owner. In these cases, we can refer to the actual building on the land as having "negative value". In a very real sense, the property would have more value if there were no buildings. In the 1950s and 1960s, the premise of "urban renewal" was to clear away older buildings in the hopes of attracting development that would generate higher returns. In many cases, the poor and disadvantaged who occupied these substandard apartments in the inner cities had no other place to go. There, destruction resulted only in displaced families and a few more vacant parking lots.

The second class of problems that developers consider is where would be a good place to construct a specific type of development. For a major big box outlet that features Swedish modern design, the question could be, "where do I locate my next store?" Again, constraints operate to limit the choice. Zoning may restrict development to only specific areas within the city. The store may also need to be located along a major thoroughfare that generates the traffic needed for business. Demographics will also pay a role: the retailer will probably want to place the development near new communities with a large number of younger families who need its homewares and furniture. Ultimately, there is the problem of finding a site that is large enough for the store and parking. The retailer may need to assemble smaller parcels into a larger site, which can increase cost and time. Going through this exercise may generate solutions that are too costly for the tenants to rent, given their business model. In those cases, the company may be forced to look in other communities for suitable development sites. Often, this search for an appropriate location can occur at a regional, and possibly at the national, level.

For many developments like the Walmarts of the world, the list of prospective sites goes well beyond a single community. The question in this case is where to locate the next store in the US or Canada. Sometimes, the decision to develop in one locale will cause loss to another. In some cases, it may be the adjacent community, where tax rates are lower and bare land is cheaper. Alternatively, the retailer might prefer to locate in a community in which the demographics are more likely to generate higher revenues.

For many businesses, the choice to locate in one city rather than another is not based on any specific attachment to a specific community. Ultimately, it is the city that offers the greatest opportunity for economic gain now and in the future that wins out.

## The supply side of the equation

The prior examples remind us that the land market is not a simple case of supply and demand. As Will Rodgers, singer and poet, stated, "Land, they aren't making any more of it" – put simply, the supply of land is fixed. Unlike manufacturers, who can always produce more widgets if the prices go up, we can do little to increase the amount of available land or where it is located. What we can change is the mix of uses from agriculture to urban. Over the last two centuries, this change has been most dramatic in North America. In 1790, 2% of the US population was living in cities and town, while in 2010, 75% of the total population was urban. The growth in urban development over two centuries was largely a response to increased productivity in the agricultural sector, the growth of basic industry, and immigration. In addition, the expansions of transportation systems, first with canals and later with rail, highway, and air travel, contributed to urban development. With the growth of regional and interregional trade, urban populations became more concentrated in large metropolitan centers. The emergence of New York, Boston, Philadelphia, and Chicago as major trading centers expanded growth of their boundaries well beyond their original plans. Where cheap land was available along the fringe of the cities, towns and villages of a few thousand expanded to cities of over 1 million during the course of the 19th and 20th centuries.

For cities like New York, Chicago, San Francisco, and Vancouver, with severe geographic constraints of rivers, bays, and mountains, development was forced to build up rather than out. In these commercial centers, land prices in the inner core were at a premium. With the invention of the skyscraper, a new building form was created to respond to these new economic realities. By relying on steel frames instead of wood and masonry, and elevators instead of staircases, developers were able to spread out the cost of land over a larger floor area. Cass Gilbert, architect of the Woolworth Building, an early skyscraper completed in 1913, is reported to have remarked, "A skyscraper is a machine for making the land pay." Chicago and New York were two of the first cities in which tall buildings emerged in the downtown commercial centers. Beginning in the late 19th century, the leading architectural firms of Burnham & Root, Adler & Sullivan, and McKim, Mead & White established American supremacy in skyscraper design.

As cities grew in commercial presence, there emerged a need for new residential development that could accommodate a growing workforce. With the construction of commuter railroads and streetcar lines that extended the ability to commute beyond the walking distance of colonial cities, the suburbs on the fringes of the city provided housing for a growing middle class. Although constrained by the Hudson and East rivers, by the mid 19th century New York's population had already expanded to the five boroughs and communities to the north, with ferry service and rail lines. With the completion of the Brooklyn Bridge in 1884, the Path Train to Hoboken in 1908, the White Stone Bridge in 1913, the Holland Tunnel Path Train in 1927, and the George Washington Bridge in 1933, New York City could increase its workforce without the burden of providing housing on Manhattan Island. After World War II, with the completion of the Garden State Parkway and the Verrazano Bridge, new communities in New Jersey and New York provided additional lands for suburban development, from where many residents still commute daily to New York City. The issue of providing affordable housing for those who work in New York City has always been at the forefront of politicians' minds and, in 2014, became a major campaign issue in the mayoral election. Finding a solution to this century-old problem continues to be difficult, given the high costs of land and construction. As cities grow in wealth, residential properties considered marginal are replaced over time by expensive condos and chic residential neighborhoods, leaving the poor to seek housing in more depressed areas.

The location of future development is constrained by a number of factors. Supply of land available on the fringes, regulations, and the cost of extending municipal services all influence the supply of land that is considered available for development. Cities such as Denver, Houston, Dallas, Fort Worth, and Calgary, with land available on the urban fringe, have been building new suburban communities since the end of World War II. In many North American cities, new opportunities for suburban growth have been provided by the conversion of farmland. What is evident from the history of urban expansion is that cities rarely place constraints on growth. Voluntary constraints on growth are a rare exception to this rule. In 1979, as part of a statewide growth management plan, Portland established a greenbelt that would limit future development. Though a goal of such programs is to encourage density and promote urban sustainability, they have also resulted in a less affordable housing stock for residents. A preference for single-family homes, along with a constrained land supply, has resulted in increased housing prices.[25]

Only in recent years has the desire to restrict suburban sprawl resulted in changes in policy and zoning that encourage greater densities within municipal boundaries. One limit on continued growth beyond the existing boundary is the cost of infrastructure. Roads, sewers, utilities, schools, hospitals, and other community services must accompany the construction of new communities. Construction of new homes on the urban fringe comes to a standstill without serviceable land to develop. To help defray the cost of providing services to new communities within a city, municipal governments require permit fees of developers. By placing a greater share of the cost of improvements on developers, inner city development becomes more attractive for redevelopment. Where funding exists for remediating brownfield sites, sites inside the city limits are now open for redevelopment where it was not possible prior to government involvement. Over the last four decades, cities such as Toronto, Philadelphia, San Francisco, Buffalo, and Boston have used a variety of vehicles to share the risk of development on largely ignored sites no long used for industry or transportation.[26] Pittsburgh, New York, Toronto, Montreal, Vancouver, and Boston are examples of cities that have removed railroad yards and highways along waterfronts, as well as demolished old steel mills, shipping terminals, and grain silos. By doing so, they have created a land supply for future development within older urban centers. In Boston, at a price of over $3 billion, the Big Dig has re-established the link to the waterfront from the older commercial centers where Nathaniel Hall and Government Center are located. In New York City, the demolition of the East Side Highway has created new opportunities for residential and commercial development with views of the Hudson River. With changes in municipal policy, we have begun to see a new mix of housing in the inner core in cities such as Portland, Seattle, Calgary, Vancouver, and Toronto. Single-family homes are being demolished to make room for townhouses and multifamily condos and apartment buildings. However, the cost of redevelopment is not insignificant, and these new developments come with a hefty price tag that often only a few can afford. In communities like Vancouver, where developers are required to create affordable housing as part of their housing scheme, there can be some hope that redevelopment will be equitable. However, given the recent escalation in real estate prices in Vancouver, finding housing will be a challenge for most.

### *Supply constraints*

Arriving at an actual number for the amount of land available for development in a community is not always simple task. Though all properties, in a sense, are available for purchase at the right price, listed properties provide only a rough approximation of what land is actually for sale. Commercial, industrial, and multifamily residential properties are bought and sold every day with little publicity or attention from the press, except for signature properties such as the Waldorf Astoria or Harrods. It is only in American films of the 20th century that you see railroad barons trying to buy all the lands adjacent to a new rail line to establish a monopoly on land pricing. A variety of constraints act as filters on the supply of land. A developer may be looking for a site to locate a fast food restaurant or a coffee-and-bake shop, only to find few sites that fit. The location may need to be on a major boulevard or be in close proximity to complementary uses. The community surrounding the site must have a demographic composition that supports their use. All of these constraints operate to reduce the number of potential sites the developer can even consider for their new project. In addition, if lots are not appropriately sized or zoning is highly restrictive, little development may occur even when there is an abundance of building sites.

#### *Geographical constraints on supply*

The land available for specific uses can be limited to a relatively small area within a region or a country. For example, land with access to rivers and lakes for transportation can be unevenly distributed. Even along a waterway, the adjacent terrain may restrict safe harbors to only a few locations. Natural deep-water ports are quite rare along many coastlines. Along the west coast of the US, there are only a few locations where a ship can safely anchor, Los Angeles, San Diego, Seattle, and Oakland being the most important. Mountains and inland waterways can restrict the supply of land adjacent to these protected harbors. Manhattan Island, San Francisco, and New Orleans are examples of port cities that have geographical advantages, not only because of their sheltered deep-water ports, but also because of their location at the terminus to a major river. The last century has seen a concentration of shipping activity resulting in fewer of these harbors being needed today; only those that can accommodate the mega-container ships and bulk handlers are still viable. Cities with harbors that are no longer suited to the demands of today's ships have transformed aging industrial areas into new waterfront communities. These harbor cities that once featured warehouses and industry have been transformed into waterfront communities with hotels, retail, and commercial office space. Once berths to tall-masted ships sending goods around the world, these harbors now are the homes to yachts and cruise ships. Tourists and business people, rather than bulk goods and manufactured items, have become the imports of choice in cities such as New York, Boston, Baltimore, San Francisco, Seattle, and Vancouver.

In the 19th century, communities in North American with access to navigable rivers and water power were choice locations for industrial development. Cities that emerged in the 19th century as major trading and manufacturing centers had logistical advantages because of their proximity to navigable waterways. Troy, located near the junction of the Eire Canal and the Hudson River, was where Burden Iron Works was established, which produced almost 60% of all iron products prior to the Civil War. It was positioned in a strategic location. Access to water power, raw materials, and New York markets made this location ideal for the production of cast iron, horse shoes, and iron plate. However, with growing competition from other producers in Pennsylvania later in the century, Troy declined as a leading iron manufacturer. Today, the warehouses along the river that once stored products heading to market are now empty. No longer serving their intended use, the mills and iron works have long disappeared, leaving only a patchwork of abandoned industrial sites. In Troy, only the fine townhouses, the banks, an opera house, and Rensselaer Polytechnic Institute, the first privately endowed engineering school in the US, stand as testimony to its former industrial prominence. The industrial activity of a former age has vanished.

Where navigable waterways did not exist, canals, railroads, and (later) highways were built to connect manufacturing centers to markets. The concentration of commercial power in cities such as Chicago, New York, and Toronto occurred in part because of the nations' transportation networks. Early in the history of these cities, the availability of land was constrained by local transit. Poor people walked, and the rich rode in carriages. However, by the late 19th century, entrepreneurs were able to expand within a relatively scarce amount of desirable land by building and operating public transportation systems. Early streetcars in cities across Canada and US were constructed to differentiate lands owned by development companies. With their completion, new suburban communities could be constructed, catering to a growing middle class seeking refuge from the bustle of city life. Though often located only a few miles from the central business district (CBD), without the advantage of public transit these lands would have remained largely undeveloped until after World War II, during the age of highways and commuting by autos.

### *Zoning, by-laws and other planning controls*

Zoning and other land use regulations, including architectural guidelines and historic district controls, limit the supply of land that can be developed under current economic conditions. These regulations can act as filters, constraining possible design solutions for a specific parcel. In effect, not all land is the same. By-laws operate to restrict the uses and also the scale of a proposed building. The underlying purpose of zoning is to reduce the impact or nuisance to third parties by limiting where industrial, residential, or commercial activity can take place in a community. A zoning map and by-laws provide a comprehensive description of both the permitted and restricted uses by type of development. In addition, by-laws establish the dimensions of any architectural solution proposed for the site. Property lines, setbacks, and maximum height and coverage all operate together to define the design solution. These constraints can be particularly difficult to overcome on small and odd-shaped lots, where setbacks may potentially leave little land for actual development. This problem can be further complicated by constraints placed by laws on the location of curb-cuts. For corner sites, finding an acceptable location for a curb-cut may make it all but impossible for a small parcel to see development. Though these corner sites are usually preferred by gas stations and fast food restaurants, making them highly desirable, the constraints of a zoning by-law may bar any development opportunities. In cases in which the shape of a lot would be just cause for applying for an exception, the process of application and review of the requested variance can take significant time.

Delays and indecisions merely add to the uncertainty attached to any proposal. With alternative sites often available in a city, it is not unusual to find many sites vacant for many years. Though not an argument for eliminating by-law restrictions, planners and elected politicians must recognize that regulations can have unintended consequences. In this case, valuable sites are often vacant eyesores for years and produce neither benefits to the community nor provide tax revenues to support public services.

Architectural guidelines and historic districts also act to limit what can be built within a community. In historic districts, usually only the interiors of buildings can be renovated without the approval of the local commission. Unless a use can be accommodated inside these older buildings, the high cost of renovation will ultimately limit their future. In 1976, the passage of the Historic Tax Credit Act was designed to offset some of the costs of historic preservation and reuse by providing tax credits for renovating buildings at least 50 years old. Architectural guidelines can also limit supply by placing restrictions on the details of a design solution for a site. When these guidelines are administered by committees that have little sensitivity to the costs and benefits of these guidelines, they serve only to discourage development and ultimately to constrain the supply of land considered viable for development. This is not a vote against historic districts. For many towns and cities, preserving their historic architecture may be their only hope. As a destination for weekend travelers and shoppers, historic streets with B&Bs, specialty shops, restaurants, and coffee shops may be the only viable strategy in an age of low-cost motels, shopping malls, and big box stores.

### *The case of the unreasonable seller*

Clearly, price matters in the negotiation of a sale. We assume that the owner has determined an asking price close to market value; otherwise, why would the property be listed? During negotiations, sellers and buyers must ultimately agree on a value that is underpinned by the state of the local market. For example, if a property is intended for condos, then the buyer cannot pay a price for the site on which, after the demolition of the existing building, and the design and construction of the new condos, the seller cannot make a reasonable profit. In some cases, the seller neither is knowledgeable of this principle nor cares about what the current market can support as development on the site. Land that is priced artificially high is outside the margins of development. This can happen when the existing property owner has operated a business at that location for many years. Without a mortgage on the property, a business is more likely to generate significant cash flow than can those with hefty mortgage payments. Not taking into account all the costs the buyer will have to endure if they are to acquire and develop the property, the market value of a property can be exaggerated in the eyes of the seller. When the present owners are looking at retirement in a few years, the anticipated windfall that may never come fuels an exaggerated price that places the property out of reach for any serious buyer.

The unreasonable seller can be a particular problem for developers attempting to assemble a large parcel. Even when multiple corporations are used to buy up the parcels so as not to alert the community, eventually an awareness of a developer trying to buy up all the parcels on a block will escalate the price in the minds of those remaining property owners, and they will attempt to hold out for a higher price. The last seller, knowing the project cannot take place unless they are willing to sell, can hold out for a higher price than those offered at the beginning of the land assembly process. Cities looking to acquire a property for a public amenity or right of way can avoid this conflict by using eminent domain. However, for elected officials, the use of eminent domain can have damaging political consequences when their action is seen as unreasonable.

### *The case for climate change*

The case that climate change is a constraint on the future supply of land is difficult to argue against. Hurricanes, fires, and rising sea levels are disasters that we now expect to see as headlines on the news or as the setting for disaster movies. In 2013, Hurricane Sandy cost over $2 billion in post-hurricane repairs and clean-up in New York City alone. The combined totals of all costs for communities that suffered destruction during the hurricane came to over $12 billion, not including the loss of business. Historically, insurance provided a measure of financial stability when facing risks such as fire. Natural disasters pose a different set of risks. Many of these communities will not receive payment for the total losses resulting from the destruction of their communities from natural disasters. The cost of rebuilding in instances in which the threat of destruction remains will bar redevelopment of these sites in the future. Federal and state disaster relief packages cannot turn back this tide.

It is estimated that over 70% of the world's population lives at elevations below 50 ft. With a projected sea level rise of 2cm a year, many of the world's great cities will either have to invest at levels that are almost impossible to justify to preserve coastal development or the result of rising sea levels may be that they abandon lands that once were important to the life of the community. With insurance companies unwilling to assume future risks, it is unlikely these communities will be rebuilt. Without national programs to manage this risk, it is doubtful that much land only a few meters above sea level will offer viable sites for development in the future.

## The demand side of the equation

In every transaction, there are both buyers and sellers. Demand for development sites is largely driven by the regional and local economies. The source of this demand is not uniform across different types of development. We must differentiate between retail, commercial, residential, industrial, and public space. The demand for each space will be driven by the unique character of the local economy. Though all communities must have buildings that are dedicated to residential, retail, and commercial space, the mix will vary. Some uses do not differ much from one city to the next. We all buy groceries, get our prescriptions filled, see our doctor or dentist, and go to city hall for a permit or a community meeting. Clearly, though, the character of every city, town, and village is reflected by the amount of space dedicated for work, shopping, and recreation.

Factors that influence the overall mix (measured by developed area) include the size of the population, the composition of employment, industrial types, transportation services, and the proximity to other cities.[27] Undoubtedly, a community that functions as a distribution hub will need more warehouse space, while a community that focuses on research and education will see more research parks, incubators, and university campuses. Communities that are isolated on the landscape will look different from those that are in close proximity to larger metropolitan centers. If you are able to drive easily to the next community for basic services, then your community need not provide as much space for shopping, health, and professional services.

### *The demographics of urban development*

Population and employment are always fundamental to understanding the direction of real estate development. As key indicators, they can be used to foretell how demand will increase for residential, commercial, and retail space. Understanding how to apply numbers to these factors requires that the population be decomposed by groups and age cohorts. How many young people are coming into the job market? What is the age of marriage (family formation)? When do families have their first child? How many children are in a household? How large is the retirement population? These statistics provide insight into the future demand for residential development: condos for young professionals, starter homes for young families, and retirement living for those who want to downsize.

For communities experiencing rapid growth, satisfying the demand for new homes can be a challenge. Even a single-family home can take over a year to build from start to finish. When housing developments are at the scale of a new community, it can be years before proposals are approved and the first homes are ready for sale. Communities that are stable in number may still have unsatisfied needs, as the age structure of the population changes. As the community ages, the need for retirement communities and healthcare facilities may put strains on other parts of the real estate market. If there are no first-time buyers able to purchase the homes of retirees, a community can experience a shortage of condos and retirement housing, while experiencing an increase in the number of available single-family homes and a decline in their value. Balancing the demand and supply of housing, retail space, and social and medical services is difficult, if not impossible, in an environment of economy in flux. Without government assistance, local economies may find it all but impossible to find the financial resources needed to make the investments in infrastructure critical to the economic health of the community.

## Demand and supply in a changing technological landscape

Over time, the economic underpinning of cities can adapt to a changing technological landscape. Cities that survive and thrive are able to build on their historic past while creating new economic opportunities for growth and development. Those cities that cannot adjust to change stagnate or decline in importance. Over the last century, industry has become much less dependent on water traffic as highways, railways, and airports give community access to markets. Communities that have been able to capitalize on advancements in technology have been able to develop new growth sectors, such as aerospace, IT, material science, pharmaceuticals, and genetic engineering.

Technological change has always played a significant role in shaping those cities that began to industrialize in the 19th century. The mill, tenement, and train station were some of the new building types associated with growing industrial centers in Europe and North America. Later in the 19th century, with the growth of service jobs in the US, architects and engineers began to create skyscrapers, department stores, apartment buildings, and tenements; all developed to accommodate a growing population of office workers, shopkeepers, middle managers, and service workers. In the last four decades, with the emergence of an IT sector, we have seen new building forms as the internet, information management, and the cloud have transformed many traditional jobs associated with an industrial economy. Centralized shipping and warehousing has created a vast number of automated warehouses often located at the confluence of airports and highways. Research parks where computer programmers create the next app or special effect are found primarily in Los Angeles and Vancouver. Server farms have been located in areas where there is cheap real estate and low energy costs. The information sector of the economy has created a need for space for back offices and call centers. In the creative industries of film, TV, and video games, many industries have located in communities with high-value amenities as a part of a strategy to attract young, talented workers. Communities such as Troy, NY, Bethlehem PA, and Halifax, NS, with older infrastructures built to service the industries of the last century, will find it difficult to update and renovate buildings. Though older warehouses and mills can be converted into work studios and space for startups, the largest concentrations of Fortune 500 IT companies are located in only a few major cities in North America.

In communities where the delivery of medical services is a major focus, hospitals, research labs, testing facilities, pharmacies, and medical supply and rehabilitation centers dominate the urban landscape. In communities with a high-tech focus, universities, incubators, and research parks are the generators of economic activity and provide the impetus for the development of retail, cultural, and recreational opportunities. More recently, universities and colleges are building retirement communities that will draw upon growing alumni who desire cultural and educational opportunities during their retirement years. Obviously, a community that is a university town will look different from that of a city devoted to manufacturing. Over the last century, the relative allocation of land by industry reflects the change in employment opportunities in North America. The cities of Raleigh, Durham, and Chapel Hill in North Carolina, which make up the "Research Triangle", look different from cities such as Detroit, Toledo, and Akron. In San Mateo County, where you find the homes of Google, eBay, Facebook, and numerous startups, it is the office parks and not the factory that dominate the landscape.

For older communities, the transformation from the last century to the present is not always simple. Buildings of the past, with their load-bearing masonry walls, are not easily adapted to these new floorplans. For example, today's average-sized retailer requires larger spaces than those created in the last century. A drug store today, with shelves devoted to cosmetics and beauty products, requires far more space than did its counterpart a century ago. Even in cases in which buildings are potential candidates for reuse, the present-day need for elevators, air conditioning, and modern electrical systems may make the cost of renovation prohibitive.

## Financial factors

### *Lending*

When buying a house, the ability to borrow is a determining factor in the purchase of a home. Given the future of long-term interest rates, lenders will establish a borrowing limit based on an individual's credit rating and income when agreeing a loan so that the borrower can buy a home. All financial lending institutions establish some criteria for a minimum debt-to-income ratio to calculate the affordability of debt payments. When interest rates are low, the actual amount that can be borrowed increases, given that less of the monthly mortgage payment (PMT) is for the actual interest payment (IPMT). In falling interest-rate markets, individuals are able to purchase more expensive homes, and the real estate market benefits as whole. However, in sustained rising interest-rate markets, the effect is just the opposite, and real estate markets take a turn for the worse. Though often following business cycles, these ups and downs in the interest rates ultimately impact the real estate market.

Given the stimulus or drag that interest rates can have on the real estate market and on the general economy, governments will use fiscal and monetary policy as a tool to stimulate or dampen investment. To the extent that interest rates are linked to global markets, capital may seem expensive or cheap from the individual's perspective. In countries that suffer from poor fiscal management, interest rates will reflect the creditworthiness of the government and its ability to borrow in global markets. Argentina, Venezuela, Greece, Spain, and Italy are countries that, in recent years, have had to pay a premium on their national debt because of depressed economies or poor fiscal management. In contrast, countries such as Japan, German, Norway, and the US have borrowing rates not seen in 60 years. Ultimately, competition for capital is both global and local. Investors will search for opportunities with the highest rates of return, though rates of return must be weighed against risk. An investor may acquire commercial office space because the forecast is higher for rental rates than for other real estate developments such as hotels or retail outlets. If the investor is a pension fund or a real estate investment trust (REIT), management can decide to purchase commercial office space in markets that show the greatest promise. Even in local markets, capital will flow to those projects with the best prospects of being profitable over the long term. Given the competitive nature of borrowing, ultimately, local lending practices will favor those projects with the best return and the least amount of risk.

### *Measuring demand*

Real estate development must always be placed within the larger context of economic growth and development. Over the last 40 years, real estate development has represented a significant percentage of GDP, ranging from 4% to 6% in Canada and 3% to 5% in the US. As an economic indicator, new construction is often a major contributor to growth. Often, by both economists and politicians alike use GDP as a yardstick for measuring the success or failure of national policies. During the economic recovery under Presidents Reagan, George H. Bush, George W. Bush, and Obama's first term in office, the US experienced a jobless recovery while GDP grew. Gains in employment were weak during this period. Employers were able to push for greater productivity, avoiding the need to hire as the economy improved. In an environment of high unemployment, recent hires are not likely to complain about long hours. This illustrates the problem of estimating the need for residential, industrial, and commercial space based on GDP. Calculating the need for new development even during periods of economic expansion can be tricky. Understanding how economic measures can be used to estimate the demand for real estate requires knowledge of structural changes in the economy. It is possible to have a rising GDP per capita without any significant gains in employment or wages. Under these conditions, economic growth may not result in increased demand for new residential development.

The IT revolution is also transforming middle management. Over the last few decades, databases and AI have eliminated jobs as smart applications replace much of the work middle managers did a decade or two ago. Like robots replacing skilled workers in manufacturing, smart applications means fewer well-paid jobs in finance, banking, and manufacturing. The hollowing out of the middle class reflected by the rising value of the Gini coefficient, a measure of income inequality, means that more wealth is concentrated, while the legions of lower-income quartiles are expanded. Under these conditions, real estate development may show a strong demand for luxury homes and high-end retail, but little demand for new communities for lower- and middle-income households.

Given that all real estate is linked to location, understanding the complexity of the economy will be critical to determining the ultimate success of a project. Conditions can vary from town to town and state to state or province to province in Canada. Differences in labor, land, and construction costs will vary from community to community. Global developments can also be a deciding factor. An economy that is critically tied to the price of oil or wheat, or to the demand for automobiles worldwide, will have a future determined by events that sometimes feel a world away. Success in real estate development, like any capital investment, is linked to both local and international markets, both of which can be difficult to predict with any certainty.

### *Competitiveness*

For anyone who is a student of business management, Michael Porter's "competitive advantage" framework offers a useful schema for understanding the changing dynamics of supply of and demand for space in an urban environment. Michael Porter proposed five factors to explain the relative advantages of one nation over another: human resources, physical resources, knowledge, capital, and infrastructure.[28] Human resources are not merely labor, but also the level of education, skill, and management needed in dynamic economies. The knowledge base that is inherent in a community represents the collective intellectual capability of a nation to advance research and solve problems. Physical resources, though now less important than they were a century ago, still play an important role in many economies. Without oil and gas reserves, neither Texas nor Alberta would have developed into world-class centers in energy and resource extraction. The Calgary and Houston skylines, with their towering office buildings, would have never materialized without the emergence of an energy industry. In addition to physical resources, access to capital is always critical to economic growth. Through banking regulations, fiscal and monetary policy will always impact the cost of borrowing, while local practices can have a direct bearing on the availability of capital. Infrastructure that includes both transportation and communication systems is also critical to businesses because it impacts a firm's ability to connect buyers to suppliers.

Within a given environment, it is important to view clustering as a critical advantage. Rather than seeing it as a zero-sum game, cities that have created an environment that supports a particular industry such as software design will have advantages over a community in which there is less competition. A new software design company locating in Silicon Valley has ready access to skilled labor educated at universities in the region, including University of California, Berkeley, University of San Francisco, and Stanford University. Being part of the community, they also have access to a professional community with a shared knowledge base that is difficult to find in any other city in North America. Being in this community also provides greater access to venture capital and potential customers. According to Porter, if cities in the 21st century are to prosper, they must develop a strategy that capitalizes on their strengths. Promoting education, providing the necessary infrastructures, and encouraging an environment free of monopoly, while protecting intellectual property, are all important aspects of planning for the future.

Estimating the demand for labor in this changing environment is critical to understanding what development projects are needed to satisfy future demand. Calgary became an important center for oil and gas development because, in 1946, a recently completed office building was available in downtown. With a single enterprise moving to Calgary in 1947, a foothold was established, attracting other companies to Calgary. Having created this cluster, Calgary now boosts 10 million sq. ft. of space devoted to oil and gas companies. Estimating the future demand for space is always tricky. We can rely on estimates based on historical trends, but during times of dynamic change this may not reveal how changes in technology, demand, and lifestyle will be reflected in the demand for a different or new kind of space.

## Building the team

Ultimately, it is the developer who seizes opportunity that is successful. A development proposal would never become a reality were it not for a risk-taker – an entrepreneur willing to create a plan, work with city officials, organize a team of experts, secure funding, and oversee construction. On smaller projects, there may be fewer players, but the tasks are the same. For example, an older home built in the 1950s might be a suitable site for building a small apartment building. For the developer who takes on this project, the building may be their only venture. The process may have begun when the developer was driving around the community, noticing older 1,600 sq. ft. bungalows from the 1950s being torn down and replaced by two 1,700 sq. ft. infill units. After meeting with a real estate agent who is well known in the area, the developer may have learned that these units are popular with young professionals and are selling quickly. Upon further investigation, the developer discovers a city website where the planning department promises a speedy approval process if basic criteria are met. There are even drawings and photos of projects similar to the one the developer saw earlier in the week.

The developer learns that the city hopes that, by cutting through the red tape on these conversions, the density of older neighborhoods will be increased. The city planning department hopes that this program will reduce the need to convert farmland on the edge of the city for new suburban growth by providing more housing options in older suburban neighborhoods. Ultimately, this increased density is part of a new sustainability program the city is promoting to developers.

Driving around the community, the developer sees a property for sale that might meet the criteria. It is an older home on a treed lot. Consulting the multiple listing service (MLS), the developer discovers the property is selling at a severe discount because of its condition. This is an ideal property for development, since the developer plans merely to tear down the existing building. For this developer, the building on this site has little value. In fact, it has negative value, since there will be a cost to have it demolished. By contacting a real estate agent, who shares his professional knowledge, the developer might learn whether interest in the property is low because of its derelict state.

During the next step, the developer may work closely with an architect or house designer on possible designs for the site, based on the city's guidelines. Given the scale of the project, a contractor is found, who may be able to provide a rough price estimate based on a sketch and photo of the proposed building. Luckily, this contractor is working on a similar project in the same neighborhood and provides a price that probably will be within 10% of the actual cost. The contractor plans to be finished with their current project in the next six months, so taking on a new project should not be an issue. In cases in which the developer and the contractor are the same individual, consideration must also be given to the availability of skilled tradespersons, who will be free to work on this project. The developer may know from experience that when development activity is high, finding available roofers, framers, electricians, plumbers, and tilers can be difficult. Like a conductor of an orchestra, the developer must see that these contributors perform their work at a specific time. If they perform out of sequence, the project could become a disaster.

With a concept plan, the developer can create a pro forma based on estimates of costs and an estimated selling price. The cost of borrowing, and fees paid to both the city and the real estate agent, are part on the expenses side of the ledger. Knowing that they will likely devote a minimum of 20–25 hours a week to this project over the next year, the developer's concern at this stage is, "Can I make a reasonable return, given the time that I will have to spend on this project over the next year?" Deciding whether to go forward with the proposal is also an issue, because concerns over an increase in interest rates before a construction loan can be secured could possibly delay completing the project by up to three months.

After contacting a banker who assures them that the rates on construction loans look like they should be stable over the next year, the developer remembers that, in the past, interest rates went up shortly before they started a project. Learning from experience can be very costly: after the expenses were paid, there was barely any profit. Working for free was not the plan. Eager not to repeat this case, the developer factors in a margin of 10% on the construction price, three months on the completion date, and a 5% reduction in the ultimate selling price – and the pro forma still looks good. There should be a reasonable return even in the worst-case scenario.

Calling the real estate agent, the developer makes an offer that is 15% under asking price. If there is any interest from the seller, the project can proceed according to plans. In the next week, the plan is to call the banker and discuss borrowing terms and rates. Also, the developer will meet with the one of the planners to discuss the approval process for this type of infill project and to secure some assurances. If the seller does not accept the offer or is unwilling to negotiate, there is always the option of looking at other properties that are for sale in the neighborhood.

Though larger projects are far more complex, the preceding scenario identifies themes common to all developments. First, there must be demand for the proposed development. Any project that is going to be successful must have buyers or tenants. If the project is being completed for a specific client with terms and a contract price, the risk of selling "on spec" has been eliminated. However, if the property is being built as a speculative venture, there is always less risk when demand is strong than when the development is being built to satisfy a niche market. Developers are sometimes criticized for not being adventurous, often going with the tried and true. Satisfying the immediate and proven demand reduces the risk of not being able to sell the property after investing significant time and money. There must be a site that is suitable for the development. For each type of development, there will be minimum size and dimensions. If there are no available sites to develop, then the developer may be required to search for sites in other neighborhoods or communities.

All developments require the approval of the municipal authorities. Having the city planning department on your side will always help. Projects that are welcomed by city hall will receive fewer objections and will take less time to secure the necessary approvals. Where projects require community input and support, the support of planning department and local politicians will certainly help. There must be capital to finance the project. Without a bank or other financial institution willing to join in as a partner, the project dies. Additionally, there must be an economic gain in pursuing the project. This requires going beyond a mere accounting of the costs and revenues of a project. There is risk associated with every project on both sides of the balance sheet. Without a critical assessment of all the risks, it is likely that something will go wrong, and the project will fail. Finally, it is important to remember that all development projects are a team effort. Negotiation and project management skills are critical to the success of all projects.

## Notes

1 Real Estate[Def. 1] (n.d.) *Oxford English Dictionary Online*: Retrieved September 18, 2011.
2 US Green Building Council, Sustainable Building Technical Manual, US Department of Energy, 1996.
3 www.paragon-re.com/3_Recessions_2_Bubbles_and_a_Baby; also see data and graphs: www.vancitybuzz.com/2015/11/real-estate-vancouver-october/; www.paragon-re.com/trend/3-recessions-2-bubbles-and-a-baby; https://datahub.io/core/house-prices-us
4 www.cbc.ca/news/canada/british-columbia/vancouver-real-estate-house-prices-1.3564528
5 Bracha and Brown, 2015, 1: "When Alan Greenspan, then Chair of the Federal Reserve Board, used the term irrational exuberance to describe the behavior of stock market investors, the world fixated on those words. 1 He spoke at a black-tie dinner in Washington, D.C., on December 5, 1996, and the televised speech was followed the world over. As soon as he uttered these words, stock markets dropped precipitously. In Japan, the Nikkei index dropped 3.2%; in Hong Kong, the Hang Seng dropped 2.9%; and in Germany, the DAX dropped 4%. In London, the FTSE 100 was down 4% at one point during the day, and in the United States, the next morning, the Dow Jones Industrial Average was down 2.3% near the beginning of trading. The sharp reaction of the markets all over the world to those two words in the middle of a staid and unremarkable speech seemed absurd. This event made for an amusing story about the craziness of markets, a story that was told for a time around the world."
6 www.cbc.ca/news/canada/calgary/statistics-canada-unemployment-calgary-alberta-edmonton-1.3836707
7 https://fred.stlouisfed.org/series/HOUS448URN (Real estate Index)
8 https://datahub.io/core/oil-prices#resource-brent-year
9 https://datahub.io/core/oil-prices#resource-brent-year
10 https://datahub.io/core/oil-prices#resource-brent-year
11 https://en.wikipedia.org/wiki/Demographic_history_of_Detroit; www.metrotrends.org/spotlight/Detroit.cfm; www.epi.org/publication/the-decline-and-resurgence-of-the-u-s-auto-industry/
12 Bomey and Gallagher, 2013 [2018].
13 http://midwest.chicagofedblogs.org/?cat=26
14 www.rita.dot.gov/bts/sites/rita.dot.gov.bts/files/publications/national_transportation_statistics/html/table_01_15.html_mfd
15 See Bomey and Gallagher, 2013 [2018].
16 http://michiganeconomy.chicagofedblogs.org/?m=201401; https://en.wikipedia.org/wiki/List_of_countries_by_motor_vehicle_production
17 Rostow, 1962.
18 www.washingtonpost.com/wp-dyn/content/article/2009/03/12/AR2009031202480.html
19 www.forbes.com/sites/jonbruner/2012/03/05/ten-american-comeback-cities-map/; www.clevelandfed.org/newsroom-and-events/publications/economic-trends/2013-economic-trends/et-20130401-gdp-growth-in-us-metropolitan-areas-during-the-recovery.aspx; www.washingtonpost.com/wp-dyn/content/article/2009/03/12/AR2009031202480.html
20 Belko, 2017.
21 *The Economist*, 2013; Chapple, et al., 2004; Barton, 2011; https://en.wikipedia.org/wiki/Silicon_Valley

22 Beitel, 2007.
23 www.zillow.com/san-francisco-ca/home-values/
24 Barton, op. cit.; Bietel, op. cit.
25 Staley et al., "A line in the Land: Urban-growth Boundaries, smart Growth and Housing Affordability." *Policy Study, n.d.*
26 Superfund was established in 1980 under the Comprehensive Environmental Response, Compensation, and Liability Act of 1980 (CERCLA).
27 Beitel, op.cit.
28 Porter, 1980; Porter, 1990.

## Bibliography


Attoe, Wayne and Logan Donn. *American Urban Architecture, Catalysts in the Design of Cities*. Berkeley, CA: University of California Press, 1989.

Babcock, Richard F. *The Zoning Game, Municipal Practices and Polices*. Madison, WI: The University of Wisconsin Press, 1977.

Barton, Stephen. "Land rent and housing policy: A case study of the San Francisco Bay area rental housing market." *American Journal of Economics and Sociology* 70, no. 4 (October, 2011): 845–873.

Beitel, Karl. "Did overzealous activists destroy housing affordability in San Francisco? A time-series test of the effects of rezoning on construction and om prices, 1967–1998." *Urban Affairs Review* 42, no. 5 (May 2007): 741–756.

Belko, Mark. "Pittsburgh downtown partnership hopes to make Golden Triangle a vibrant retail destination." In *TCA Regional New.* Chicago, (15 March 2017).

Benevolo, Leonardo. *History of Modern Architecture, The Tradition of Modern Architecture*, Vol. 1. Cambridge, MA: The MIT Press, 1989.

Bomey, Nathan and Gallagher John. "How Detroit went broke: The answers may surprise you – and don't blame Coleman Young". *Detroit Free Press*, originally published September 15, 2013, updated July 18, 2018, www.freep.com/story/news/local/michigan/detroit/2013/09/15/how-detroit-went-broke-the-answers-may-surprise-you-and/77152028/

Bracha, Anat and Donald J. Brown. (Ir)rational exuberance: Optimism, ambiguity and risk.

Burg, David F. *Chicago's White City of 1893*. Lexington, KY: The University Press of Kentucky, 1976.

Chapple, Karen, John V. Thomas, Dena Blezer, and Gerald Autler. "Fueling the fire: Information technology and housing price appreciation in the San Francisco Bay area and the twin cities." *Strategic Economics; Strategic Economics Source: Housing Policy Debate* 15, no. 2 (2004): 347–383.

Haar, Charles M. and Jerold S. Kayden, eds. *Zoning and the American Dream*. Chicago, IL: Planners Press, 1989.

Collins, Richard C. "Changing views of historical conservation in cities." *Annals AAPSS* 451 (September): 86–97.

Cooke, Jason. "Compensated taking: Zoning and politics of building, height regulation in Chicago 1871–1923." *Journal of Planning History* 2, no. 3 (September, 2016): 207–226.

Cuthbert, Alexander R. "Urban design: Requiem for an era – Review and critique of the last 50 years." *Urban Design International* 12 (2007): 177–223.

Davis, Morris A. and Francois Ortalo-Magne. "Household expenditures, wages, rents." *Review of Economic Dynamics* 14, no. 2 (April, 2011): 248–261.

Kaiser, Edward J., David R. Godshalk, and F. Stuart Chapin. *Urban Land Use Planning*, 4th ed. Urbana and Chicago, IL: University of Illinois Press, 1995.

Ellickson, Robert C. and A. Dan Tarlock. *Land-Use Controls, Cases and Materials*. Boston and Toronto: Little, Brown and Company, 1981.

Forester, John. *Planning in the Face of Power*. Berkeley, California: University of California Press, 1989.

Forester, John. *The Deliberative Practitioner: Encouraging Participatory Planning Processes*. Cambridge, Massachusetts: MIT Press, 2000.

Hirt, Sona. "Home sweet home: American residential zoning in comparative perspective." *Journal of Planning Education and Research* 33, no. 3 (2013): 292–303.

Grant, Jill, ed. *A Reader in Canadian Planning*. Toronto, ON: Thomson. Nelson, 2008.

Kwartler, Michael. "Legislating aesthetics: The role of zoning in designing cities." in *Zoning and the American Dream* (187–220). Chicago, IL: Planners Press, 1989.

Levy, John M. *Contemporary Urban Planning*. Upper Saddle River, NJ: Prentice Hall, 2000.

Meck, Stuart and Rebecca Retzlaff. "The emergence of growth management planning the United States: The case of *Golden v. Planning Board of Town of Ramapo* and its aftermath." *Journal of Planning History* 7, no. 2 (May, 2008): 113–157.

Moore, Aaron A. "Decentralized decision-making and urban planning: A case study of density for benefit agreements in Toronto and Vancouver – Decentralized decision-making and urban planning." *Canadian Public Administration* 59, no. 3 (September, 2016): 425–447.

Papayanis, Marilyn Adler. "Sex and revachist city: Zoning out pornography in New York." *Environment and Planning D: Society and Space* 18 (2000): 341–353.

Porter, Michael E. *Competitive Strategy*. New York, NY: Free Press, 1980.

Porter, Michael E. *Competitive Advantage of Nations*. New York, NY: Free Press, 1990.

Rostow, Walt W. *The Stages of Economic Growth*. London: Cambridge University Press, 1962.

Shiller, Robert J. *Irrational Exuberance*. Princeton, NJ, and Oxford: Princeton University Press, 2015.

Staley, Sam, Edgens, Jefferson G. and Gerard, C. S. Mildner. *A Line in the land: Urban-Growth Boundaries, Smart Growth, and Housing Affordability, Los Angeles*: Reason Public Policy Institute, no 263, 1999.

Steven, Moga. "The zoning map and American city form." *Journal of Planning Education and Research* (June, 2016): 1–15.

Tarlock, Dan A. "Zoned not planned." *Planning Theory* 12, no. 1 (2014): 99–112.

*The Economist*. San Francisco, Growing Pains. 8865 (Saturday, December, 7, 2013: 48.

ULI, Urban Land Institute. *Multifamily Housing Development Handbook*. Washington, DC: ULI, 1999.

US Green Building Council, Sustainable building technical manual. US Department of Energy, 1996.

Wilensky, L. Harold. "The professionalization of everyone?." *American Journal of Sociology* 70, no. 2 (September, 1964): 137–158.

## Web resources

Information and technical assistance on the Americans with Disabilities Act of 1990
www.ada.gov/regs2010/2010ADAStandards/2010ADAstandards.htm#c1

# 2 An introduction to financial analysis

## Risk and reward

Every project begins with a financial analysis. Even a modest real estate development project requires a considerable investment in land, construction, and professional fees. To assure investors and financial lenders, a financial analysis based on best practice provides a measure of security. Fundamental to this analysis is the return and the associated level of risk. Logic would support the principle that the greater the reward, the greater the risk. All financial analysis must abide by this principle. Risk and reward have been a subject of interest by mathematicians since the 16th century, when Cardano, the gambling scholar, wrote his book *Liber de Lude Alea or Book on Games of Chance* in 1564.[1] Published almost a century later in 1663, his treatise provides anyone who plays games of chance with a set of probabilities of expected gain and associated risks. Investments, like games of chance, have a probability of gain associated with each state in a game. For example, anyone familiar with a deck of cards knows that there is a 1 in 52 chance of getting an ace of hearts at the first draw. Having this knowledge, we would probably not bet our entire savings on a 2.1% chance of seeing a positive outcome. However, if you are now down to the last card in the deck and the ace of hearts has not been played, you might actually bet your life savings on a sure thing – as long as you are confident this is a fair deck that contains an ace of hearts.

Most financial decisions are more complex than merely selecting a single card from a deck of cards. Often, predicting success in life is dependent on more than the outcome of a single event. For example, what is the probability of picking a queen of hearts from two separate decks in succession? The probability of this event is 1 in 52 multiplied by 1 in 52 – or 1 in 2,704. Clearly, if we need to predict the outcome of multiple events, we face a more difficult challenge. Real estate development is a complex process. Success may require that certain financial, economic, and political outcomes occur simultaneously, each of which is difficult to predict. We should not expect every investment to pay off as predicted, though, when we are presented with charts, tables, and projections showing favorable financial prospects, it is hard not to feel optimistic about the future. If a project's success requires low borrowing rates, low rental vacancies, rising rents, real estate taxes and to be solvent throughout the entire duration of the project, we might as well be asking for alignment of all the planets in our solar system. With such stringent requirements, it is unlikely our investment will generate the projected rate of return stated in the prospectus.

The level of risk someone is willing to accept varies based on their unique situation. Knowledge of the risks and rewards may change the decision on whether or not to invest. There are many factors that can influence the actual risk associated with a development project. Understanding the local economy or having the political skill to negotiate a contract can reduce risk.

There may also be government programs that share the risk, reducing the investor's exposure in the project. Diversification is also a strategy that we can use to reduce our risk. An institution such as a pension fund with a portfolio of properties located in different regions will be able to assume more risk than an individual who has all of their life savings invested in a single venture in their home town. Provisions in the tax code may also reduce risk by providing tax credits and deductions. The property being located in a community with good economic prospects can also reduce the risk of failure. Finally, to the extent that insurances can be used to offset potential losses from catastrophic events, our risk can be reduced; however, this is predicated on the existence of insurance that will provide protection from a catastrophic event. Flood insurance may not be available in areas prone to flooding, and, of course, the clause "act of war" exempts insurance companies from payment in areas of conflict. Even when insurance companies are willing to assume the risk of an adverse event, the high cost of insurance may force the decision to do without because paying premiums will bankrupt the venture. Like life, taking chances cannot always be avoided.

### *Risk-free rate*

In theory, the risk-free rate would be the rate of return an investor would be willing to accept if there were zero risk. Knowing the risk-free rate allows us to determine the premium an investor would accept to make an investment with a higher level of risk. Although all investments have some risk, we can use a class of investments that are considered by the financial

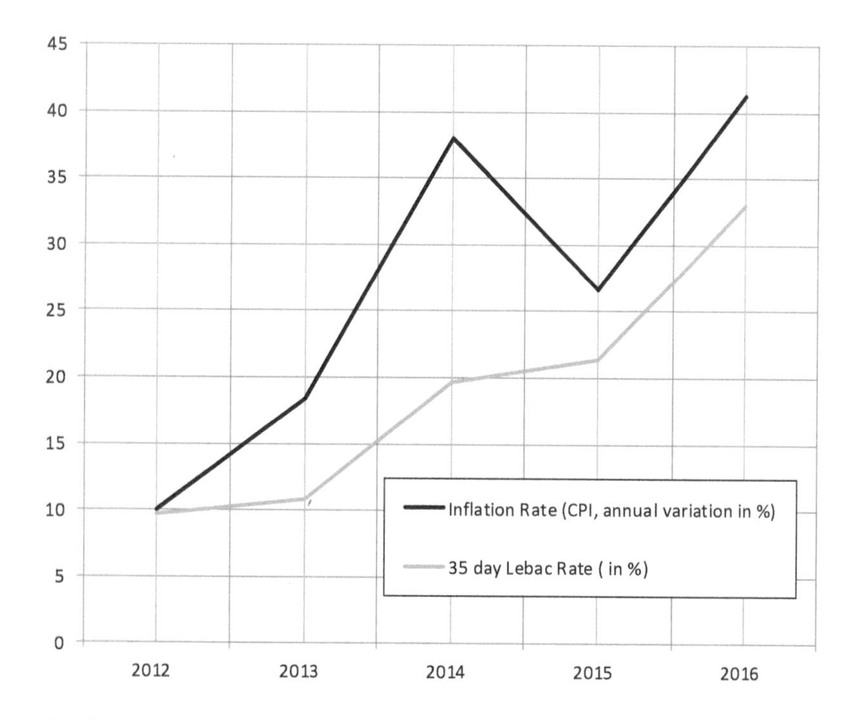


*Figure 1.2* Inflation rate and short-term interest rates in Argentina, 2012–16.

Data Sources: Focus Economics, n.d., www.focus-economics.com/country-indicator/argentina/inflation; Trading Economics, https://tradingeconomics.com/?ref=ieconomics.com/argentina-interest-rate

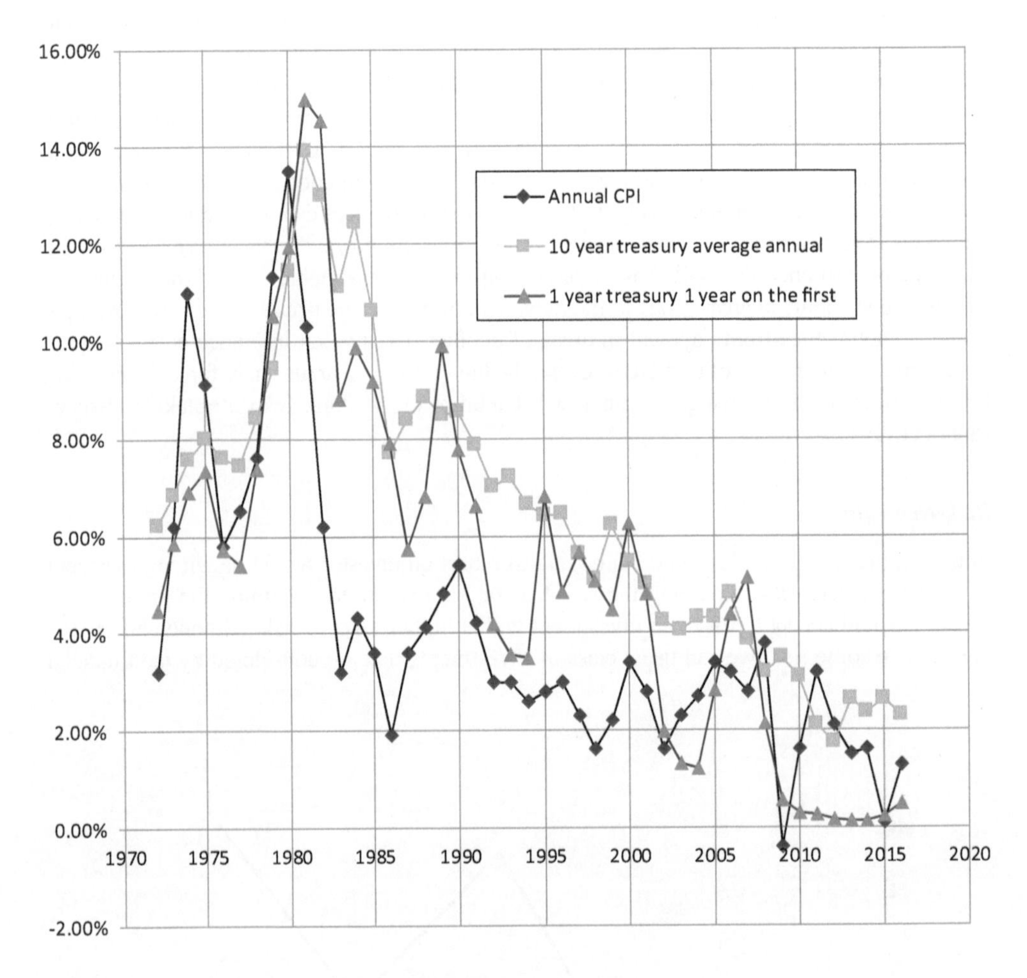


*Figure 2.2* Interest rates and inflation in US, 1972–2015.

Data Source: Bureau of Labor Statistics, US Department of the Treasury

community to have the least risk as our baseline for this measurement. The rate on short-term US government bonds is often used as a proxy for this measure. Though similar investments exist in all of the G20 countries, in times of economic stress and international political turmoil a flight to US Treasury bills often occurs, revealing that the world financial markets still considers these to be an investment with minimum risk. Of course, real estate is not an investment that we would consider buying and selling during a short period of time. Much of real estate is owned by insurance companies and pension funds that have ten-year or longer holding periods prior to their sale. Given this time horizon, a comparable risk-free return could be offered by the 10-year Treasury rate. Given the longer holding period, the investor is making assumptions about the future of interest and inflation rates. The risk of inflation can destroy any economic gain from the interest earned on a fixed-rate investment. Logic would inform us that as inflation rate goes up, investors will want higher returns to offset the devaluation of the underlying asset. Consider the yields on US vs Argentina bonds: the relatively higher rates paid by Argentina's central bank reflects a higher inflation rate (Figures 1.2 and 2.2).

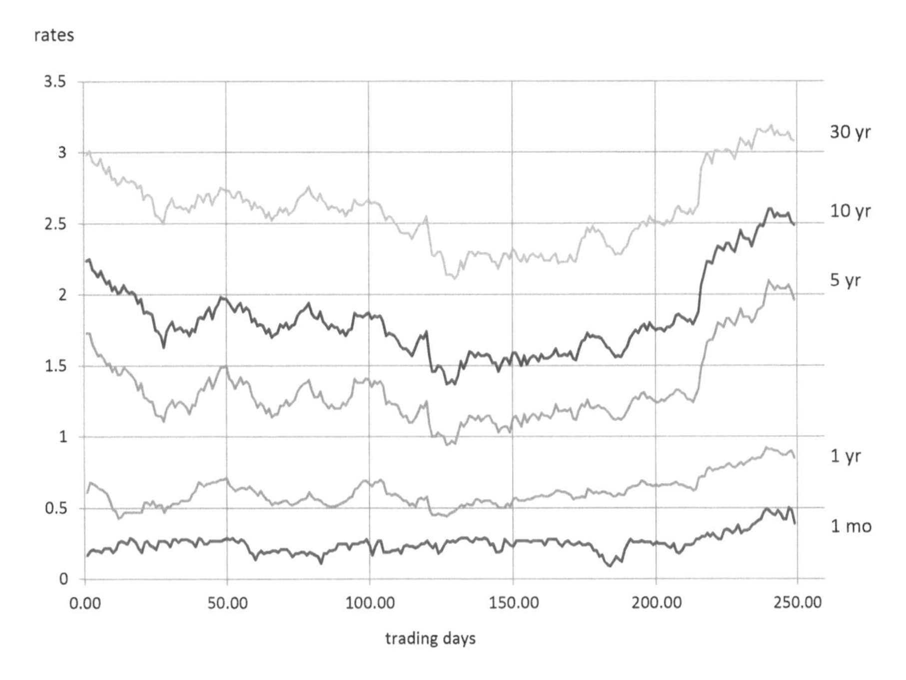


*Figure 3.2* Daily treasury yield curve rates, 2016.

Data Source: US Department of the Treasury

When deciding to invest in real estate, investors are assuming that interest rates will not go up substantially in the future, making their investment less valuable. When the Bank of Canada or the Federal Reserve raises interest rates, existing debt instruments such as bonds will have to be discounted to levels that will provide a return equivalent to the new issues. Likewise, investors will be willing to pay a premium for the same bond when there is a decline in interest rates. Inspecting the interest rates of Treasury bonds of different duration will reveal a clear differential between each period of repayment (Figure 3.2). For each period, whether 90 days, or two or ten years, this differential is almost a constant, reflecting the premiums that must be paid and the risk associated with future interest rates. Predicting interest rates ten years into the future will always be more difficult than a forecast of the next 90 days. For this reason, we would expect to receive a premium for tying up our money for a longer period of time. For any point in time, this difference between rates paid as a function of holding period is given by the "yield curve". During periods of stability, the yield curve will have a positive slope without any sharp breaks (Figure 4.2).

Using our knowledge of yield curves and risk-free rates, it is possible to consider the financial return on a particular class of investments. Acknowledging that the risk associated with real estate varies according to property type and location, it is possible to determine the premium that will be paid over the risk-free rate. For example, industrial properties with long-term tenants will exact a lower premium than will apartments, for which the turnover is greater and the vacancy rates are higher. In both cases, this difference is based on the history of returns for the specific class of properties in the community. One issue faced by investors

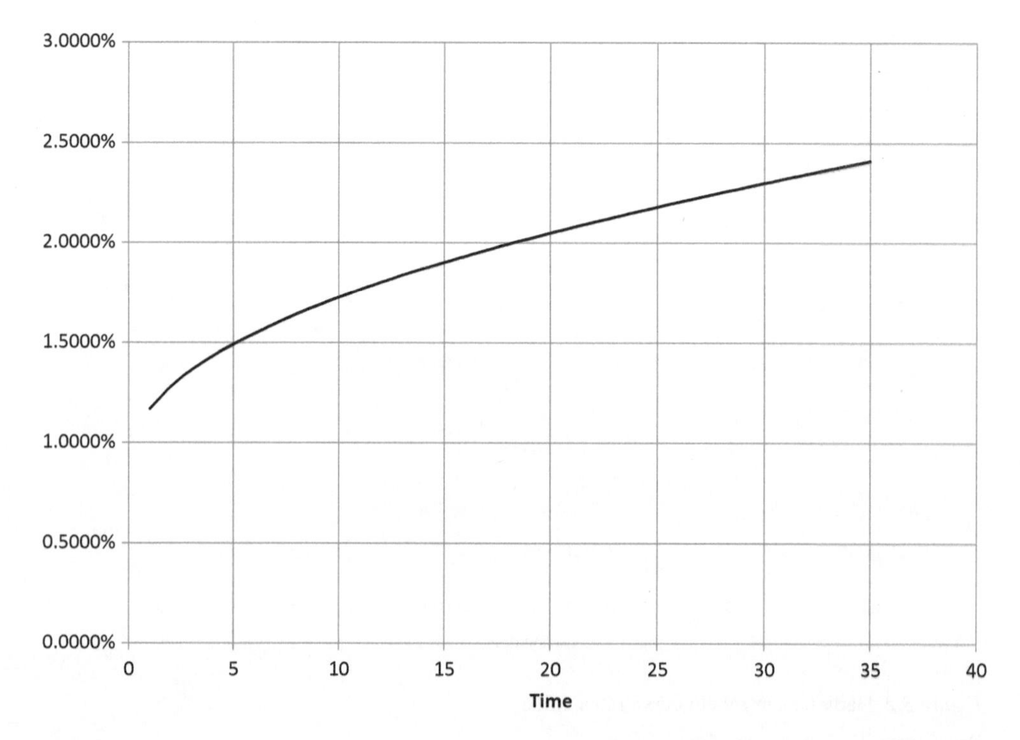


*Figure 4.2* Normal yield curve.

Data Source: Author

is acquiring the information needed to make this judgment. Obtaining rates of return for a specific city or region of the country may be complicated by the lack of published information that is both accurate and timely. Agency reports published in the US by the Department of Housing and Urban Development (HUD) and the Department of Commerce, and in Canada by Canada Mortgage and Housing Corporation (CMHC), will provide some guidance in making this evaluation. Non-governmental sources of data and trends can be found in National Association of Realtors (NAR) and Chambers of Commerce publications; these can provide valuable insight into the future of real estate development in a community.

One issue faced by investors seeking higher rates of return is the choice either to go long on the investment or to assume more risk. Investors looking for higher rates of return may be willing to accept the higher risks associated with lower-quality bonds. Ignoring the fact that high rates are being paid because there are actual real risks is foolish. In 2010, a US Treasury bond with 90 days' duration paid considerably less interest than a bond issued by a government with economic difficulties with the same duration. Defaults of bonds, though rare, have occurred, and when that happens, investors have received only a partial payment of their principal. Chasing higher return is normal in times of falling interest rates. Investors who have been acclimatized to higher yields in their search to maintain cash flows sometimes seek out higher returns on investments without discounting sufficiently for the associated risk.

### *Expected return and the evaluation of risk*

Expected return allows us to consider the potential gain and risks associated with different investments. Consider two cases: the first is an investment in government bond, and the other is an investment in a publicly listed company. In the first case, a government-issued bond with a guaranteed 4% per year presents us with little risk of default. In our hypothetical case, the annual rate of return on equities will be either +20% or –10%. A coin flip (50/50 chance) determines whether you have a 20% gain or a 10% loss. Clearly, depending on your time frame, you may elect for one investment over another. If you can wait several years before you need your cash, the investment in equities is a better choice. After all, if we are looking at multiple coin flips, our average return can be calculated assuming the probabilities of each event:

$$\text{Rate of return for equities} = (50\% \times 20\%) + (50\% \times -10\%) = 5\%$$

$$\text{Rate of return onTreasury bonds} = 4\% \times 100\% = 4\%$$

However, if you need your cash in one year's time, the thought of a potential loss of 10% would probably keep you from making this choice. In that case, the government bond looks like your only choice.

One assumption underlying this analysis that is often ignored is that investment outcomes and their associated probabilities are based on historical knowledge. Rates of return can fluctuate over time; it is difficult to predict exactly when peaks and troughs will occur (Figure 5.2). In the period 1955–2018, the prime rate in the US reveals

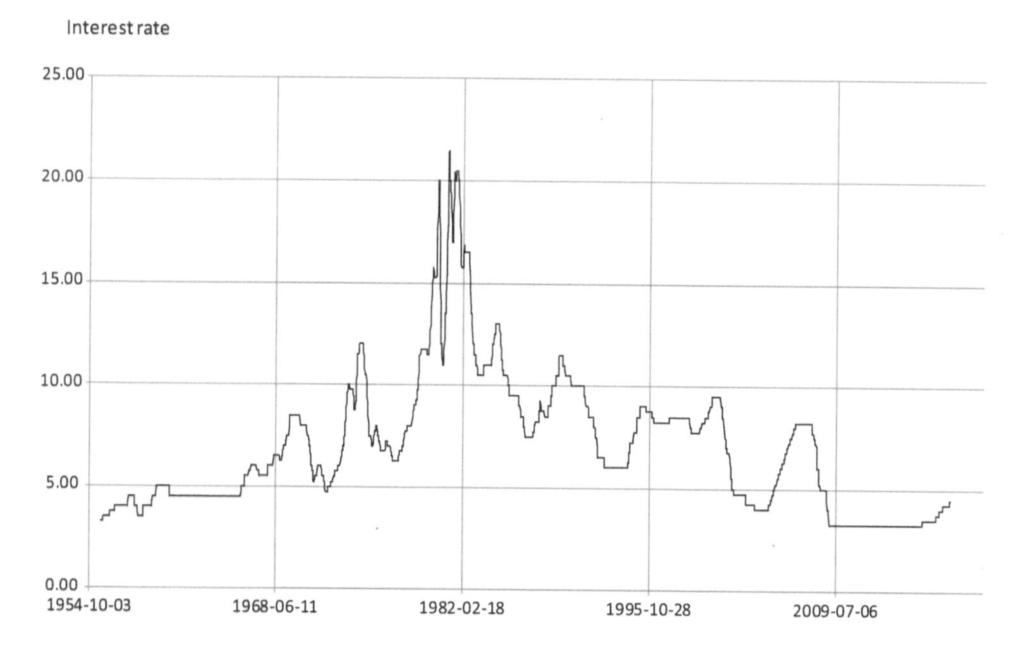


*Figure 5.2* Prime rate, 1955–2018.

Data Source: Federal Reserve, Economic Research

a long-term rising interest with rates peaking in the mid-1980s, and with several significant peaks and troughs occurring every ten years.

On several occasions, rates have spiked or have fallen dramatically due to external events and changes in domestic monetary and fiscal policy. Given the complexities of our political economy, it is very possible that the future will not look like the past, placing our projections and predictions in jeopardy. Often, dramatic changes in the economy can be brought on by a shock to the system due to oil embargos, wars or massive unemployment. A collapse of the world's largest banks can wreak havoc with the financial markets. Sometimes, these causal relationships will be apparent only to future historians. The recessions of 1929, 1986, and 2008 were not predicted, though some individuals certainly warned the financial community of impending financial collapse. In the recession of 2008, interest rates were kept low to stimulate the economy. For years after the initial crisis, property values continued to fall in many regions of the country because of a general weakness in the overall economy.

Changes in economic circumstances outside the initial projections can have catastrophic repercussions on a project. For example, planning for a condo development project may have begun when the cost of borrowing was low. If interest rates suddenly were to increase prior to securing a construction loan and property values were to fall because fewer buyers can afford the higher cost of borrowing, the results could be devastating. Of course, the converse is also possible: falling interest rates and rising price values will only improve our profit margins. Predicting the swings in interest rates and property values are all part of the game of chance. Historical knowledge can only provide some guidance in making our financial projections.

Clearly, the time horizon for any investment will influence our attitudes towards risk. As an investor, you can feel confident that today's economic climate will not change much in the next few days or weeks. However, we will have much less faith in any projection that goes beyond a year, much less 10 or 20 years. Developers are often criticized for not taking on projects that are designed to satisfy a new market or demographic., For the investor, the choice is between an investment in which there is much risk and less certainty about the financial returns and another that might appear to guarantee a definite rate of return. Like the investor choosing between equities or government bonds, if I am looking for a sure bet – one with little risk associated with the potential reward – I will favor investments with fewer risks even though there is an associated lower return. If I am the buyer of a property who plans to hold the property for 15 years, I can withstand some fluctuation in the return as long as the project remains profitable over the lifetime of the project. Like the investor in equities, as long as the return over the lifetime of the project is commensurate with the risk, I will invest. However, if I am the bank that is providing financing on the project, I will need some assurance that year-to-year fluctuation in cash flow will not impact the borrower's ability to make mortgage payments.

### *Measuring return*

The goal of any financial analysis is to ascertain the total return on a project. Real estate return can be generated from either the appreciation of the underlying asset or positive cash flow. In simple terms, total return is a function of both growth and income. Growth in the asset is realized at the time of sale, while net operating income (revenue minus expenses) is accrued on a periodic basis: monthly, quarterly or annually. Holding period return (HPR) is a simple measure of return. Commonly used to determine the overall return of bonds, it can also be applied to real estate. In the case of bonds, our face value is the amount the corporation or government will pay at maturity. The coupon rate will determine the actual interest paid each month, quarter or year, which payment is made until maturity. Because many bonds are purchased at either premium or discount rates, we actually may pay an amount greater or less than the face value of the bond depending on the prevailing interest rates. If, for example, I purchased a $10,000 bond yesterday paying 3.75% annually and now rates suddenly go up to 4%, I will have to offer a discount price on my bond to account for the difference in interest. An investor would not purchase a bond with a 3.75% coupon rate when 4.0% is the prevailing rate. If I need to sell my bond in the current markets, the price will have to fall to an amount that offsets the coupon rate of 3.75%. We can use the formula for HPR to calculate the simple rate of return. Using HPR, we are calculating the return based only on the price paid, the price the bond sold for, and the dividend received during a single period[2]:

$$\text{HPR} = \text{Holding period return} = (P_1 + D - P_0)/P_0$$

Where:

$$P_0 = \text{Beginning value of investment}$$

$$P_1 = \text{Ending value of investment}$$

$$D = \text{Dividend or cash flow}$$

Example 1

Determine the HPR for a bond with a face value of $10,000 that was purchased for $9,800 and sold for $10,200 with a dividend of 4%:

$$\text{HPR} = [(10{,}200 - 9800) + .04(10{,}000)]/9{,}800 = (400 + 400)/9{,}800 = 8.163\%$$

In this example, part of our return is from the appreciation of the bond. During our holding period, interest rates fell, making our bond more valuable in the marketplace.

Example 2

If, in this same period, the value of the bond were to fall to $9,200, the HRP would be negative:

$$\text{HPR} = [(9{,}200 - 9{,}800) + (.04(10{,}000)]/9{,}800 = (-600 + 400)/9{,}800 = -2.0408\%$$

Example 3

Of course, if, during this same period, my bond matures, I will receive my payment of $10,000 and my dividend payment of $400. My HPR is consequently as follows:

$$\text{HPR} = [(.04(10{,}000) + (10{,}000 - 9{,}800)]/9{,}800 = (400 + 200)]/9{,}800 = 6.1224\%$$

In real estate, we purchase a property for one price and hope to sell it at a profit in the future. During that period, we hopefully will have a positive cash flow.

In evaluating a real estate investment, it is important to consider its economic performance over its expected holding period. During that time, a series of positive cash flow payments will hopefully accrue. Using the formula for yield to maturity (YTM), we can calculate the approximate yield over which periodic coupon payments (dividend payments) are made. Applying the formula requires some trial and error by making a best guess for the rate until the price matches the actual price of the bond. Most financial calculators are designed to make this calculation.[3]

$$ApproxYTM = \frac{C + \frac{F-P}{n}}{(F+P)/2}$$

Where:

$$C = \text{Coupon/Interest Payment}$$

$$F = \text{Face Value}$$

$$P = \text{Price}$$

$$n = \text{years to maturity}$$

Example 4
If a bond is priced at $9,000 with a face value of $10,000, has a coupon rate of 5%, and matures in ten years, then we can calculate YTM as follows.

$$YTM = [500 + (10{,}000 - 9{,}000)/10]/(10{,}000 - 9{,}000)/2 = (500+100)/9{,}500 = 600/9{,}500 = 6.316\%$$

Often, we are making a choice between several potential investments. Having a formula that would help us make the better choice would be useful. With some knowledge of the markets and some best guess about future values, we could determine what we should pay for the investment. Using the formula for yield to maturity, book value or redemption value, we can calculate the present value stream of future payments. The formula for YTM can be used either to establish the present value of the investment or, if we solve for "r" (rate), to determine the rate of return, given values for the current purchase price, selling price, and dividends. The YTM formula allows us to determine if the asking price (P) is reasonable, given our assumptions about the future of the economy.[4]

$${}^{n}_{t=1}P = \sum C/(1+r)^{t} + F/(1+r)^{n}\ n$$

Where:

$$P = \text{price of the bond}$$
$$n = \text{number of periods}$$
$$C = \text{coupon payment}$$
$$r = \text{requiredrate of return on the investment}$$
$$F = \text{maturity value}$$
$$t = \text{time period when payment is to be received}$$

Example 5

a) What should I pay for an investment that is expected to sell for $10,000 in three years that is generating a 3% positive cash flows? In this example, I will assume that the investor would be satisfied with an overall rate of return of 2.5%.
b) What should I pay for the investment if my expected overall rate of return is 3.5.%?

a) 2.5% *overall return*

$$P = 300/(1+.025)^1 + 300/(1+.025)^2 + 300/(1+.025)^3 + 10{,}000/(1+.025)^3$$
$$P = 10,143$$

b) 3.5% *overall return*

$$P = 300/(1+.035)^1 + 300/(1+.035)^2 + 300/(1+.035)^3 + 10{,}000/(1+.035)^3$$
$$P = 9,860$$

Using this approach, I would not expect to pay more than $10,143 for the investment if I am satisfied with a 2.5%, but I should pay $9,860 if a rate of return of 3.5% is required.

Looking closely at this formula, if we assume there is no growth in the asset and our P = F, then our return merely equals the discounted coupon divided by the price we paid for the investment. This makes sense if we consider a case of real estate where the local economies are experiencing little growth in price. We might still buy the property because the cash flows may be significant when compared to the purchase price. Likewise, in fast-growing markets, we may have paid a premium for the property. After paying our expenses, there is little cash flow. In fact, it may be zero. Then, it is our hope that the increased value (F) will be significantly higher than our price (P) so as to generate a higher rate or return, offsetting the lack of yearly cash flow from the investment.

From this formula, it is easy to visualize that our return is merely a function of the purchase and selling price when the cash flow is zero. In the next chapter, we will consider other measures, including net present value and internal rate of return, both commonly used to measures the rate return on real estate investments.

### *Sharpe ratio*

In the case in which we have two investment portfolios with similar returns over the same period of time, which would be the better investment? The Sharpe ratio, named after William Forsyth Sharpe, a Nobel Laureate in Economics, can provide us with a number that allows us to compare two or more investments.[5] In simple terms, the Sharpe ratio is as follows:

Sharpe Ratio = (Mean portfolio return − risk free rate)/standard deviation of portfolio return

$$\text{Sharpe Ratio} = S = r_p - r_f/\sigma_p$$

Where:

$$r_p = \text{Expected Return on the investment portfolio}$$

$$r_f = \text{risk free rate}$$

$$\sigma_p = \text{Standard deviation on the investment portfolio}$$

As an example of how the Sharpe ratio would apply to real estate, consider a pension fund assessing two real estate investment portfolios: Which would make for a better investment, given the volatility of the market?

Example 6

*Portfolio 1*

$$r_p = 8\%$$
$$r_f = 3\%$$
$$\sigma_p = .15 \text{ or } 15\%$$
$$\text{Sharpe ratio} = (.08 - .03)/.15 = .333$$

*Portfolio 2*

$$r_p = 8\%$$
$$r_f = 3\%$$
$$\sigma_p = .1 \text{ or } 10\%$$
$$\text{Sharpe ratio} = (.08 - .03)/.1 = .5$$

In this case, the choice is obvious. If two investments have the same return, the preference would be for an investment that has less volatility or the Sharpe ratio with the higher value. However, sometimes the choice is less obvious. Consider the following example.

Example 7

*Portfolio 1*

$$r_p = 7\%$$
$$r_f = 3\%$$
$$\sigma_p = .1 \text{ or } 10\%$$
$$\text{Sharpe ratio} = (.07 - .03)/.10 = .4$$

*Portfolio 2*

$$r_p = 8\%$$
$$r_f = 3\%$$
$$\sigma_p = .15 \text{ or } 15\%$$
$$\text{Sharpe ratio} = (.08 - .03)/.15 = .333$$

In this case, Portfolio 1 would be given preference because of the substantially lower volatility over the expected lifetime of the investment. Even with a slightly lower rate of return, a Sharpe value of 0.4 makes it a better choice than Portfolio 2 (0.333). However, there are some important caveats with the Sharpe ratio. If calculations are based on daily returns, the standard deviation will be higher than those based on monthly returns where peaks and troughs are averaged out. In addition, the Sharpe ratio may not differentiate between periods of upswing and periods of decline. Risk and volatility are not the same. Using this calculation, large and sudden increases over a short period of time can produce large standard deviations. That said, the Sharpe ratio is another tool for assessing relative risk.

### *After-tax rates of return*

Whether it is better to have growth or income is not a simple question and will ultimately depend on the tax policies, the structure of the investment, and the desire for cash flow or future appreciation. Tax considerations will weigh heavily on the decision to structure the investment to produce positive tax cash flow prior to its sale. In the US and Canada, income from rental properties is taxed at different rates, affecting the gains made on the sale of a property (see Figure 9.2). In both countries, when the property is sold, capital gains exclusion of 50% cuts the effective tax in half. Though tax policy can change, clearly, if minimizing tax is the only concern, it would be preferable to structure the investment to produce no taxable income while it is being rented and see the profit as long-term capital gain at the end of the project. This does not necessarily mean the investment is not producing any free cash until the end of the project. In both the US and Canada, it is possible to use depreciation to offset a positive net income. Though electing to use depreciation will effectively increase your taxable long-term capital gain, this may be preferable to paying tax on rental income during the life of the project.

A climate in which changes in the tax code are being considered often dampens business investment since any prospective changes can impact the returns of investments with long holding periods. Any prospective change in the tax law can have a negative impact on investments in real estate as investors become more reticent about making substantial investments. Each investment has a unique structure. Interest rates, cash flow, appreciation, and tax considerations will set the rate of return required by any individual investor.

## Pro formas

Pro formas are statements that detail the financials of an individual or business and are essential tools in understanding the inner workings of any investment. Pro formas generally include a balance sheet and an income or cash flow statement.[6] For the individual, a pro forma is a summary of a person's wealth and income. For a business, the profit-and-loss statement (P&L) will reveal the current cash position (positive or negative), while the balance sheet states the net worth of the enterprise after all outstanding debt obligations have been paid. All businesses must prepare a P&L and balance sheet for their shareholders and tax authorities as part of their annual responsibilities. They can also be prepared in an abbreviated format for perspective investors to review as part of a solicitation. For anyone considering the purchase of a property, reviewing the pro forma is essential. Usually prepared by the seller's accounting firm for this purpose, the buyer's accountants should always scrutinize them carefully for errors and omissions. An experienced accountant with a Certified Public Accountant (CPA) or Chartered Accountant (CA) designation is an essential member of the decision-making team. A saying attributed to Mark Twain and Charles H. Grosvenor, among others, that "Figures don't lie, but liars do figure" should be the rule of the day prior to making any investment.[7]

### *Cash flow, or profit and loss*

The cash flow statement, commonly referred to as a profit-and-loss statement or simply P&L, provides a snapshot in time of a business's current financial health. Think of it as a budget for a specific period of time. A P&L is are important to the management of all businesses. Any business today needs to know more than how much money is in the cash register at the end of the day and must have accurate P&Ls generated on a regular basis to support management decisions. Some of the earliest accounting records date back to Mesopotamia, and document the purchase of goods and services. By the 14th century, double-entry bookkeeping allowed the merchants in Venice, Padua, Florence, and Constantinople to track complex financial transactions. With the growth of railroads in the mid-19th century, the use of cash flow statements became part of a manager's operational toolkit. Given this long history, it is noteworthy that P&Ls became mandated by the Financial Accounting Standards Board (FASB) as part of a company's audited financial statements only in 1987. Statements can be prepared monthly, quarterly or on an annual basis and are divided into two parts: income and expenses. Today, with our dependency on financial databases, financial statements, including P&Ls, can be updated in real time and used as tools in daily business operations.

#### *Income*

The income portion of the statement for managed rental real estate will begin with the line entry "Gross Potential Rental Revenue" (Figure 6.2). This is the amount collected from tenants for renting an apartment, an office, an industrial space or a store location in a mall. Rents can include utilities or the tenant may be responsible for all or some of these expenses. It is possible to include escalators (increases as a percentage of rent) on long-term leases to offset inflation. On retail locations, a participating lease also includes a percentage of gross sales. With participating leases, both tenant and landlord are partners. The landlord might receive a share of 5–10% of the gross receipts. In return, the tenant would receive a slightly lower base rent, reducing their cash flow demands and risk. Under this arrangement, both parties have a stake in improving the retail sales of the location. As the landlord, I can improve sales for my tenants by maintaining the common space and marketing, and by hosting special events, which benefits everyone.[8]

Finding an appropriate rent must take into account current and future market conditions. Web-based rental services, government-published data, and leasing agents offer sources of information useful in establishing rents prior to lease negotiations. In determining rents, the judgment of the direction of the market will always be a factor. In a rising market, the tendency is to escalate rents out of concern that rents will be only higher in the future. Rents may be raised based on the assumption of a continued rising market, reducing the number of potential renters for any particular class of properties. Similarly, when the market is in decline, the concern of finding a tenant from a shrinking pool of potential prospective renters may push the landlord to offer rents that are lower

| **Income** | |
|---|---|
| Gross Potential Rental Revenue | 140,000.00 |
| Vacancy Factor 10% | -14,000.00 |
| Interest on Cash | 2,500.00 |
| Reimbursements for Utlities | 5,000.00 |
| **Total Income** | 128,500.00 |
| | |
| **Operating Expenses** | |
| Utilties | 6,000.00 |
| Maintenance | 4,000.00 |
| Real Estate Taxes | 12,000.00 |
| Interest on Mortgage Payments | 42,000.00 |
| Management Fees | 6,000.00 |
| Legal and Accounting Fees | 4,500.00 |
| **Total Expenses** | 74,500.00 |
| | |
| **Net Income (NOI)** | 54,000.00 |

*Figure 6.2* XYZ Apartments cash flow statement.

than the average. The fear of having no rental income for an extended period encourages landlords to create incentives, such as a free month's rent as a signing bonus. Such measures may help the landlord avoid having vacant apartments that generate no revenue.

In developing a pro forma for a potential project, it is important to take into account the time period over which the space is vacant. Though it would be ideal if our tenants never moved, it is unrealistic to expect them to remain for the life of the project. Apartment tenants may move to another city because of better employment opportunities, a business may leave because their growth requires more space than you can offer or a retailer may be forced to close their doors because of a downturn in the economy. In each case, there will be a time when space is vacant. You may need time to paint, replace carpets, and make general improvements between each lease. Then the property will need to be advertised and shown, and new leases negotiated. Eventually, the space will be leased again, but only after some time has passed. To account for this lost revenue, it is common to list the income from rent and apply a vacancy allowance. Subtracting the vacancy allowance from the gross scheduled rent income will give you a net value often referred to as the gross operating income.

When rents for a proposed development are stated for a pro forma, they should be treated as mere projections. Once the project is sold and rented up, the actual rents may vary from those stated in a pro forma. Rents can go up or down, depending on economic circumstances during the construction period. Unless the properties are 100% preleased, it is not possible to know the revenue once the property is occupied. Even for properties that have been rented for years, if leases are due for renewal, rents can change making a forecast difficult. If markets are improving for space, you should be able to increase rents in the future when leases are renegotiated. However, if the market for commercial space is seeing a number of new office buildings near completion, tenants may leave for newer space. Special concessions, including allowances for renovation on long-term leases made by your competitors, may also empty your building of tenants as they leave for these new locations. Looking at the history of rents for a property can be helpful in identifying trends useful in forecasting future revenues, but again only to a limited degree.

Other income that can be listed in a pro forma includes any charges above the rent. It can include interest and dividends on money held in reserve for future capital improvements. Capital improvements must be part of a long-term improvement plan to stay current with the market. Properties can begin as quality locations with high rents. Wear and tear, changes in architectural style, and new technological advances can make an "A-class" office space look tired and old after a few decades. As neglect takes its toll, the property will no longer maintain its "A" class designation. Now a "B-class" space, you will find the rents are lower than those of newer buildings in the area. After a few more decades of decline, a property that was once considered of high quality and in a good location may become a C-class space requiring extensive renovations. If the property is now situated in an undesirable area of the city, the needed improvements will be difficult to justify.

### *Expenses*

Expenses accrued during the reporting period include all fees and expenses associated with operations (Figure 6.2). Like a household budget, this will include utilities. These expenses itemized show the cost of gas, oil, electricity, telephone, cable TV, telephone, trash removal, water, and sewer. It is also possible, under a triple net lease, for tenants to pay for utilities, taxes, and maintenance. In these cases in which the tenants pay the utilities directly, charges will accumulate during vacancy periods for utilities not paid by tenants. If specific expenses are to be passed along to the tenant, it is useful to show these expenditures with categories for both reimbursed and unreimbursed expenses. It is always beneficial to organize the income and expense portion of the "cash flow statement" under headings that can aid in understanding the exposure of the operations to fluctuations in utility rates.

Where properties are managed and maintained by paid staff, cash flow statements will list expenses for property management, including onsite janitors and a superintendent. For smaller operations, it may be more cost-effective to use third-party contracts for basic maintenance services. It is also possible to hire a property management company to collect the rents and maintain the property. Though the basic fee can be based on a percentage of the gross scheduled rent, other charges, including those of plumbers, electricians, gardeners, snow removal contractors, and marketing, will be added to the monthly fee. Higher-than-normal expenses for services such as painters, plumbers, and electricians should be a warning sign. Higher-than-normal maintenance expenses in residential properties may suggest that tenants are placing undue wear and tear on their apartment unit. Buildings that are poorly constructed, with cheap finishes and cheap appliances, will show their wear in just a few years. If the contractor was cutting corners during construction, you may be forced to fix problems that are only now coming to the surface. Close examination of the cash flow statement may suggest areas that need closer examination by a qualified energy consultant, an engineer or an architect prior to purchase.

Rental properties will also include fees paid to lawyers, accountants, and other professionals, as required. Many professionals who provide services will require a retainer, from which charges are deducted on a monthly basis. Higher-than-normal legal fees could indicate a high number of evictions, indicating a number of bad tenants.

Finally, taxes must be paid to the municipality for local services. As a charge to expenses, real estate taxes and special assessments can represent a significant expense. In communities in which real estate taxes have been going up quickly, there may be a systemic underlying decline in the local economy. As businesses and residences move, the cost of maintaining the municipal infrastructure must be shared by ever fewer taxpayers. Having a history of tax payments to review may reveal trends that can help in the negotiation of a sale price for the property.

*Rates of return*

Subtracting the total expenses from total income gives you the net operating income (NOI) for the property. For a project that is being built or is currently under management, the NOI provides a measure of overall cash flow. For properties purchased with cash, the NOI is the profit of the operation. In these cases, figuring the return on equity (ROE) is a simple calculation that involves first deducting taxes paid on income, then dividing income after taxes by the investment made in the property. For example, on a property that was purchased with cash for $1 million and which generated a $100,000 NOI, the ROE before taxes would be 10%:

$$\text{ROE} = \text{NOI/Equity} = \$100,000/1,000,000 = 10\%$$

A simple measure, ROE allows us to compare this investment's rate of return against that of other properties and investments. We may find that, given the risk, the 10% ROE before taxes is not enough to justify the purchase price, and we may want to negotiate for a lower price. It may also be possible to use debt to increase my leverage and ultimately to lower my equity in the project, thus increasing my rate of return. However, when interest rates are high, the cost of borrowing will only increase the operating expenses and lower the total income.

In evaluating the relative value of real estate investments, ROE is one measure. For two properties of equal value, the one that produces more income will be favored, if all other factors are equal. This ratio of NOI to current market value is defined as the capitalization rate (or simply cap rate). In arriving at an appraisal value for income property, the cap rate will be a useful measure. Though cap rates will vary by region and property class, they can provide some basis for determining the value. If, for example, it is expected that commercial space in the downtown will produce a rate of return of 6% of current market value, then anyone selling a property with a return of less than 6% should expect the buyer to offer less than the asking price.

Cap rates can also be used to help define rates of return relative to an individual's risk tolerance. If an investor in commercial office buildings requires a premium of 6% above the 10-year Treasury bond and if the 10-year Treasury pays 4.5%, then a commercial building would have to return at least 10.5% before the individual would consider it.

In our case, if the building generated $200,000 in cash flow in the year and the cap rate is 10.5%, then the value of the property is $1,904,761. In this case, we would make an offer of $1,904,761 if the return on our investment were our only concern. However, if the risk-free rate should fall by 1%, then the value of the property would increase to $2,105,263:

$$\text{Capitalization rate} = \text{Net operating income/Current market value}$$

We can rearrange this to solve for current market value:

$$\text{Current market value} = \text{Net operating income/Cap rate}$$

$$\text{Current market value} = \$200{,}000/10.5\% = \$1{,}904{,}762$$

$$\text{Current market value} = \$200{,}000/9.5\% = \$2{,}105{,}263$$

### *Debt*

For most real estate purchases, debt represents a significant portion of the purchase price. Mortgages use the land and building as collateral for the loan. In cases in which the owner is unable to make their payments, the property is forfeited. The legal history of this arrangement dates back to the Middle Ages, if not to Roman times. It allows the investor to acquire a significant real estate investment without making the entire payment in cash, which would be a king's ransom for most investors. Without debt financing, our economy would be hamstrung by the severe constraint of paying for capital investments out of savings. Economic expansion during periods of population growth would be difficult to accommodate if real estate development and municipal improvements were all-cash transactions. It would be all but impossible for a country, province, state, county or city to build roads, transportation systems, schools, universities, hospitals, and water and sewer systems without the ability to borrow from financial markets. Taxpayers may be able to support maintaining the existing infrastructure out of current tax revenues, but building and extending services such as sewer and water for future residents would certainly be difficult, if not impossible. For this reason, debt is a positive for an economy when there is responsible investing and a solid plan for the repayment of debt.

Government programs and guarantees can play a significant role in ensuring that investors are not at risk in making capital available for home financing and other projects. The rationale is that, by putting the faith of the US government behind mortgages for those with modest credit, national social objectives can be supported, including home ownership and related investments in community infrastructure. The Government National Mortgage Association (GNMA, or Ginnie Mae), Federal National Mortgage Association (FNMA, or Fannie Mae), and Canada Mortgage and Housing Corporation (CMHC) all provide guarantees to financial institutions for loans made to home buyers. Until recently, the activity of GNMA and FNMA were viewed as programs with little associated risk. In the aftermath of the global financial crisis of 2007/08, high levels of default among borrowers have put these loan programs in question. As a consequence of the risk associated with these government-backed loans, these federal government agencies have come under closer scrutiny. While the legal aspects of debt borrowing will be discussed in Chapter 7, in this chapter the focus is on the financial aspects of debt in real estate acquisition.

### *Borrower and lender*

There are at least two parties to every loan: the borrower and the lender. The lender provides the money for an investment and, in return, receives repayment of the loan with interest at a later date. In the interim, the borrower has the cash to invest, during which time they receive payments from their tenants to both pay the loan and cover expenses. If all goes as planned, the investment will generate a positive rate of return after taxes. When these loans are secured or collateralized by real property, we refer to these loans as mortgages. Lenders traditionally finance less than 100% of the value of the property to reduce their risk. Bankers and insurance companies who make most of their loans on real estate in Canada, the US or Europe are not in the business of managing real estate. Foreclosing on a property after a borrower does not make payments is a costly affair. Legal fees, court delays, and expenses associated with the sale of the property can add significant cost to the lender when selling a foreclosed property. Besides, if the property went into foreclosure because of an economic downturn, the property may now be worth far less than when it was purchased several years earlier. To reduce this potential risk, borrowers are required to have an equity position by making a down-payment when purchasing the property. The percentage varies, but it will generally range from 5% to 50%, depending on the type of property, location, associated risks, and practices of the country and locale. Sometimes, debt may be shared by more than a single lender. In these cases, lenders in a subordinate position to the primary lender will expect a higher interest rate given that their financial position is less secure in the event of a default.

*Loan mechanics*

Historically, loan payments of principal and interest were made at the end of the term. This term was fairly short, and loans could be extended if the lender was agreeable. Today, most mortgages are amortized, with payment schedules that are usually either 15, 25 or 30 years. With amortized loans, the payments are a fixed amount usually made on a monthly or biweekly basis. For each loan payment (A), part of the payment is the interest on the outstanding balance and part is for the repayment of the principal:

$$A = P\frac{r(1+r)^n}{(1-r)^n - 1}$$

Where:

$A =$ payment amount (per period)

$P =$ initial principal (loan amount)

$r =$ interest rate (per period)

$n =$ total number of payments or periods

Before the use of desktop computers, printed tables provided the amount of the principal payment, interest, and principal for each a mortgage payment for a specific interest rate and term. Using Microsoft Excel, a spreadsheet program, or an online application, it is now easy to calculate and explore the relationships between interest rates, the number of payments, and mortgage payments.

Consider the case of borrowing $1 million over a ten-year period. If I were able to borrow the money at 0% interest, with payments made annually, I could divide the $1 million by 10 to determine the amount due at the end of each year ($100,000). Now consider a more realistic case of a mortgage rate with a 5% per annum interest rate. I would have to pay an additional $50,000 for interest due at the first year. In this particular case, my interest in the first year represents a significant portion of the payment due. Though no one likes to pay interest, as a tax-deductible expense the payment of interest can provide a significant tax benefit in the early years of the project. The question is: "How much would I have to pay at the end of each year if the loan were paid off in ten years?" Using the formula above or the financial functions provided in Microsoft Excel (PMT, IPMT, PPMT), we can arrive at the solution: $129,504 (Figure 7.2). What is important is that, after making each payment, the principal balance is reduced, and the actual interest payment will be a smaller proportion of the next

| PV (amount borrowed) | $ 1,000,000 |
|---|---|
| No of Periods | 10 |
| Current Period | 1 |
| Interest rate | 5% |
| | |
| PMT | -$ 129,505 |
| PPMT | -$ 79,505 |
| IPMT | -$ 50,000 |

(a)

| PV (amount borrowed) | 1000000 |
|---|---|
| No of Periods | 10 |
| Current Period | 1 |
| Interest rate | 0.05 |
| | |
| PMT | =PMT(B4,B2,B1) |
| PPMT | =PPMT(B4,B3,B2,B1) |
| IPMT | =IPMT(B4,B3,B2,B1) |

(b)

*Figure 7.2* Mortgage payment calculations with Microsoft Excel: (a) calculations; (b) Excel formulas.

payment. The mortgage payment would reduce our cash flow by both principal and interest payments, though only the interest portion of the payment would be a tax-deductible expense.

*Debt, depreciation, and cash flow statement*

Accounting for the debt payment made on the ABC Apartment appears as adjustments to the net operating income (Figure 8.2). In this tabulation, it is important to distinguish between deductions that reduce cash flow and those that reduce taxable income. Investors ultimately will want to know what is the amount of income subject to taxation. This will be an important number when calculating the after-tax return for the investor. Differentiating between what is and what is not a taxable deduction is critical to this calculation.

To arrive at the bottom line of what is taxable, the mortgage payment is subtracted from the NOI. This gives a number for the pretax cash flow. Mortgage interest paid on an investment property is a deductible expense in the US and Canada, as well as many other G20 countries. Mortgage repayment of principal, though reducing your pretax cash flow, is not treated as a tax-deductible entry, but appears as an adjustment to pretax income. Another adjustment to taxable income is depreciation. Depreciation refers to the decrease in value of a capital investment from wear and tear and technological obsolescence. It is an important number because of its tax implications. In the US and Canada, depreciation on real estate and capital equipment used in a business is deducted from taxable income. It can include a computer or a car, truck or heavy equipment required in the operation of a business. Every country's taxing authority will have a list of rules for depreciating each type of asset. The rationale for this deduction is that, at some time in the future, the property's value will have been reduced, sometimes to zero, and is no longer useful. Providing some tax benefit for each year over which the capital investments are held is done partly to

| Market Value Increase/year | | 1<br>2.5% | 2<br>2.5% | 3<br>2.5% |
|---|---|---|---|---|
| **Operating Income** | | | | |
| | Gross Income (net vacancy) Note this is an end of year calculation | $868,073 | $889,774 | $912,019 |
| **Operating Expenses** | | | | |
| | Operating Expenses | $75,402 | $77,287 | $79,219 |
| | Maintenance | $75,402 | $77,287 | $79,219 |
| | Management Fees | $75,402 | $77,287 | $79,219 |
| | Real Estate Taxes | $18,158 | $18,612 | $19,077 |
| | *Short term Loan Interest ( Interest can be positive or negative) | $0 | ($5,116) | ($10,753) |
| **Total Expenses** | | **$244,363** | **$250,472** | **$256,733** |
| | | | | |
| **Net Operating Income** | | **$623,710** | **$639,303** | **$655,285** |
| | | | | |
| **Debt Service** | | | | |
| | Interest Paid | ($311,784) | ($303,564) | ($295,073) |
| | Amortization (Principal) | ($249,091) | ($257,311) | ($265,802) |
| | Total Debt service (PMT) | ($560,875) | ($560,875) | ($560,875) |
| | Principal Balance | 9,198,908.99 | 8,941,597.98 | 8,675,795.70 |
| **Pre tax Cash Flow** | | | | |
| | Net Operating Income - Total Debt service | $62,835 | $78,428 | $94,410 |
| | | | | |
| **Taxable Income** | | | | |
| | Pre tax Cash Flow | $62,835 | $78,428 | $94,410 |
| | Plus : Amortization (+) | $249,091 | $257,311 | $265,802 |
| | Less : Deprecation ( - ) | ($428,000) | ($428,000) | ($428,000) |
| | Taxable Income | ($116,074) | ($92,261) | ($67,787) |
| | Tax Credit(-) or Paid(+) [personal tax rates] | ($45,965) | ($36,535) | ($26,844) |
| | After Tax Cash Flow (Net spendable) = Pre TaxCash Flow - taxes | $108,800 | $114,963 | $121,254 |
| | | | | |
| | **Market Value** | $12,105,250 | $12,407,881 | $12,718,078 |
| | Line of credit Short: Short Term Loan Account Before Tax (Principal + Interest at the end of year) - If your net spendable is less then 0, you will need to borrow money at current rates | $113,696 | $238,949 | $376,412 |
| **Ratios and Measure of Financial Success** | | | | |
| | Cummulative Principal | ($249,091) | ($506,402) | ($772,204) |
| | Total Equity = Down payment + Accumulated Principal + Cash Account | 2,724,787.28 | 3,107,351.12 | 3,510,616.63 |
| | Cash on Cash = Pre Tax Cash Flow/Equity | 2.31% | 2.52% | 2.69% |
| | Return on Equity after Tax = Cash flow after Tax/Equity | 3.99% | 3.70% | 3.45% |
| | **IRR AND NVP CALCULATIONS** | | | |
| $2,542,479 | NPV Before Tax | ($2,299,165) | $78,428 | $94,410 |
| **17.51%** | IRR Before Tax | ($2,299,165) | $78,428 | $94,410 |
| ($1,303,681) | NPV After TAX( Includeds gain in 15th year and Cash Account in 15th yr) | ($2,253,199.74) | $114,963 | $121,254 |
| **16.48%** | IRR Aft Tax | $(2,253,199.74) | 114,963.16 | 121,254.09 |

*Figure 8.2* ABC apartment pro forma.

encourage reinvestment in business operations. On this basis, a rapid depreciation would encourage an enterprise to make investments sooner, making it more productive as it fights off obsolescence. When reinvestment is not made, there may be significant deferred maintenance, ultimately reducing the value of the property.

Accounting practices reflect the taxing authority rules in calculating depreciation. There are often several methods that can be employed to depreciate a capital asset, including straight-line depreciation, declining balance, and sum of the digits. In our example, we used the straight-line method to calculate depreciation. Using this method, the asset value as established at the time of purchase is divided by the years of service given in the tax code by the Internal Revenue Service (IRS) or Canada Revenue Agency (CRA). For example, in the US in 2010, the value was 27.5 years for residential properties and 39 years for commercial properties. In Canada, the value was 29 for both residential and commercial. Note that, in making this calculation, only the building and structures on the property are depreciated; land does not depreciate in value over time. Clearly, being able to write off the depreciation more quickly results in a greater benefit on a net present value (NPV) basis – though the total, the accumulated depreciation at the end of the property's useful life, is 100% of the price paid for the asset. For this reason, changes in the tables used for depreciation can spark a politically charged debate. In Figure 8.2, the completed profit-and-loss statement shows the importance of depreciation in the tax calculation.

For the ABC Apartment Complex, we deduct the $428,000 of depreciation against income under adjustments to pretax cash flow (Figure 8.2). The taxable income number is then used to calculate the tax liability for this investment. How this rental property is owned will bear upon its treatment for tax purposes and ultimately the actual after-tax income for the year. In the US and Canada, property can be owned personally or by a corporation, a partnership, a limited partnership or a real estate investment trust (REIT).

Finally, when the property is sold, we will need to account for the accumulated depreciation. The basis of the property will be an important part of this calculation. The basis is defined as the purchase price *less* any depreciation *plus* the value of any improvements. Depreciation reduces the basis of a property from its original purchase price. If no improvements are made in the property, then, at the time of sale, capital gains are calculated by subtracting the selling price from the basis. However, if substantial capital improvements have been made, the value of these improvements will be added to the basis. In figuring the taxes for long-term capital gains at the end of the project, only 50% of the gain is taxed in both the US and Canada.

$$\text{Capital gains} = \text{Selling price} - \text{basis}$$

$$\text{Basis} = \text{Purchase price} - \text{accumulated depreciation} + \text{improvements}$$

$$\begin{aligned}\text{Capital gains} = {} & \text{Selling price} - (\text{Purchase price} - \text{accumulated depreciation} \\ & + \text{improvements})\end{aligned}$$

For our ABC Apartment Building, if we sell the property in the 15th year and receive $17,104,401 for the property, and if we have made no substantial improvements, our capital gain would be $11,714,401 and only $5,857,200 would be subject to taxation:

$$\text{Capital gain} = 17{,}104{,}401 - (11{,}810{,}000 - 6{,}420{,}000) = \$11{,}714{,}401$$

If we made substantial improvements of $1million that we are now accounting for in the year the property is sold, the capital gain will be $430,000:

$$\begin{aligned}\text{Capital gain} &= 17,104,401 - (11,810,000 - 6,420,000 + 1,000,000)\\ &= \$10,714,401\end{aligned}$$

For a comparison of tax rates used in the US and Canada, see Figure 9.2. Given the after-tax rate of return is used to compare one investment opportunity against another, it is important that the form of ownership used to hold real estate be given careful consideration by a qualified accountant (a CPA in the US or a CA in Canada).

In the P&L shown in Figure 8.2, it is assumed the property is owned outright, and any gain is reported on a personal tax return. In the US, if a property is owned by an individual, rental income will appear on a Schedule E and is taxed at personal rates. In Canada, income from a rental property is considered passive income and is taxed (at time of writing in 2019) at a rate of 37%. If the investments were acquired by a C corporation, then corporate rates would apply. By placing the property in an S corporation, losses and gains will pass through to the individual interest. In the US and Canada, investors will have to consider their personal tax situation to determine whether a specific rental property is an appropriate addition to their portfolio.

### *Leverage*

Given the significant demands on cash flow for the repayment of a mortgage, it may be difficult to visualize how increasing levels of debt can actually increase the rate of return for a property. In the pro forma for the Elmwood Apartments, we compare two cases: one, a cash purchase; and the other, a substantial mortgage at 6%. At the end of year 1, the mortgage balance will be 87% of the property value. The resulting pro forma shows the rate of return at the end of the first year of the project based on the rents and expenses in the previous year. In comparing these two options, the rate of return is approximately double when there is long-term financing (Figure 10.2). If a mortgage is secured, then a much smaller amount of the investor's money is at risk, while at the same time the rate of return increases to 18.3% from 7.6%. This is because the equity position is only 13%, while the cash flow is reduced by about half to cover the debt payment. In this case, using debt to purchase the apartment building provides an investment opportunity of $5million, even though I may have much less than the $5million to purchase this property. Of course, to cover the added expense of a mortgage, cash flow is reduced significantly from the case of an all-cash deal ($381,000 vs $119,000 before

| Country | Corporate | Individual Min | Individual Max |
|---|---|---|---|
| USA | 21% + state & Local tax | 0% | 37% + state tax & local tax |
| Canada | 15%-26% | 0% | 58.75% |
| Germany | 29.65% | 14.00% | 47.45% |
| England | 19.00% | 0.00% | 45.00% |
| Australia | 28.5%-30% | 0.00% | 49%+state(territory) |

*Figure 9.2* Tax tables, Canada and the US.

Data Source: Wilson and Parlapiano, 2017

Interest rate on Debt = 6.0%
Tax rate = 28.0%

| | No Debt | Debt 87% |
|---|---|---|
| **Income** | | |
| Gross Revenue | 500,000 | 500,000 |
| | | |
| Expenses | | |
| Utilities | 50,000 | 50,000 |
| Management Fee | 10,000 | 10,000 |
| Maintenace | 30,000 | 30,000 |
| Real Estate Taxes | 30,000 | 30,000 |
| Interest on Debt | - | 261,000 |
| Total Operating Expenses before taxes | 120,000 | 381,000 |
| NOI before taxes | 380,000 | 119,000 |
| NOI after tax | 273,600 | 85,680 |
| Property Value | 5,000,000 | 5,000,000 |
| Debt | | 4,350,000 |
| Equity | 5,000,000 | 650,000 |
| ROE | 7.60% | 18.31% |

a

Interest rate on Debt = 6.0%
Tax rate = 28.0%

| | No Debt | Debt 87% |
|---|---|---|
| **Income** | | |
| Gross Revenue | 350,000 | 350,000 |
| | | |
| Expenses | | |
| Utilities | 50,000 | 50,000 |
| Management Fee | 10,000 | 10,000 |
| Maintenace | 30,000 | 30,000 |
| Real Estate Taxes | 30,000 | 30,000 |
| Interest on Debt | - | 261,000 |
| Total Operating Expenses before taxes | 120,000 | 381,000 |
| NOI before taxes | 230,000 | - 31,000 |
| NOI after tax | 165,600 | - 22,320 |
| Property Value | 5,000,000 | 5,000,000 |
| Debt | | 4,350,000 |
| Equity | 5,000,000 | 650,000 |
| ROE | 4.60% | -4.77% |

b

*Figure 10.2* Leverage and ROE: (a) gross revenue = \$500,00/yr; (b) gross revenue reduced by 30% to \$350,000/yr.

taxes). Given that the equity position is less, the ratio of income to investment will be higher (rate of return) for the more leveraged case.

It would appear from this simple illustration that debt is desirable. This is certainly the case when there is sufficient cash flow to meet all expenses and the future economic prospects are favorable. But consider the case of acquiring the property when cash flows were positive, but now a sudden downturn in the local economy has reduced the demand for rental apartments. Some of your tenants have vacated the property earlier than expected. Some have even left before the termination of their leases. Your attorney has advised you that, given that many of these tenants have left for other parts of the country, the legal costs of suing would outweigh any potential gain. Besides, without jobs, your delinquent tenants may have no ability to make restitution. To fill the space, you had to make special concessions to new tenants, and still there are still many vacant apartments. With a gross revenue reduction of 30%, you are now showing a loss for the year. Now, the situation looks very different. In this case, without any debt, the loss of rents has resulted in a lower rate of return, but you are surviving (4.6% vs –4.77% with a debt level of 87%). If reserves are insufficient to carry you through this difficult period, you may need to use your line of credit to help defray future expenses. In 2008, at the lowest point of the credit crunch, many investors walked from their recently purchased properties, leaving the bank with properties they could not sell and investors without their equity. Leverage is a double-edged sword, cutting both ways.

In considering this example of how leverage impacts return on investment in the Elmwood Apartments, one factor not considered was the market value of the property. Though, in the case presented, there were no immediate plans to sell the property, it is likely that a rapid decline in rents would also be reflected in the market value of the property. If difficult times continue for any period of time and the owner is forced to sell the property, it is likely that they will receive less than \$5million. Under this scenario, it is unlikely that, after they have paid the lawyers and the real estate agent's fees, there would be anything left over for the investors.

In appraising the creditworthiness of real estate investments, simple ratios are often used to measure the level of risk associated with debt. The debt coverage ratio and breakeven ratio are two simple measures of indebtedness. Many financial institutions will use these ratios as a test during the loan approval process. The first of these measures, the debt coverage ratio, is simply the annual net operating income divided by the annual debt service. In our ABC Apartment Project, this number is 1.112 for year 1:

$$\text{Debt coverage} = \text{NOI/Annual debt service} = 623,710/560,875 = 1.128$$

The breakeven ratio is equal to the debt service and operating expenses divided by gross operating. As a measure of solvency, when this number equals 1, the business is able to cover expenses and debt payments. When the breakeven ratio is greater than 1, the business is making a profit. A breakeven ratio equal to 1 means that there is sufficient cash flow to pay the bills, but there is nothing left in reserves at the end of the year. Submitting this statement to your bank for a loan will probably result in a rejection. Banks and financial institutions will want to see at a margin of safety before feeling comfortable making a loan.

### *Balance sheet*

Cash flow statements are only one part of the any pro forma. Cash flow statements, also referred to as profit-and-loss statements (or P&Ls), show the income and expenses over the year, while balance sheets provide a statement of assets and liabilities at a specific point in time. Like P&Ls, balance sheets are produced at least once a year as part of annual reports produced for investors, financial institutions, and tax authorities. Depending on the purpose and the accounting practices, they can vary in style, but all balance sheets are governed by one principle: assets *minus* liability *equals* shareholder equity. Equity, in theory, is the value of the investment after all outstanding liabilities are paid. It is a useful number in comparing two or more real estate purchases. A balance sheet will reveal the assets and liabilities acquired as part of the real estate purchase. Equity calculations are also important in foreclosures, divorce settlements, and bankruptcies. Although this number does provide some guidance in the valuation of a property, it should be treated as a starting point. If an actual liquidation of the asset were required, there would be expenses not shown on the balance sheet, including real estate fees, legal fees, and improvements required to make the property saleable. Most importantly, until a property is sold, it is often difficult to determine its actual value.

#### *Assets and liabilities*

Figure 11.2 is a typical balance sheet for an apartment building. Under the heading "Assets" is "Current assets". Cash in bank accounts, money markets, and other liquid assets will appear under current assets. This number will fluctuate from day to day as expenses and rents are paid. Accounts receivables will track any bill payments made on behalf of the operation. Also included under assets is the building, land, and other structures that are fixed to the land or investments that are generating income. There are a number of approaches that can be used to arrive at the

**Balance Sheet for XYZ Apartments**

| | | | |
|---|---|---|---|
| **Assets** | | | |
| | Current | | |
| | | Cash | 50,000 |
| | Non Current | | |
| | | Land | 1,200,000 |
| | | Building | 3,000,000 |
| | Good Will | | 50,000 |
| **Total Assets** | | | 4,300,000 |
| | | | |
| **Liabilities** | | | |
| | Short Term (Current) | | |
| | | Line of Credit | 24,000 |
| | Long Term | | |
| | | Mortgage | 2,500,000 |
| **Total Liabilities** | | | 2,524,000 |
| | | | |
| **Equity (Assets - Liabilities)** | | | 1,776,000 |

*Figure 11.2* XYZ apartments balance sheet.

current value of a building. In appraising the value of a property, one approach is to use the cap rate as the basis of this valuation. Recent sales of similar properties or comparables can also provide the basis for this calculation. Finally, the cost replacement method calculates the cost to construct the building in today's dollar value *less* any deduction for obsolescence and wear and tear. Needless to say, depending on the approach, the calculated current value can vary. Other assets can also include values for goodwill and other miscellaneous assets. How these numbers were determined will require research and scrutiny by any potential buyer. This is particularly true for line items such as goodwill, which can certainly evaporate during a liquidation.

Liabilities include short-term obligations. Short-term obligations can include balances on credit cards, lines of credit, and accounts payable for services and repair, while long-term obligations include any mortgages and longer-term debt. When these liabilities are deducted from assets, the net asset value of the company can be determined.

The current ratio is one measure that can be used to determine the health of an investment. Calculated by dividing the current assets by current liabilities, it reveals whether current obligations can be met by cash on hand. Any number greater than 1 indicates that you can meet current obligations. In our example, this number is 2.083, indicating that even after debts are paid there is cash on hand. The debt/equity ratio indicates the proportion of equity to debt that the company is using to finance its assets. High debt ratios can place an enterprise at risk, especially when interest rates are rising. This is especially true when the budget is operating close to its breakeven point. Though debt/equity ratios depend on the type of enterprise, this number can vary from 0.5 to 2, on average, for most businesses. For highly leveraged real estate, this value can be higher than 5 at the start point for a project, but as the the debt is repaid over time, this number would be reduced significantly.

In summary, there are several simple measures that are used to assess the profitability of a project, as follows:

$$\text{Current Ratio} = \text{Current Assets/Current Liabilities}$$

$$\text{Break Even Ratio} = \text{Debt/Equity}$$

$$\text{Return on Assets} = \text{Net Income/Total Assets}$$

$$\text{Cash on Cash} = \text{before tax cash flow/total cash invested}$$

$$\text{Return on Equity} = \text{Net income/Shareholder's Equity}$$

$$\text{Return on Equity after tax} = \text{Net income after tax/Shareholder's Equity}$$

$$\text{Cash flow before tax} = \text{Net operating income} - \text{debt service}$$

Depending on your concern, each ratio can be used to reveal a different aspect of a project. Return on assets is a simple measure of income against assets, but does not take into account the impact of borrowing. Cash on cash, or equity dividend rate, provides a simple measure of cash flow before tax over the amount of money you have invested into a project. If the balance sheet was prepared prior to the property sale, it provides a quick assessment of projected cash flow in the first year over what you will invest in the project. A useful number when comparing similar projects in an area; however, it does not consider issues of taxation that may be of concern to the investor. In cases in which after-tax returns are of concern, return on equity after tax will be a measure of greater interest. Cash flow before tax is calculated by subtracting the debt service from the net operating income. An investment that does not have a positive cash flow before tax will not be able to meet its obligations and probably will not be in business for very long. Even those that are slightly positive, if faced with a serious financial crisis, may find it difficult to survive. In using any of these measures, it is important to consider how changes in the balance sheet and P&L can impact the future health of the investment. Can the real estate investment withstand a reduction in rents or an increase in real tax assessments and borrowing rates? What if a major repair or upgrade to the building is required? Applying these measures to the worst-case scenario can help us lessen the financial difficulties that result from the unplanned and the unexpected, but which are often unavoidable even for the most seasoned investor.

## Cautions

When reviewing any cash flow statement or balance sheet, the devil is in the detail. Often, summaries prepared for investors may hide some of the negative operational aspects of an investment. There is no substitute for independent research and expert advice. All purchases and proposals must be preceded by a close inspection of the numbers presented in a pro forma. Are the numbers representing the true state of the asset? Were the improvements accounted for in the pro forma actually made? Where these improvements were done to the building "up to code" or will they need to be redone because of shoddy workmanship? Is the city going to require costly changes to the existing electrical system in the near future? Are service contracts currently being renegotiated, resulting in higher fees in the future? Are the leases due to come up for renewal, and are fewer of the existing tenants planning to renew? Accountants with experience in real estate investments are critical to this investigation. Only an inspection of the actual site by experts with knowledge of construction and design will reveal the property's true value. Most importantly of all, pro formas are a snapshot in time. Realizing that the pro forma provides only one step in the analysis of a project is perhaps the first lesson in real estate financial analysis.

## Notes

1 Cardano, 1965.
2 Brueggman and Fisher, 2001, 591.
3 The yield to maturity (YTM) formula is often used to calculate the yield on a bond. It is based on the current market price. The formula is based on compounding the rate as opposed to the simple yield using the dividend yield formula. Note that the formula calculates only an approximate value for the yield to maturity. To calculate the actual yield requires a trial-and-error process. There are many online resources capable of calculating the yield to maturity. Holding period return (HPR) is the return realized by an investor during a specific period of time. It is calculated as income plus price appreciation during a specific time period divided by the cost of the investment. See: www.investopedia.com/
4 The yield to maturity is calculated by dividing the annual cash flows by the market price. The yield to maturity accounts for the future payment of cash flows. See: www.investopedia.com/
5 See: www.investopedia.com/terms/s/sharperatio.asp
6 Lesheret al., 2017; Beresford, 1988.
7 See: http://quoteinvestigator.com/2010/11/15/liars-figure/
8 Brett and Schmitz, 2009

## Bibliography

Beresford, Dennis R. “The ‘Balancing Act’ in Setting Accounting Standards”, *Accounting Horizons* (March 1988): 1–7.

Brett, Deborah and Adrienne Schmitz. *Real Estate Market Analysis, Methods and Case Studies*. Washington, DC: Urban Land Institute, 2009.

Brueggman, William and Jeffrey Fisher. *Real Estate Finance and Investments*. Boston: McGraw-Hill /Irwin, 2001.

Cardano, Gerolamo. *The Book on Games of Chance*. trans. by Sydney Henry Gould, published in Cardano, *The Gambling Scholar*, by Oystein Ore, New York: Dover Publications, Inc, 1965.

Gallinelli, Frank. *Mastering Real Estate Investment, Examples, Metrics and Case Studies*. Southport, Connecticut: Real Data, Inc, 2008.

Haugen, Robert A. *Modern Investment Theory*. Englewood Cliffs, NJ: Prentice Hall, 1986.

Havard, Tim. *Financial Feasibility Studies for Property Development, Theory and Practice*. London & New York: Routledge, 2014.

Krugman, Paul. *The Return of Depression Economics and the Crisis of 2008*. New York & London: W.W. Norton & Company, 2009.

Lesher, Dale L., Gary John Previts, and William D. Samson. “Working on the Railroad: Public Accounting Talent in the United States – The Case of Haskins & Sells”, *ABACUS – A Journal of Accounting, Finance and Business Studies* 53, no 1 (March, 2017): 133–157.

Linneman, Peter. *Real Estate Finance & Investments: Risks and Opportunities*, 2nd ed. Philadelphia: Linneman Associates, 2008.

Poorvu, W. J. and J. Cruikshank. *The Real Estate Game, the Intelligent Guide to Decision-Making and Investment*. New York: The Free Press, Co, 1999.

Reed, Richard and Sally Sims. *Property Development*, 6th ed. New York: Routledge, 2015.
Squires, Graham and Erwin Heurkens, editors. *International Approaches to Real Estate Development*. New York: Routledge, 2015.
Staiger, Roger. *Foundations of Real Estate Financial Modeling*. New York: Routledge, 2015.
Wilson, Andrew and Alicaia Parlapiano, "What's in the Final Republican Tax Bill?", *The New York Times*, updated December 18, 2017.

## Web resources

American Accounting Association
http://aaahq.org/
American Institute of CPAs
www.aicpa.org/Pages/default.aspx
Bank of Canada
www.bankofcanada.ca/
Canada Mortgage and Housing Corporation
www.cmhc-schl.gc.ca/en/
CFA Institute
www.cfainstitute.org/Pages/index.aspx
Chartered Professional Accountants of Canada
www.cpacanada.ca/
Department of Housing and Urban Development
https://portal.hud.gov/hudportal/HUD
Federal Reserve Banks
www.federalreserve.gov/aboutthefed/federal-reserve-system.htm
Statistics Canada
www.statcan.gc.ca/eng/start
Urban Development Institute
http://udi.bc.ca/
Urban Land Institute
https://uli.org/
US Census Bureau
www.census.gov/
US Department of the Treasury
www.treasury.gov/resource-center/data-chart-center/interest-rates/Pages/Legacy-Interest-Rate-XML-Files.aspx

# 3 Model building and real estate development

## Models and decision making

Anyone who invests in real estate or other financial assets implicitly relies on a model to calculate future gains and losses. Prior to writing a check or transferring funds, we may run through a multitude of scenarios using a balance sheet or spreadsheet on our computer desktop. We may consider what will happen if consumer demand changes if there are lower operating costs, higher taxes, and so forth. By evaluating each possible scenario, we assure ourselves that the likelihood for gain is better than it would be if we were simply to leave our cash in the bank. Ultimately, when we write that check, although we may have considered the prospect of losses, we have decided that they are unlikely and have not been swayed against making the investment. With advances in computing and real-time data feeds, mathematical models have become indispensable tools in financial decision making, though models can take many forms. Mathematical models are used by fund managers to trade currency, stocks, and bonds. Models are used by governments to determine when to change monetary and fiscal policy to either stimulate the economy during a recession or dampen inflationary pressures during a boom period. Although you may never build a model, you have probably reviewed output from an economic model before making a financial decision. From the direction of interest rates to the future price of real estate, knowing something about building models is essential if you are to make informed decisions. After all, economic predictions that are based on models must all make assumptions about the financial universe they are simulating. And, given that the world is far too complex to include every factor that may have an impact on the predictive outcome of a model, all models are flawed merely by being simple constructs. Knowing something about the limitations of these models will hopefully make us more prudent and savvy investors.

Recognizing their usefulness, we must also acknowledge that all predictions of models are accompanied by noise or statistical errors. Furthermore, it becomes more difficult to predict with any certainty events that are to happen in the more distant future. A mathematical expression, the Weibull distribution, defines the upper and lower boundaries for our predictions. When viewed on a graph, this pair of curves on either side of our predictive line function reminds us that the certainty of any prediction dramatically diminishes with time. When predictions are viewed without the presence of these upper and lower constraints, we may become overly confident in the forecasts generated by the model.

### *What is a model?*

Models are constructs or representations of reality. Models can take on many forms: physical, graphical, logic or mathematical. When most people think of models, what probably first comes to mind is a physical three-dimensional (3D) scale model created to impress an architect's client, or possibly a scale model of an airplane or sports car assembled from a kit of parts purchased from a hobby outlet. These types of physical scale model can be useful in understanding the 3D qualities of objects that are not easily grasped from viewing drawings on paper or a computer display. For scientists, models are an essential part of their research activity. Models written in the language of mathematics give researchers a framework for presenting and testing a particular phenomenon or behavior. As part of normal scientific activity, data can be collected and compared to the predictions given by the accepted model. When the data disagrees with the predictions of the model, we may either modify the model or entertain using other competing models. Models can also be didactic, allowing us to explore, examine, and present new concepts and ideas. In an urban planning context, models can be used to simulate how a city will grow or how individuals will commute to work or where a prospective buyer will purchase a new home, given a multitude of choices. Perhaps the most commonly watched of outputs from simulations or models are the animated weather maps showing the weather over the next 48 hours. As with all models, we are cognizant that, with any prediction, there is a degree of error. The weather, like many phenomena, is difficult to predict depending on the locale. For that reason, bringing an umbrella to work might be good precaution even when there is only a slight likelihood of rain.

### *Issues in model building: the practice of model building*

Predicting the future is always based on historical data. During periods of economic, political, and social stability, models are particularly good at predicting the future. Not surprisingly, when business environments are stable, when yesterday was pretty much like today and tomorrow, the predictive powers of models are quite good. A sudden change in the economy is what models have a more difficult time predicting. The sudden collapse of the banking system in 2008 was a shock to the economy requiring the concerted actions of banks and governments around the world. Though a few analysts foretold of this coming economic disaster, it is important to remember that in every period there always a few prognosticators who envision doom and economic collapse.[1] Forecasting is as much an art as it is a science. We must never be too confident about our predictions. In *The Inferno*, Dante reminds us that the eighth circle of Hell has been reserved for fortune tellers, astrologers, soothsayers, and false prophets, forced to spend eternity with their heads facing backward as punishment for having asserted the power of seeing into the future.

Model building is always constrained by resources, the availability of data, and the complexity of the problem.[2] Lee Douglas, in his classic paper "Requiem for Large-Scale Models", though published in 1973 still presents what is a valuable precautionary set of concerns that should be considered before beginning a modeling exercise. First, models that are complex and broad in scope, and which serve a number of purposes, will ultimately have a great number of variables. Though complexity would seem to be an asset by giving the appearance of greater sensitivity to the complexity found in real systems, ultimately it may result in models that are incomprehensible, making it more difficult to recognize when these models are producing values that are just wrong. Ultimately, models with a strong theoretical underpinning are needed to avoid the errors made when correlation becomes a substitute for understanding. Ultimately, models with a great number of variables will require data that may not exist or lacks the fineness to be useful. For example, spatial models used by urban planners and economists are dependent on data tied to specific locations. When this data is aggregated to the level of community or city, it is not possible to predict events for a single city block or parcel.

### *Model building and statistics*

Models predicting the behavior of economic systems can take many forms. Economic models can be used to describe the behavior of an economic sector or the action of investors. In creating these models, the approach is often statistical. Models using the statistical technique of regression, sometimes referred to as "curve fitting", are constructed to determine relationships between the dependent variable (what we are trying to predict) and a set of independent variables. For example, we may want to determine the relationship between household income (I) and the price of home purchase (P) in a specific city. The purpose may be to help a developer make decisions about a proposed marketing plan for a new subdivision. Understanding the relationship between income and potential purchase price would be the first step in gauging the number of potential buyers in the area for a home within a specific price range.

Arriving at this relationship requires data on the purchases made by actual buyers. As a first step, we may simply obtain some recent sales data. If we know the price the owner paid for their home and their annual income, we can graph this relationship using a spreadsheet program (Figure 1.3). If we draw a straight line through the points so that as many points are above as are below the line, we have begun the process of curve fitting or fitting a line to a set of points. In this example, K represents the slope of the line (change in Y divided by change in X) or rise to run. This value could be estimated by taking two points on the line a, b, measuring the change (distance) along the Y-axis, and dividing by the change (distance) along the X-axis. In this example, the slope is 1.44. Now, given a household income of $50,000, we can predict the price of the household's home purchase to be $175,826 (Figure 1.3).

The actual process of fitting the line to a set of points uses a method of least squares where the distance between the line and the sum of distances to each point squared is minimized (least squares linear regression). Statistical programs, such as SPSS and SAS, and spreadsheet programs, such as Microsoft Excel, can perform this calculation. One measure of fitness is the $R^2$ correlation coefficient. If all the points were on the line, the $R^2$ value would be equal to 1, while if points fall far from the line, a more distant relationship between income and price is indicated. In the case of no relationship between income and house purchase price, the points would be scattered and their $R^2$ value would be 0. In this example, $R^2$ equals 0.906. A second measure, the adjusted $R^2$, takes into account the number of observations in the dataset. In our case, it gives a value of 0.902. Another way of thinking about adjusted $R^2$ is that our model explains 90.2% of the variation in the data, given the number of observations used in constructing the model. For our study, that might be good enough. If the adjusted $R^2$ were only 0.50 (50%), then we might need to look at variables other than income to determine the price of home sale by prospective buyers.

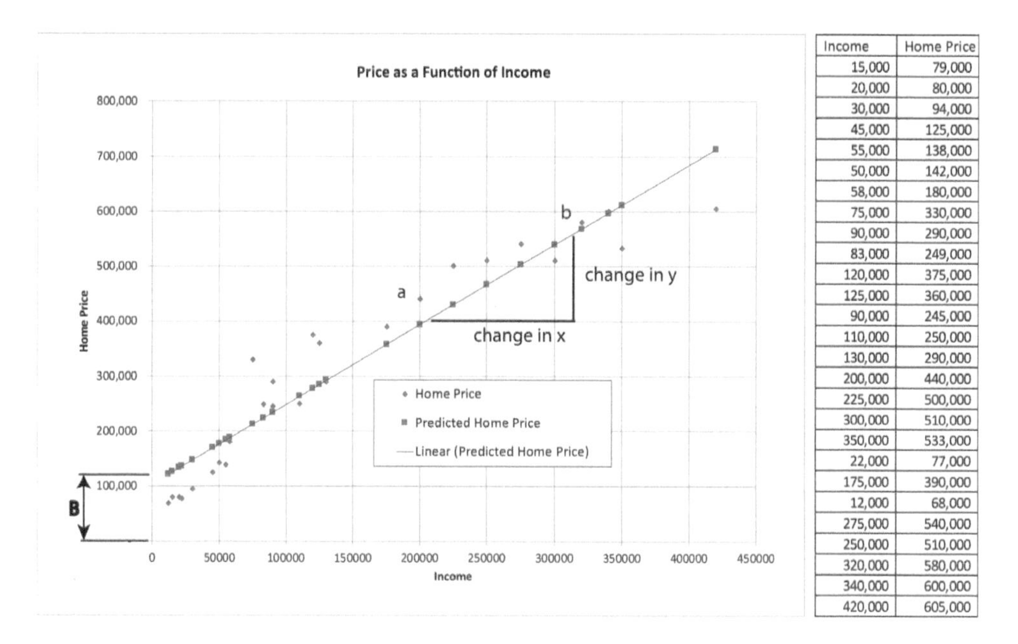


| Income | Home Price |
|---|---|
| 15,000 | 79,000 |
| 20,000 | 80,000 |
| 30,000 | 94,000 |
| 45,000 | 125,000 |
| 55,000 | 138,000 |
| 50,000 | 142,000 |
| 58,000 | 180,000 |
| 75,000 | 330,000 |
| 90,000 | 290,000 |
| 83,000 | 249,000 |
| 120,000 | 375,000 |
| 125,000 | 360,000 |
| 90,000 | 245,000 |
| 110,000 | 250,000 |
| 130,000 | 290,000 |
| 200,000 | 440,000 |
| 225,000 | 500,000 |
| 300,000 | 510,000 |
| 350,000 | 533,000 |
| 22,000 | 77,000 |
| 175,000 | 390,000 |
| 12,000 | 68,000 |
| 275,000 | 540,000 |
| 250,000 | 510,000 |
| 320,000 | 580,000 |
| 340,000 | 600,000 |
| 420,000 | 605,000 |

*Figure 1.3* Home purchase price as a function of income.

In our example, the expression can be written as an equation:

$$P = K(I) + B$$

Where:

P = the price an individual is willing to pay for a home (sale price)
I = the household income
K = a constant = slope = Change in Y/Change in X = Delta Y/DeltaX
B = line intercept
P = K (I) + B

P = 1.44 (I) + 103,826
R2 = 0.906; Adjusted $R^2$ = 0.902
Example:
I = $50,000
Then
P = 1.44(50,000) + 103,826 = $175,826

In developing models that are useful in practice, we may include additional variables. Clearly, our simple model cannot determine why two households with the same income would have different purchasing preferences. To make our model a better predictive tool, we may want to include the number of family members, the age of the home buyers, their wealth, and other independent variables that may help explain the purchasing behavior of our home buyers. Each additional variable may give us greater accuracy in our predictions and provide greater insight into the relationship between price income, family size, age of the purchaser, and the amount they are willing to pay for a place to live. Having arrived at a profile of our ideal buyer, we can estimate the potential number of buyers for a particular type of development in a community.

*Data and curve fitting: when relationships are non-linear*

A visual inspection of our model reveals that the data does not perfectly fit a straight line (Figure 1.3). The fact that our $R^2$ is not equal to 1 tells us there may be a better model. (Figure 2.3). Looking at the graph, households in both the lower and higher income brackets spend less proportionally on their homes than those who are in the middle-income brackets. Perhaps upper-income households are saving more or using their income to pay for expensive vacations and cars. For lower-income households, the cost of food, energy, and transportation takes greater priority. In a linear model, each variable contributes proportionally to the behavior of the dependent variable. However, for many phenomena, we find that their relationship of the independent to the dependent variable is non-linear.

How best to treat each variable in the model is never a simple problem. Our model of income vs selling price could take several forms:

**Linear relationship**

$$Y = B_1X + B_o$$

**Second-order relationship**

$$Y = B_1X + B_2X^2 + B_o$$

**Third-order relationship**

$$Y = B_1X + B_2X^2 + B_3X^3 + B_o$$

**Logarithmic relationship**

$$Y = \ln X$$

In this example in which the price is a function of income, the fit is improved when we consider a second-order relationship (Figure 2.3). In this example, we may also want so test a second- or third-order polynomial or even a logarithmic relationship. In the practice of curve fitting, we may examine several different relationships in our search for the best fit. Comparing the correlation coefficients for each trial allows us to contrast the ability of each model to explain the variation in our data (Figure 2.3). In this case, if we believe that as wealth increases, a smaller proportion of income goes toward housing, then a second-order model is a better choice over a simple linear model. It is also possible to test out other forms for the model: polynomial or logarithmic functions may describe even better how home buyers allocate their income.

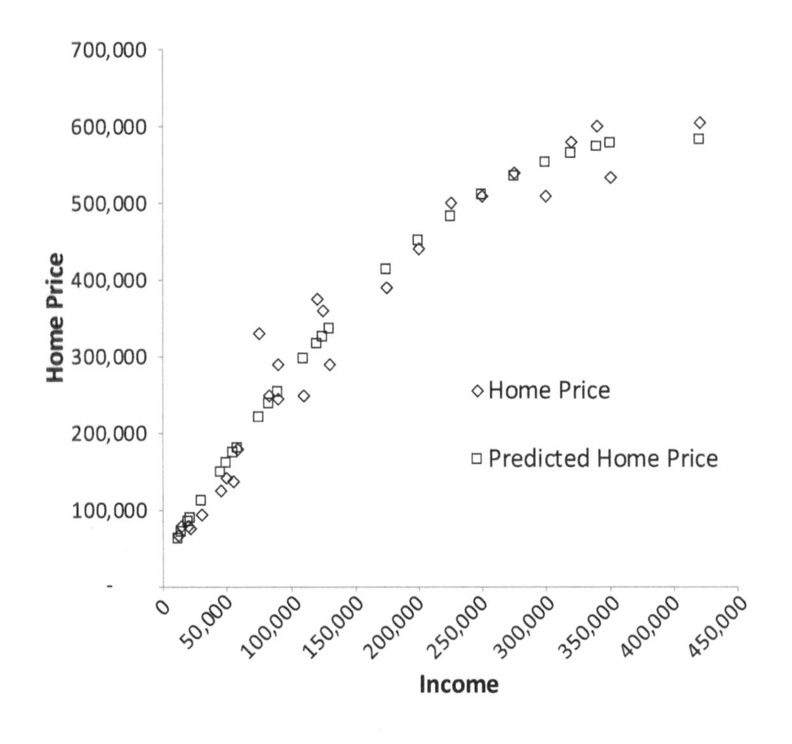


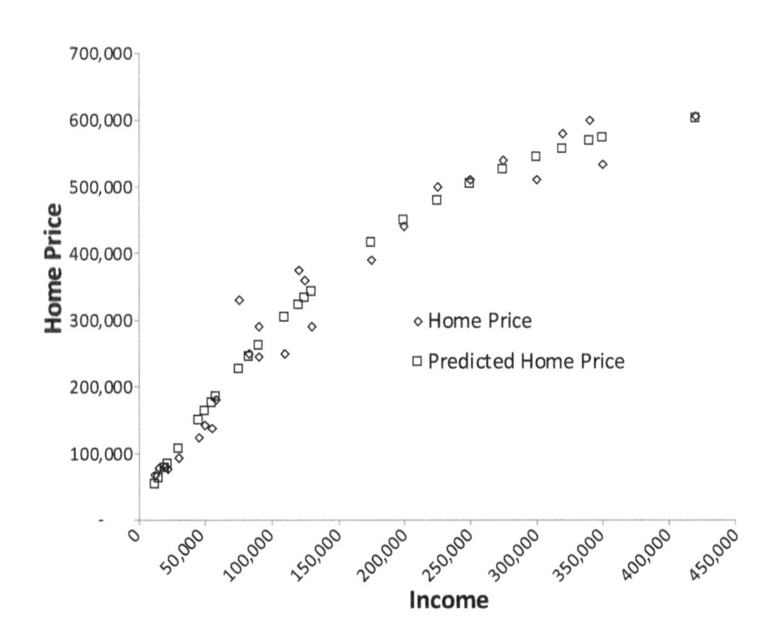


*Figure 2.3* Home purchase price as a function of income: second order (a) adjusted $R^2$ = 0.9610; third order (b) adjusted $R^2$ = 0.9615; and natural log (c) adjusted $R^2$ = 0.9111.

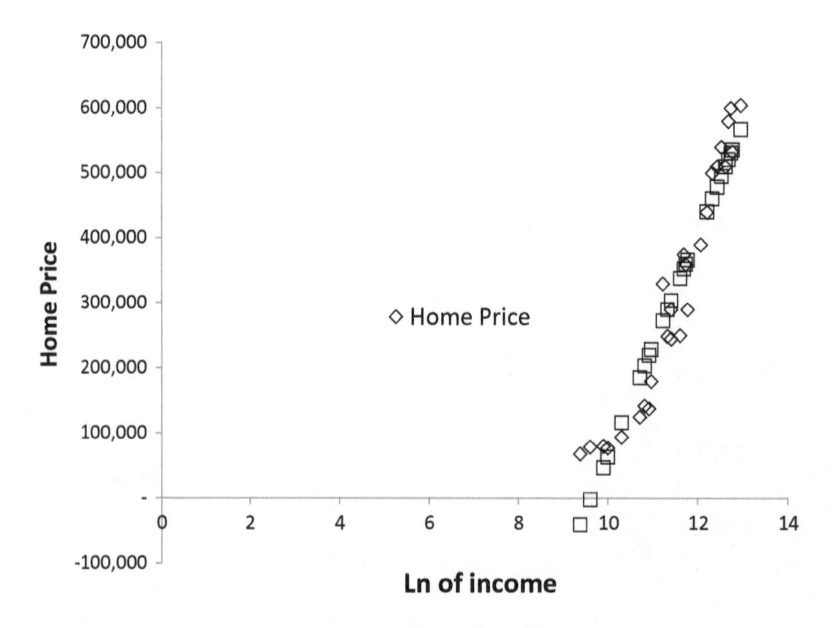


*Figure 2.3 (continued).*

In our example, the second-order function, a parabola, makes for a better fit than a straight-line function, while the $R^2$ for natural log function is close to the linear fit (adjusted $R^2$ = 0.902 vs adjusted $R^2$ = 0.911). In our example, using the third order is a slight improvement over the second-order function (adjusted $R^2$ = 0.9610 vs adjusted $R^2$ = 0.9650). Ultimately, selecting the appropriate model must be supported by both the data and our knowledge about consumer behavior.

### *Appraisal models*

In real estate appraisals, a model that can fairly assess the price of a home is needed by both financial institutions and taxing authorities. For the tax authority, having a basis for the equitable distribution of the municipal tax burden is essential to good government. Property owners need to be assured that their tax is based on calculations that are reasonable. Two properties that are similar in age and physical attributes should not vary in their tax assessment. For the tax assessor, it is important that the assessments are fair and defensible. Financial institutions want assurances that, in the event of foreclosure, there will be sufficient value to cover any mortgage on the property.

To begin the process of building an appraisal model, it is important to select variables that reflect the largest component of value.[3] A model based on the interior gross area would be the simplest model, where P is price, A is the total livable area, and K is a coefficient (price/sq. ft.). Using regression as our tool, data on selling price and size could be used to calibrate the model. Once calibrated, the model predicts the value of the home based on its gross square footage (Figure 3.3):

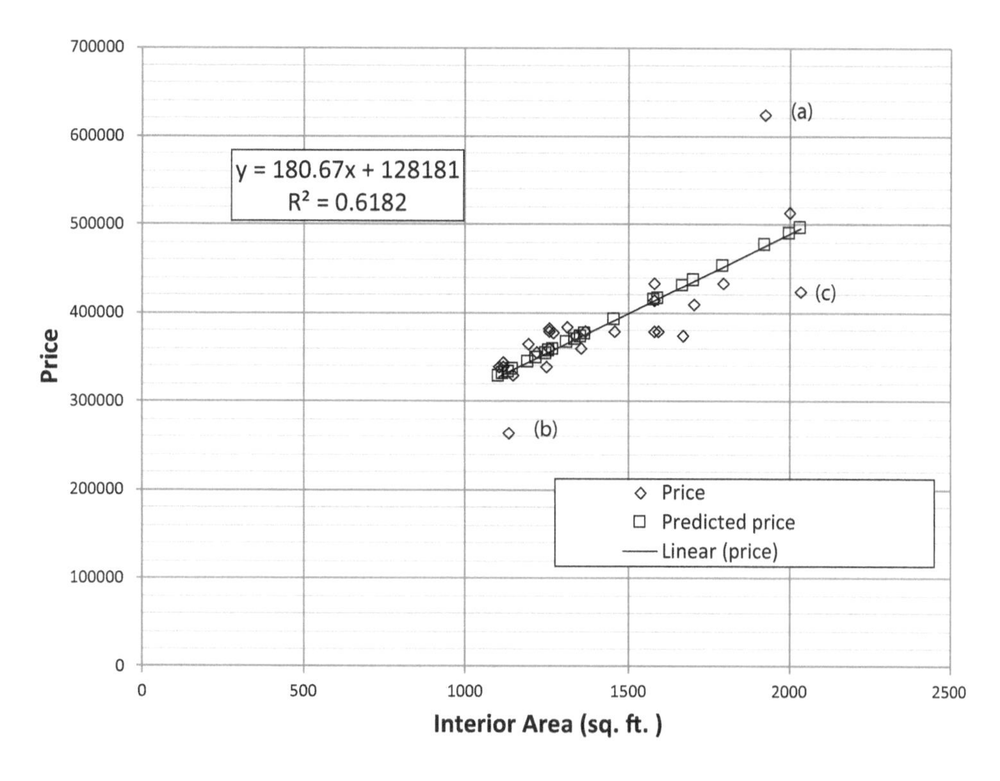


*Figure 3.3* Real estate price as a function of interior area.

$$P = K(A) + B$$

Where:

P = sale price
K = cost per sq. ft.
A = gross living area
B = Y intercept, value of Y when X = 0

Clearly, square footage cannot capture all the variance in a price of a home. Our initial model has an $R^2$ of only 0.6182. Other factors clearly matter in determining the price of a house, including age, lot size, and the number of bedrooms, bathrooms, and garages, for example. Inspecting the output from our first attempt at building a model reveals several outliers. These are points that clearly fall far from the line.[4] Investigating these data points further may provide some insight into the addition of variable that explains the deviation. For point (a) which is over $100,000 greater than the predicted value, we may find that this house property has a lot size much larger than the average home, while for points (b) and (c) the poor condition of the properties may be reflected in a lower-than-predicted sale price. Quantitative data describing the physical attributes of residential properties, for example from the Multiple Listing Service (MLS), which services realtors, can begin to help fill in the gaps of our knowledge. A sample appraisal form lists many of the variables that are used to determine the relative value of similar properties. Real estate agents and appraisers will commonly compare recently sold properties, or "comparables", to determine a reasonable value for a property. Within this form, the attributes of recently sold properties are listed, with corresponding adjustments made to offset for features not found in the appraised property. For example, if one recently sold property has central air conditioning and this is not found in the appraised property, then some deduction is made to compensate for the lack of this feature. Building a database that can be used to predict the price of homes that differ in size, quality, and features require a sample of properties large enough to capture the variation that exists in the housing stock. What makes this a challenging exercise is the need for current data. In markets with little activity, it may be difficult to build a database that reflects the variation in housing stock.[5] One solution is to use historical data and apply an adjustment based on the consumer price index (CPI) to arrive at an estimate of its current value. The problem with this approach is the need to compensate for the changing preferences in taste that will be reflected in the buyer's price.

To begin the modeling process, we can use data on the recent sale of properties to determine the expected price for each potential project. A simple model could take the form:

$$P = B_1(AR) + B_2(L) + B_3(A) + B_4(N) + B_5(BA) + B_6(G) + B_o$$

Where:

P = selling price
AR = gross area
L = lot size
A = age
N = number of bedrooms
BA = number of bathrooms
G = number of garages
Coefficients are $B_1$, $B_2$, $B_3$, $B_4$, $B_5$, $B_6$, $B_o$

For this list, there are clearly many more variables we could add. Quality of the finishes, materials, and fixtures can vary significantly among houses. Granite countertops or a jet tub in the master en suite can result in a premium price for a house. Even if the model includes an extensive list of physical attributes, we have yet to capture all the factors that determine price. Finally, the quality of the community can have a direct bearing on the choice of a buyer. Neighborhood schools, access to health, retail, and commercial services, the location of neighborhood amenities, and the levels of crime in the area are all factors that home buyers consider before purchasing a property.[6] Clearly, some of these factors will be more difficult to quantify than others. Taxes and utilities are numbers that are readily available, whereas qualitative variables such as the excellence of the neighborhood schools can pose a challenge. In our model, we may need to use proxy measures to capture these differences. For example, in finding a measure for the quality of the schools in a community, we could use test scores or the percentage of students who graduate from high school and attend college. Of course, each new variable will require additional effort in finding the necessary data. As we build a longer list, several problems emerge. A single equation or holistic model can become very unwieldy. Even in our example, it is possible to imagine a list of more than 100 variables that could be used to determine the price of a single-family house.[7]

$$P = B_1(AR) + B_2(L) + B_3(A) + B_4(N) + B_5(BA) + B_6(QK) + B_7(QB) + B_8(QA) + B_9(QF) + B_{10}(QT) + B_{11}(QE) + B_{12}(QL) + B_{13}(G) + B_{14}(S) + B_{15}(R) + B_{16}(H) + B_o$$

Where:

P = selling price
AR = gross area
L = lot size
A = age
N = number of bedrooms
BA = number of bathrooms
QK = quality of kitchen
QB = quality of bathrooms
QA = quality of appliances
QF = quality of lighting fixtures
QT = quality of bathroom fixtures
QE = quality of exterior
QL = quality of landscaping
G = number of garages
S = quality of local schools
R = access to retail
H = access to health services
Coefficients are $B_1$, $B_2$, $B_3$, $B_4$, $B_5$, $B_6$, $B_7$, $B_8$, $B_9$, $B_{10}$, $B_{11}$, $B_{12}$, $B_{13}$, $B_{14}$, $B_{15}$, $B_{16}$, $B_o$

A major issue with models that contain qualitative measures is their scoring and interpretation. Getting consistent results can be a challenge with variables that rate the subjective quality of a property. Subjective scoring of the property can vary considerably depending on the individual. It must also be recognized that a Likert scale is not a measured value like square footage. With Likert scale data, the values are not numerical quantities, but nominal categories, such as presence of view, water frontage or a swimming pool. To control for different classes within a data set, "dummy variables" can be included in the model. Dummy variables are not less intelligent variables than other variables, as the name would suggest, but independent variables that take on a value of either 1 or 0. In this way, we differentiate between different classes in the model. In practical terms, each measure you are trying to account for would receive the value of 1 when it is included in the property. By reviewing the statistical measures for each of the dummy variables in a model, it will be possible to know if there are differences between class types. For example, by including dummy variables in a consumer preference model, it will be possible to know if having a view of the ocean makes a difference in the valuation of a property.

In building complex models, a useful approach is to break down the model into components or partial models. In our example, we could construct a model for neighborhood crime or the quality of the education offered by neighborhood schools. These partial models could then be assembled into a larger model that could be used to predict property values. As part of this framework, we could also have models that determine the price of condos or even estimate the rent tenants will be willing to pay for an apartment. By including rental properties in the model, it may be possible to simulate the choices made by households whether to buy or rent. A model that simulates the choices of prospective renters and buyers must account for changes in mortgage interest rates and terms. Creating a model that can accurately predict the actual price a house will sell for can be a daunting task that requires dedication to data collection. Interestingly, online real estate information marketplaces such as Zillow boast of having more than 100 million active entries in their databases. Through extensive use of statistical data, they are able to make estimates on home values, making their website an important part of a prospective home buyer's research.[8]

For someone who is looking to undertake or invest in a real estate development project, a model estimating the selling price of a property would be part of a larger model that could also predict the profitability of each prospective proposal. Looking at the land acquisition, construction, design, and approval costs for each possible project, it will be possible to predict which site may be the most profitable to develop. If the model were to contain a spatial component, competition from existing projects could also be part of the equation. In this case, the model could consider the balance of supply and demand for each housing type in their respective neighborhoods.

### *Dynamic models*

In the previous examples, our predictions were based only on the current state of the market. But, in reality, many outcomes are the product of events that emerge over time. Consider the urban development process. Land must be purchased, plans approved, and financing arranged, and only after the construction process is complete is the project ready for the market. For example, following a period of high lending rates that depressed construction activity, it may take months or years before a recovery is in progress. Similarly, lowering tax rates or offering abatements on real estate taxes to stimulate commercial and industrial development may take time to result in any measurable activity. With these time-dependent relationships, dynamic models are particularly useful in predicting the future state of the real estate market. A dynamic model can be invaluable when forecasting both the supply and demand for real estate. Just consider, for a moment, events leading up to the purchase of a home by first-time buyers. Traditionally, couples fell in love, got married, and looked for a place to live. If they could afford the down-payment or if they were lucky enough to have parents who could help with the down-payment, our couple would be looking to purchase a "starter home", a small house on a minimum-sized lot. In this idealized scenario, if you could track the number of engagements in a community, you could probably predict the number of potential first-time home buyers. If you knew something about the distribution of income and wealth of these engaged couples, you could create a more detailed model of demand to predict the types, quality, and number of homes this group would purchase in the future. And although the definition of what constitutes a family has changed over the last 70 years, family formation is still an important factor in predicting future demand for housing.

Models that predict the future based on the current state of things are known as Markov models. Named after the Russian mathematician Andrey Markov (1856–1922), the Markov process has been used by modelers of the urban economy.[9] In applying the Markov process, you need to define the current state variable and have some knowledge of how that state will change in the future. For example, if we know a population is growing by 4% per year, we would merely take the current population and multiply it by 1.04 to get the population a year from now. If there is confidence that this growth rate will continue for some time into the future, this process could be repeated to calculate the population of the community next year and the year after for the next 30 years (Figure 4.3).

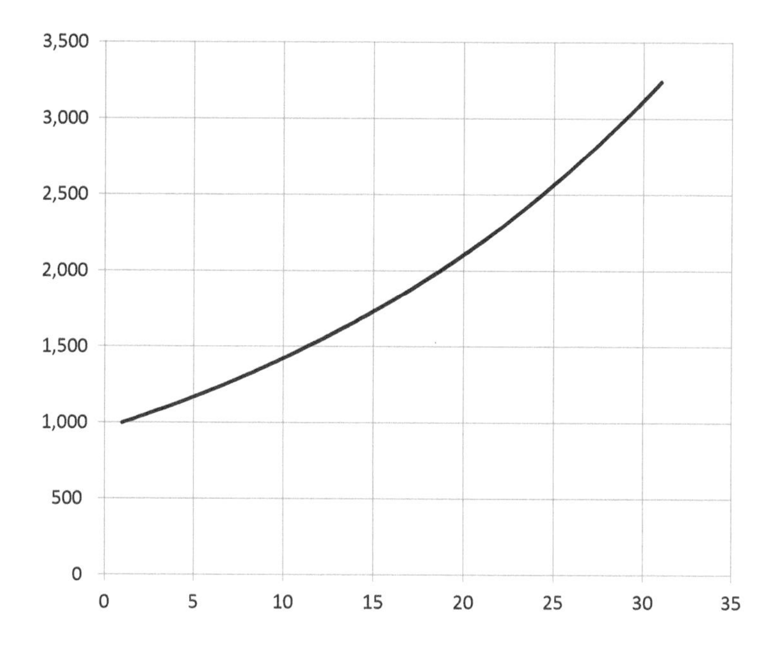


*Figure 4.3* Population growth.

$P_{(t+1)} = K\ P_{(t\ o)}$
$P_{(t+2)} = K\ P_{(t+1)}$
$P_{(t+3)} = K\ P_{(t+2)}$
Growth rate is 4%
***K = (1+.04)***

$$P(t) \text{ is the current population} = 1000$$

Then:
Population at

$$P_{(t+3)=}P(1+.04)(1+.04)(1+.04) = P(1+.04)^3 = (1000) * (1.04)^3 = 1092$$

Population at $P_{(t+30)=}P(1+.04)^{30} = 1000 * 3.243 = 3243$

Applications of Markov processes have been used to explain many urban processes.[10] For example, the future quality of housing stock in a community can be predicted by applying a set of probabilities that will tell us what percentage of our existing housing stock will be improved, stay the same or deteriorate in the next year. In constructing our model, we could use past data to determine the values for this transition matrix. If we have detailed historical knowledge, we should be able to forecast the condition of our housing stock in the future, which could be a useful exercise for a city government estimating tax revenues based on assessments. To begin this exercise, I need to know the probability a house will improve in value, fall in value or have no change in its assessed value. In using historical data, I could assemble this set of probabilities and apply them to our existing housing stock. If for example, I know that, over the

| Initial State | Probability | Initial Value | N | Value of all Homes |
|---|---|---|---|---|
| | 100% | 50,000 | 1000 | 50,000,000 |
| | | | | |
| Future State | 15% | 20% | 150 | 9,000,000 |
| | 35% | -25% | 350 | 13,125,000 |
| | 50% | 0% | 500 | 25,000,000 |
| total | | | 1000 | 47,125,000 |
| | | | | |
| | Difference in value over five years | | | 2,875,000 |

*Figure 5.3* Markov housing model.

next five years, 1,000 houses with a current assessment in the lower quartile and an average value of $50,000 have a 15% probability that their value will increase by 20%, a 35% chance their value will be reduced by 25%, and a 50% chance it will remain the same, we could then apply this set of probabilities to determine the total value of all houses in the future. In this example, we can expect a $2,875,000 reduction in the total value of tax assessments of properties in the lower quartile from $50 million to $47,125,000 (Figure 5.3).

Using dynamic models can also be helpful in predicting future additions to the housing stock for a community.[11] Often, models that predict new home construction are based on leading economic indicators – factors that predict future events. Housing permits are just one of many indicators that could be used for this purpose. Government agencies such as the US Department of Housing and Urban Development (HUD) and Canada Mortgage and Housing Corporation (CMHC) will customarily publish these statistics in their quarterly reports. For many communities, this information is a matter of public record. With knowledge of the average time period needed to complete a single-family home or apartment building, it should be possible to predict the supply of new housing. A simple variant of this model would predict future housing on the basis of current building permits. In building this model, we need to account for those projects that are left uncompleted. Some may never get built or may be delayed until some time in the future.

NH = New housing created in the next period
P = No of housing permits filed for new homes

Then:

$$\mathrm{NH}_{(t+1)} = (K)P_{(t)}$$

If I want to estimate the total supply of housing in the future at (t+1), I would need to add a few more variables to the models, such as existing housing stock (H) and housing slated for demolition (HD):

H = Total number of units
HD = Houses demolished and condemned

Then:

$$H_{(t+1)} = H_{(t)} + (K)P_{(t)} - (HD)_{(t)}$$

Many statistical packages allow you to create and test these time-dependent models. If supplied with a reasonably long series of data, it is possible to analyze not only the long-term trends, but also the cyclical nature of these data points. These time-dependent models are referred to as time series models. Time series models are particularly useful when the data exhibits strong seasonal patterns, as is the case for real estate, where home buyers tend to be more active during the spring and summer months and less active in the winter. Creating a model that reflects seasonal variation will be important, looking at the volume of futures sales. Many economic data series exhibit cycles.[12] Some are seasonal and others are longer than a year, like the business cycle. Still others are even longer, like the Kondratieff (sometimes spelled Kondratiev) cycle, which is associated with new periods of economic expansion tied to the emergence of new technology. Using economic and population data, models can be built that help explain the dynamic nature of real estate development. For example, in looking at the price and supply of new housing units, it is possible to construct models that take into account a range of factors, including lending rates, cost of materials, land prices, and labor costs. In cases in which the impact of change on the outcome may take several quarters to impact the construction industry, variables can be lagged. As in the case of our Markov model, the impact of a change in the present will impact the growth of housing stock in the future.

One precautionary note, however: lack of data can be a limiting factor in building time series models. If the need for quarterly or even monthly data does not exist for the variable, then it becomes all but impossible to apply theory to practice.

### *Two big questions in real estate development*

In real estate development, two questions emerge in most modeling discussions. The first question, simply stated, is: what is the most profitable development for an existing location? The second question addresses the issue of location: with plans for a particular type of development, where would be the best location for this proposed enterprise? Modeling techniques can help us answer both questions. To answer the first question, we look to the guiding principle of "highest and best use":[13]

> Fundamental to the concept of value is the theory of highest and best (or most profitable) use. Briefly it can be defined as that use which at the time of appraisal is most likely to produce the greatest net return to the land and/or building over a given period of time.
>
> (The Appraisal Institute[14])

First, understanding the zoning constraints will be key to determining what will be most profitable. Zoning places limitations on the use of the site. If single-family homes are the only permitted use, then there are no other choices. However, if the zoning allows for commercial, residential, and mixed use, then we must analyze a range of building options before deciding which is the most profitable. Although it might be possible to secure a variance to pursue an option not permitted under the current zoning, the additional costs and delay often makes choosing this route prohibitive. Evaluating each alternative must also take into account the constraints of the site on building form. Setbacks, maximum height restrictions, and building coverage will ultimately limit the total floor space. Where sites are vacant, the calculation involves estimating the cost of purchasing the land, the cost of construction, and the future revenues and expenses once the building is completed. If there is already a building on the site, the cost of demolishing the existing structure becomes another expense that enters into the calculations. Knowing the cost of demolition also gives us an estimate of the value for the land independent of any building. For each scenario, we can compare potential rates of return, deciding which scenario will be the most profitable, giving us an answer to the question of what is the highest and best use. In performing these calculations, we may also want to conduct a sensitivity analysis. Asking the questions "what if mortgage rates go up?" or "what if rents do not go up with inflation?" will help us decide which scenario should be the most profitable and, with any luck, steer us away from development options that have no chance of success.

The second question of where, given a specific development, would be the best location requires us to consider the potential profitability of building on suitable sites throughout the city. In each case, we will need to consider the cost of land acquisition and the cost of construction. If the plan is to build a grocery store, the constraints can include minimum parcel size, being on a major road, and locating in a neighborhood with specific demographic characteristics. We will also need to examine how existing competition will impact the cash flow of our new venture. Having arrived at a list of potential sites for this new development, the most profitable will be our first choice. For a major retailer like Walmart, making this choice can require a survey of numerous potential sites across a city or a region. Spatial modeling techniques can assist in making the decision of where to locate.

### *Spatial modeling of land values*

If you have ever driven from the downtown core out to the suburbs, you will know that it is not surprising to find taller buildings at the center of the city, where land prices are higher. As you travel further out from the central business district (CBD), the density diminishes, reflecting lower land costs. Scarcity and demand will increase land prices, referred to as bid–rent theory (Figure 6.3). In 1826, Heinrich Von Thunen (1783–1850) developed a model to explain the pattern of agricultural practice around a central market place. With transportation costs going up as you move out towards the center of the settlement, land that is further from the center would be used for activities such as growing grain. There are cheaper rents farther from the center reflecting higher costs per trip out to the fringe. Those lands closest to the center of the community would then be reserved for market gardening and other high-valued crops that need constant tending, but which fetch a high price in the central market. According to Von Thumen's model, there are a series of concentric rings, each devoted to an activity that corresponds to the rent charged and the cost of transport. In building this model, it was assumed there was only one means of transport, the land topography was flat, and the quality of the farmland was uniform. We also assume that farmers are rational and that maximization of profit is their primary motive.[15]

In the 20th century, the work of Walter Christaller, Alfred Weber, August Loesch, Ernest Burgess, and Homer Hoyt would extend this study of land use to the growth of cities. The "concentric zone model" developed by Burgess in 1925 saw the city as a series of radiating rings, with the CBD at the center and residential areas located on the urban fringe of the city. Using this model and applying the "bid–rent curve", you would predict that high-intensity developments such as skyscrapers and department

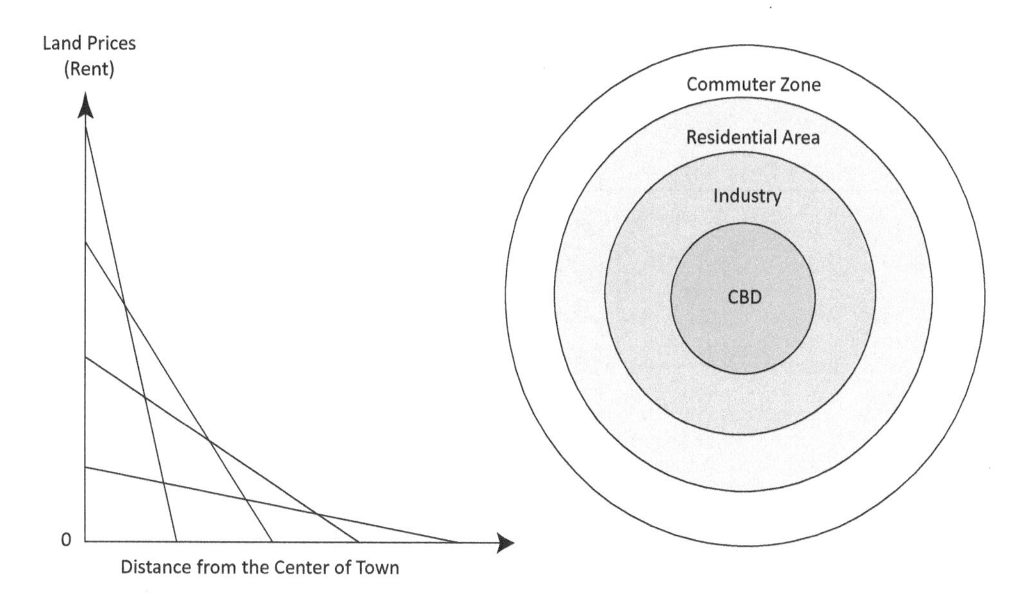


*Figure 6.3* Bid–rent function.

stores would be reserved for the center of the city, while low-density housing would be reserved for the urban fringe (Figure 6.3). Though many towns and cities organize themselves as economic units, geography, politics of land development, and the location of early activities will have a strong influence on the specific location of commercial, retail, industrial, and residential uses. The sector model first proposed by Homer Hoyt in 1939[16] can be seen as a modification of the concentric ring model, introducing sectors that grow out from the city core. Introducing these sectors helps explain the growth of industrial development and working-class neighborhoods established during the early history of North America. Developments grow out from the center of the city as the population increases. In the period following World War II, the construction of highways and mass transit encouraged a flight to the suburbs and bedroom communities, which distorted this basic relationship between distance from the CBD and land values.

In the study of urban form, the work of Walter Christaller and August Loesch on central place theory provides an explanation for the differentiation of goods and services found across a region. If you were to rank villages, towns, cities, and major metropolitan centers, you would find that population and the number of kinds of specialized services are highly correlated. Smaller towns and villages, though numerous, would contain fewer services. The trade areas that support a drug store, grocery store, and a bank might only contain a few thousand individuals. Having a natural monopoly, the services of these small towns and cities would be evenly distributed over the countryside, so as not to compete directly with one another. Residents in each of these trade areas would frequent these local businesses as they go about doing their daily chores. Larger cities containing services frequented less often would form the next level of service. Though found in cities substantially larger in population, more specialized businesses would be fewer in numbers and positioned strategically to service a number of smaller communities. This placement pattern creates a network of smaller and larger communities. In these locations, you might find medical facilities, government

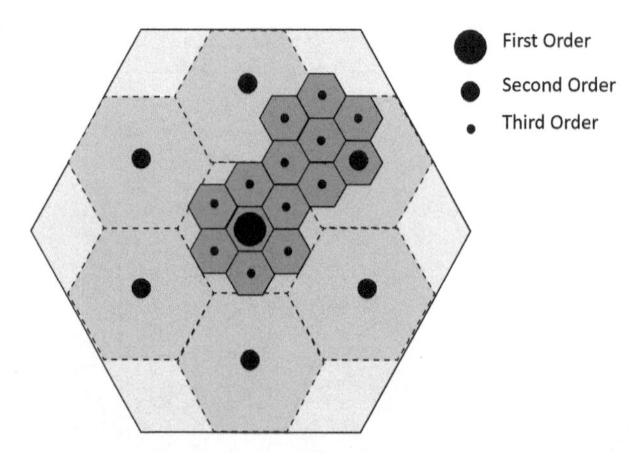


*Figure 7.3* Hierarchical arrangement of cities with an undifferentiated landscape.

offices, and a more varied shopping experience. This hierarchical arrangement has been described spatially as an integrated set of hexagons across the landscape: from the small shopping center up to the metropolitan center (Figure 7.3). Government policy, geography, history, natural resources, and the location of transport systems will shape the character of each community, distorting this geometric arrangement across the landscape. In general, as a city's population grows, we find a larger number of services and firms operating in a particular sector. Graphing these relationships, we find a strong correlation between population and the number of locations providing specific specialized services. For example, consider the location of corporate head offices. When the number of headquarters in each city is plotted against the population for each city, we find a strong correlation between the number of headquarters and a city's population (rank size rule). Though it is not possible to use central place theory to predict the exact number of specific types of business you should find in a city of a specific size, it can reveal which communities are underserved. It is not surprising that a study of Walmart locations in the US reveals a pattern predicted by central place theory.[17]

Clearly, having a model that can predict the price of land would be useful to a developer. By using a model that can determine the value of the land, it is possible to estimate what you should be paying for the building. Using comparative and replacement value approaches to the valuation of the building would allow you to determine if the asking price is fair. It would also allow an investor who is reviewing a number of properties a means of comparing the land value component of one potential investment to another. In cases in which buildings are to be demolished, land value models can help the developer arrive at a price for only the site.

In building a model, simplicity should always be a guiding principle. Mathematical relationships used in a model should always have a theoretical underpinning. The market value of a property as a function of distance to the downtown or CBD is a good example of this case. Given that the amount of land increases by the square of the radius from the center of the city, it is customary to use a non-linear function to describe how the distance to the CBD impacts its property value (Figure 8.3). The Gompertz equation describes this relationship between land values and distance to the CBD:

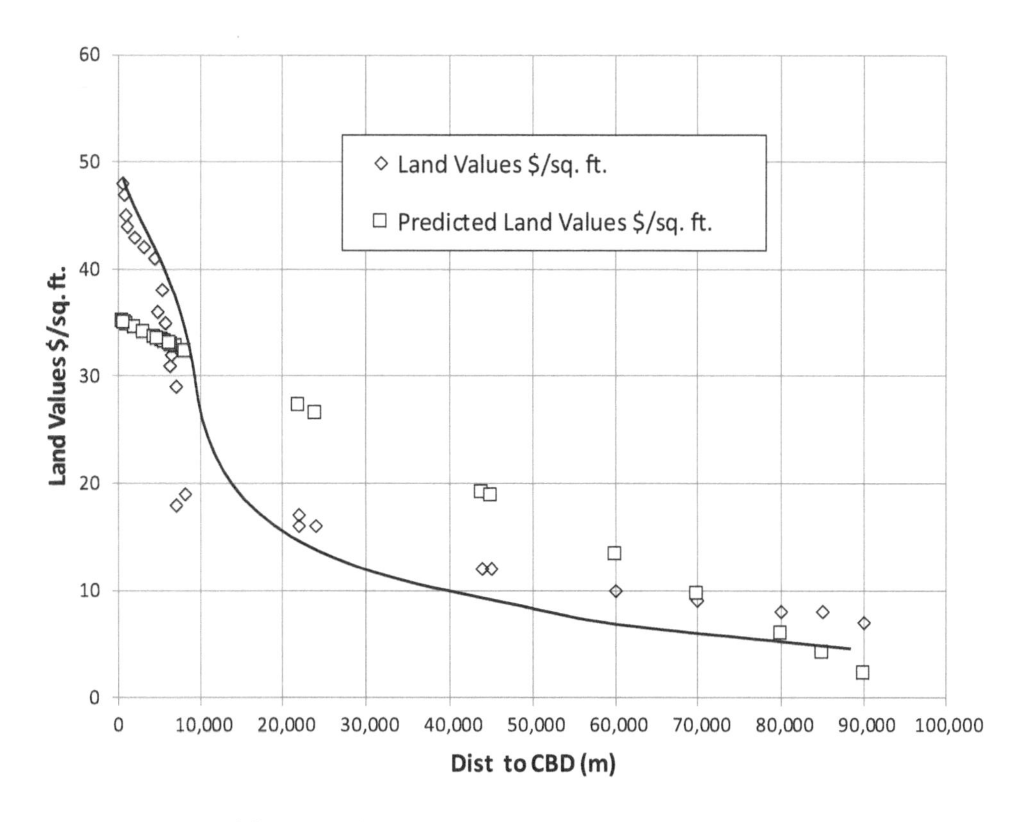


*Figure 8.3* Impact of distance to the CBD on property values (3rd order equation).

$$y(t) = ae^{-bx-ct}$$

Where:
$a$ is the upper asymptote, since $ae^{be^{-\infty}} = ae^0 = a$
$b$ sets the displacement along the $x$-axis (translates the graph to the left or right)
$c$ sets the growth rate ($y$ scaling)
$e$ is Euler's Number ($e = 2.71828 \ldots$)
Note: $b$, $c$ are positive numbers

In applying this equation to solve the problem of land values, we can use the parameters "$a,b,c$" and "$e$" to fit our data to the curve. Though a simple model, the Gompertz equation can be used to predict the distribution of land values in a city that has a prominent CBD. These cities are considered monocentric, in contrast to those that are multinodal. Consider the case of cities such as Los Angeles vs Calgary. For Calgary, where there is a single CBD, we would expect our distribution of land values to conform more closely to this idealized model. For Los Angeles, which has several major commercial districts, we would expect the pattern of land values to be more complex.[18]

### *Gravity models and retail*

Finding the ideal location for a new business in a community requires some knowledge of the location and size of the competition. Estimating the importance of a store, shopping district or mall begins by examining the draw of customers to each location. It is not uncommon to find numerous women's shoe and clothing stores at a single regional shopping mall. Each may cater to a slightly different clientele. Some are for teens; others focus on children clothing; others, on coats; some, on wilderness and hiking attire; still others, on exercise outfits or formal attire. In each category, there may also be clothing stores at different price levels. It is not unusual to find retail stores clustering in close proximity to each other. Many large cities have an established district for clothing, shoes, cameras, and electronics. By clustering in this way, the outlets know that they will attract more shoppers on average, because the knowledgeable shopper understands that they will find what they are looking for by visiting this part of the city. This phenomenon of clustering has been part of urban life for centuries. Today, retail analysts can use a gravity modeling approach to estimate this draw to a shopping mall or downtown shopping district.

Consider two shopping centers located within an hour of each other. The larger of the two shopping centers may be clearly more attractive because it offers a greater variety, even if the other is closer. Estimating the probability that a shopper will go to one location over another is possible using the Huff gravity retail model, which adapts Newton's equation predicting the behavior of celestial bodies to the problem of solving retail shopping locations:

$$P_{ij} = \frac{\frac{S_j}{T_{ij}^{\lambda}}}{\sum_{j}^{n} \frac{S_j}{T_{ij}^{\lambda}}}$$

Where:

$P_{ij}$ = probability of a consumer at a point *i* traveling to retail location *j*
$S_j$ = size of the retail location
$T_{ij}$ = travel time (or distance) from the consumer at a point *i* to travel to location *j*

By applying this model to retail, we can estimate how shoppers will distribute themselves. Larger centers will attract more shoppers over a larger distance. This equation explains why a shopper will travel farther to go to a large regional shopping center, while a local strip mall will attract shoppers only from the immediate neighborhood. This equation, which looks at travel time, can also help explain why shoppers will travel a greater distance to reach a regional shopping mall. For shoppers who own an automobile, it is travel time and not distance traveled that weighs heavily on their decision of where to shop. The actual mechanics of calculating how many potential shoppers will come to a location can be processed by many geographic information systems (GIS) applications. Within these applications, it is possible to consider a range of factors, including multiple competing shopping centers. Transportation planning and retail consultants are usually called upon to help analyze these types of problem. In particular, they can assist a developer in discriminating between several locations and finding the one that provides the best opportunity for their particular retail outlet. This approach to locational analysis is also relied on by big box retailers and fast food restaurants to decide where the best location is for their next outlet.

### *Transportation modeling*

Critical to any development is access to transportation. The transportation planner must know where people live and work to determine the demands on public transportation, and on roads and highways for truck and auto traffic. The process of forecasting demand requires extensive data on the travel behavior of their citizens. Demographic, income, and survey data on travel behavior are needed for models that look at origins and destinations of commuters. Though rush hour commuting can place severe pressure on any transportation network, many trips are made each day that are unrelated to work. In communities where public transit is limited, auto traffic dominates. This is particularly true in suburban communities, where mass transit is often limited to a few routes and may be infrequently available, except during rush hours. In every city, numerous trips are made daily: grocery shopping, errands, picking up the children after school, and taking them to activities such as soccer and band practice. It is not unusual for one adult in a family to devote a significant portion of the day to being a chauffeur and delivery person. Recent developments in transportation modeling have allowed modelers to incorporate the behaviors of every commuter into large-scale model. Using agent behavior theory, we can simulate the activity of every citizen, commenting in their daily life. Using massive data sets built from the US Census data on population, family structure, and income, it is possible to build a detailed model of travel behavior.[19] Using this approach, it is possible to evaluate how well a city's transportation system is performing throughout the day. When land use patterns or the availability of transit are altered, it is possible to see how individual travel behavior will change in response.

In designing a transit system, planners must consider more than only the trips made by individuals to and from work and home. Much of transit analyses focuses on the delivery of goods and materials by truck, rail, and air – even by bicycles and draft animals, which, in some parts of the world, still make a large contribution to the economy. Planning for transportation infrastructure depends in part on where people live and work. It is for this reason that transportation planners should collaborate with policy and land use planners when making decisions about zoning and the location of future development. For example, by zoning higher-density residential near a mass transit station, we can reduce the need to drive by providing public transport. Similarly, by having residential development close to work, commute times may be reduced in theory. Of course, in two-income households, it is less likely that the household will find a residence in close proximity to both workplaces. Often, at best, the household can locate so that one individual is close to their work location, while the other will have a modest commute time.

### *Urban land use models*

Urban land use models offer policymakers, as well as the developer, an opportunity to explore the impact of new municipal policy on future development. For example, if we provide grants for construction of affordable housing in single-family neighborhoods, will the shortage of rental units for lower-income households be alleviated? To explore even simple policy changes, models need to simulate the actions of all the players involved in property development. We may want to know how many units will be produced over the next year and who will rent them once they are constructed. This will require a model that simulates the behavior of all those actively involved in the real estate industry. Questions a model might consider include the following. When will a developer consider it profitable to build an apartment building in a residential neighborhood? What will be the response by mortgage lenders who make capital available for new ventures? Once the project is completed, will there be enough renters willing to occupy this new apartment building? To simulate the behavior of renters, investors, developers, and lenders, the model will need to incorporate decision rules that can simulate the behavior of all the parties that contribute to the success of a new real estate development. For the city planner, models that can predict the impact of change in public policies are particularly useful. Being able to predict the future state of the housing market will have a direct bearing on the need for health, education, transportation, and social services.

Creating urban models for evaluating growth policies requires a sensitivity to the assumptions that underlie the projections. Variables, including employment growth, may be difficult to predict beyond a few years. Without employment opportunities, cities and communities stagnate. However, many land use models assume that job growth and employment opportunities increase, with population growth at the same rate as that of the near past. Experience has taught us that many cities that emerged as major industrial centers in the 19th and 20th centuries went through periods of growth, maturity, and decline. Planners and developers with time horizons measured in years, not decades, may overlook the factors needed to accurately predict the future.

One of the goals of land use modeling is to determine where development is most likely to occur. An early example of urban modeling is the California urban futures model (CUFM), created by John Landis at the University of California, Berkeley. Like many models, it is useful not only as a forecasting tool, but also as a means for evaluating the impact of policy on development. The underlying principle in this model is that developers will seek maximum profitability in their choice of projects. The model has two components: demand and supply. The demand model is based on population projects based on Census data. Sites that are too difficult to build on or have environmental restrictions are eliminated from the supply. The remaining vacant sites are sorted by profitability and then population statistics are used to fill these sites until the demand is satisfied. Matching supply with demand looks at the profitability of building on vacant sites. The assumption is that the most profitable project will have the greatest opportunity for success. A list of potential projects sorted by profitability is then used to determine which projects will be built first. The obvious advantage to these models is that they can show where future development will most likely occur.

There are several caveats to all these models. First, the measure of profitability is based on a general cost model that may have to estimate the cost of land acquisition, construction, and financing. In reality, these inputs vary from site to site depending on the details of the design. There is also some variance in construction costs and profitability. On a per square foot basis, a large expensive home will cost more to build, but still may be a more profitable venture than building affordable housing. Also, the assumption that demand is an extrapolation of a past trend is good only for the short term. There is also considerable variance in demand functions. When population growth falls off because of a lack of employment opportunities, residents are forced to move to other cities, resulting in a fall-off in the demand for residential development.

Urban land use models assume that the actors or agents have perfect knowledge of their environment. In reality, perfect knowledge of the real estate development market is an economic assumption. Many developers operate with only limited or partial knowledge of their industry. A developer may decide to pursue a project because they already own the property and a marginal profit is better than no profit at all. In some cases, a project keeps key employees busy who otherwise might have been dismissed because of a lack of work. Other problems faced by these models is that they do not consider the competing development opportunities in nearby cities. A neighboring community offering better overall value may see greater activity and interest among developers and future home buyers.

### *Building models and the need for data: a practical approach*

One central issue in building any model will be the availability of data on every aspect of a community. In some cases, there may be high costs to acquire data. In other cases, the data is collected only on an infrequent basis and may be out of date. Also, given that we may be building a model that is a function of having data that is specific to a location, data summarized for an entire neighborhoods or city will be of little value. Once data has been aggregated to the level of a town or city, it is impossible to say that any particular value reflects what is happening on your block. The most widely used and readily available is Census data, which exists at the parcel and block levels. A vast array of data is contained in the Census: not only population by cohort (five-year age groups), but also information on businesses, housing stock, and employment are a few categories that are available to the public. It is now possible to acquire this data in electronic formats compatible with GIS and spreadsheet software, making it easy for anyone to build a model. Cities will also assembly population and business data by community or neighborhood. When Census data is collected only on a five-year basis, the information can be out of date after a few years, though often it is better than no data at all. Cities also have data on attendance in school, student success on national tests, crime, and local businesses. Though sometimes this data is available only as reports, as more cities adopt open data access policies this data will become available as downloads to the public.

Government agencies, including housing, labor, transportation, commerce, and the Treasury, also publish data that is useful in building models. Though, in the past, data collected on population, agriculture, industry, and commerce served government planners, today, in addition to the federal government, many cities, states, and provinces recognize that open data policies can be the basis of new mobile applications designed to serve the public and businesses.[20] One example of how this open data policy can benefit the public and the business community is the San Francisco Enterprise GIS Program. You can download from its website data maintained by the Program's corporate GIS. Those without GIS software can access the data via a web interface, including data on crime, education, zoning, building permits, and even the trees planted along the street. As part of the Open City Agora Project, the public can use a web browser to access GIS data online. There are even mobile applications based on this data that allow the user to provide residents with useful information on services. Referred to as the "Neighborhood Score", this data allows the user to provide the city with resident feedback on what they feel needs to be improved in their community. Based on crowd-sourcing as the means of acquiring data, this application will be used to help correct problems and provide the community with the location of problem areas. What is interesting about the San Francisco initiative is the recognition that open sources can actually stimulate economic development by providing software developers with the data needed to drive new apps for the web and mobile devices.[21] The website even provides some assistance on how to get started. Ever wonder how Zillow makes its "Zestimate" for a showcased property? The answer in part is that it analyzes data from both public and commercial sources.[22]

In building any spatial model, it is essential that the data exists for the study area. Knowing the boundaries of the actual study area are just as important. In most cases, time and resources will limit the focus to a particular city. Where communities are spaced far apart from one another, boundary conditions will not be a major factor. But in regions in which cities and towns are close together, actions taken by one community will impact its neighbors. If, for example, there is a sudden increase in property and school taxes, buyers may start looking in adjacent communities for housing. Similarly, if a municipality changes its zoning, barring a big box retailer in their community, an adjacent community can become an attractive location for this development. In this age of globalization, actions taken halfway across the world can impact local development, as has been seen most dramatically in the industrial sector, where lower-cost production overseas has shut down or stifled the development of whole industries, including clothing, shoes, and electronics.

## Modeling human behavior: models are rational, people are not

One of the challenges in model building is that the actors in the model are not always rational. Buying real estate is part investment, home, and expression of ourselves. Economic models assume that all decisions are rational decisions. Research in economic behavior has revealed that we are not rational when it comes to many life decisions, including economic and financial matters. We now even have terms such as "retail therapy" describing the action of shoppers who make purchases not for their value or use, but for a sense of pleasure or a desire to feel better. Many life decisions, including those that seem rational, have an emotional component.

The study of economic behavior has enlightened us to the importance of human behavior in saving for future retirement. Research has shown that when individuals were given the opportunity to save for the future, a majority were unable to carry through even when financial resources were available. Evaluating financial risk is another factor that is subject to human emotion. In Chapter 2, expected return was used to calculate the likely rates of return for two investment choices. In a classic experiment conducted by Kahneman and Tversky, subjects were presented with choices between two sets of outcomes. Each subject was asked to select their preferred option. In these experiments, participants were asked to choose between a pair of outcomes: one positive; the other negative. In tallying the results, it was found that the respondents had a strong preference for minimizing loss over maximizing gain, which attests to our risk-averse nature in making a financial decision. This aversion towards risk was also demonstrated by an experiment in which doctors were asked to make choices between various cancer treatments when death was imminent. Again, the preference for options when death was less likely were favored over those options with better long-term statistical survival rates, but in which death was more likely in the first year of treatment.[23]

Taking account of human behavior in our modeling of economic choices will always present a challenge that is simply easier to overlook. In classical economics, rational behavior is assumed to underpin our economic decisions. However, upon closer inspection, human behavior does not mimic these predictions. Individuals, when faced with choices, may not always select those that would have been decided by a rational choice model. Research in the field of economic behavior can help us understand better the attitude of all the players involved in real estate development. Developers, when considering possible development options, may not always select those proposals that generate the greatest reward. Instead, they may select proposals that generate more modest rates of return because they have a lower risk quotient. Similarly, home buyers who must consider the resale potential of their homes may opt for more conservative designs in well-established neighborhoods over houses that feature bold design statements in marginal neighborhoods, even though the marginal neighborhoods could see major increases in property value in the future. We all know individuals who bought properties in communities in transition. At the time, we may have considered their decision a very risky proposition with no guarantee that homes in this area would ever appreciate. However, in retrospect, their choice was very fortuitous, the neighborhood having since seen significant appreciation in real estate values. How much risk an individual is willing to assume is constrained by their experiences and knowledge. In a world dominated by social and mass media, it is often the negative news events that not only are more prominent, but also are more influential. It is not surprising that our tendency is to avoid risk over maximizing economic gain.

## Conclusions

Model building can be useful in estimating both the supply and demand for new and existing space in each sector: residential, commercial, retail, and industrial. Though it is not possible to predict the future with absolute certainty, models are still important in that they allow us to test our assumptions and consider alternative scenarios. All models are constrained by their underlying assumptions. Even with their shortcomings, models can be designed to improve our understanding of spatial questions of development or macroeconomic events, or to explain the decisions made by the individual. Models require data if they are to be applied to real-world problems. With economic and demographic data available in the public domain, model building has become a mainstream activity pursued by economists, as well as urban and transportation planners. Though data sources have become more generally available, data may still be lacking to answer the specific questions posed by a project. Model building is a creative activity, which allows us to explore an economic landscape. Acknowledging that real-world events can alter the predictions of any model is both instructive and precautionary. Even with these inherent drawbacks, models are important for understanding the development of a landscape. A quote attributed to Dwight D. Eisenhower, "In preparing for battle, I have always found that plans are useless, but planning is indispensable", should be applied to the effort to build models that can assist in planning for the future.

## Notes

1 Taleb, 2010.
2 Lee, 1973; Lee, 1994; Harris, 1994.
3 Kummerow, 1997.
4 Caples et al., 1997; Isakson, 2001.
5 Wilson, 1997.
6 Fisher et al., 2009.

7 Detweiler and Radigan, 1999.
8 Zillow, n.d., http://investors.zillowgroup.com/index.cfm
9 Catanese, 1972.
10 Ibid.; Arsanjania et al., 2013.
11 Weber and Devaney, 1996.
12 Pyhrr et al., 1996.
13 Wilson, 1993.
14 Wendt, 1972.
15 Brooks, 2006, 30.
16 Hoyt, 1939.
17 DeMarco and Matusitz, 2011.
18 Wadell et al., 1993.
19 Miller and Salvini, 1998.
20 San Francisco Enterprise GIS Program, see: https://data.sfgov.org/
21 The Socrata Open Data API, see http://dev.socrata.com/
22 Zillow and MLS Listing Syndicators, Realtors Push Back, www.ultimateidx.com/blog/realtors-push-back-against-zillow-and-mls-listing-syndicators/

## Bibliography

Abraham, John, John Douglas Hunt, Robert A. Johnston, Michael McCoy, Shengyi Gao, and Eric Lehmer. "Incremental cooperative development of land use models in California." (April 2010): 1–29.

Abraham, John E., John Douglas Hunt, and Allan T. Brownlee. "Incremental modeling developments in Sacramento, California: Toward advanced integrated land use-transport model." *Journal Transportation Research Record* 1898, (2004): 108–113.

Alonso, William. *Location and Land Use, Towards a General Theory of Land Rent*. Cambridge, Massachusetts: Harvard University Press, 1964.

Arsanjania, Jamal Jokar, Marco Helbichb, Wolfgang Kainza, and Ali Darvishi Bolooranic. "Integration of logistic regression, Markov chain and cellular automata models to simulate urban expansion." *International Journal of Applied Earth Observation and Geoinformation* 21, (April 2013): 265–275.

Ason, Okoruwa. A., Hugh O. Nourse, and Joseph V. Terza. "Estimating Sales for Retail Centers: An Application of Poisson Gravity Model." *The Journal of Real Estate Research* 9, no. 1 (Winter 1994): 89–97.

Batty, Michael and Yichun Xie. "From Cells to Cities." *Environment and Planning B: Planning and Design* 21, (1994): 531–548.

Brett, Deborah L. and Adrienne Schmitz. *Real Estate Market Analysis, Methods and Case Studies*. 2nd ed. Washington, DC: ULI, 2009.

Brooks, S. Michael. *Canadian Real Property Theory and Commercial Practice*. Toronto, Ontario: Canadian Real Property Association, 2006.

Caples, Stephen C., Michael E. Hanna, and Shane R. Premeaux. "Least Squares Versus Least Absolute Value in Real Estate Appraisals." *The Appraisal Journal* 65, no. 1 (January 1997): 18–24.

Chen, Yimin, Xiaping Lui, Li Xia, Yilun Liu, and Xu, Xiacong. "Mapping the fine-scale spatial pattern of housing in the metropolitan area by using online rental listings and ensemble learning." *Applied Geography* 74, (September 2016): 200–212.

Demarco, Michael and Jonathan Matusitz. "The impact of Central Place Theory on Wal-Mart." *Journal of Human Behavior in the Social Environment* 21, (2011): 130–141.

Detweiler, John H. and Ronald E. Radigan. "Computer-assisted real estate appraisal: A tool for the practicing appraiser." *Appraisal Journal* 3, (July 1999): 280–286.

Fisher, Lynn, Henry Pollakowski, and Jeffrey E. Zabel. "Amenity-based housing affordability indexes." *Real Estate Economics* 37, no.4 (2009): 705–746.

Goddard, Bryan L. "The Role of Graphics Analysis in Appraisals." *The Appraisal Journal* 67, no. 4 (October 1990): 429–435.
Harris, Britton. "Quantitative models of urban development: Their role in metropolitan decision-making in decision-making." in *Urban Planning*, Ira Robinson eds., (Beverly Hills, California: Sage Publications Inc., 1972): 115–138.
Harris, Britton. "The real issues concerning Lee's requiem." *American Planning Association Journal* 60, (Winter 1994): 31–34.
Hodges, McCloud B., Jr. "Three Approaches?" *The Appraisal Journal* 61, no. 4 (October 1993): 553–564.
Hoyt, H. The Structure and Growth of Residential Neighborhoods in American Cities (Washington, Federal Housing Administration, 1939).
Huang, Qingxu and Dawn C. Parker. "Tatina Filatova and Shipeng Sun. "A review of urban residential choice models using agent-based modeling." *Environment and Planning B* 41, (2013): 189–661.
Huff, David L. "A Programmed Solution for Approximating and Optimum Retail Location." *Land Economics* 42, (1966): 293–303.
Huff, David L. "A Probabilistic Analysis of Shopping Center Trade Areas." *Land Economics* 39, no. 1 (1963) (February 1981): 81–90.
Hunt, Douglas. "Modeling choice behavior of non-mandatory tour locations in California – An experience." *Travel Behaviour and Society* 12, (May 2017): 122–129.
Isakson, Hans R. "Using Multiple Regression Analysis in Real Estate Appraisal." *The Appraisal Journal* 69, no. 4 (October 2001): 424–431.
Jacobs, Erich K. "Appraising the appraisal: A developer's guide to appraisal techniques." *The Journal of Real Estate Development* 4, no. 4 (1989): 37–44.
John Douglas, Hunt and John Abraham. "Modeling logistics and supply chain with an integrated land use transport model: PECAS." *Ninth International Conference of Chinese Transportation Professionals (ICCTP)* (August 2009) Harbin, China.
Kahneman, Daniel and Amos Tversky. "Prospect Theory, An Analysis of Decision Under Risk." *Econometrica* 47, no. 2 (March 1979): 263–291.
Kahneman, Daniel and Amos Tversky. "Choice values and frames." *American Psychologist* 39, (April 1984): 341–350.
Landis, John. "Imagining Land Use Futures, Applying the California Urban Futures Model." *American Planning Association Journal* 61, no. 4 (Autumn 1995): 438–457.
Landis, John. "Modeling urban land use change: Approaches, state-of-the art, prospects *Socio economic Benefits Workshop: Defining, measuring, and Communicating the Socio-economic Benefits of Geospatial Information* (June 2012): 1–7.
Landis, John and M. Zhang. *Modeling Urban Land Use Change: The Next Generation of California Urban Futures Model.* Berkeley, California: University of California Berkeley (May 15, 1997).
Lee, Douglas. "Requiem for large scale models." *American Planning Association, Journal* 39, no. 3 (May 1973): 163–178.
Lee, Douglas. "Retrospective on large-scale urban models." *American Planning Association, Journal* 60, no. 1 (Winter 1994): 35–40.
Lin, Jian, Bo Huang, Min Chen, and Zhen Huang. "Modeling urban vertical growth using cellular automata – Guangzhou as a case study." *Applied Geography* 53, (2014): 172–186.
Linne, Mark R. and John Cirncione. "Integrating Geographic Information and Valuation Modeling for Real Estate." *The Appraisal Journal* 78, no. 4 (Fall 2010): 370–378.
McCloud B. Hodges, Jr. "Three Approaches?" *The Appraisal Journal* 61 no. 4 (October 1993): 553–564.
Miller, E. and Paul A. Salvini. The Integrated Land Use, Transportation, Environment (ILUTE), "Modeling System: a Framework," Prepared for presentation at the *77th Annual Meeting of the Transportation Research Board* (January 1998).

Moeckel, Rolf and Kai Nagel. "Maintaining Mobility in Substantial Urban Growth Futures." International Conference on Mobility and Transport Transforming Urban Mobility *Transportation Research Procedia* 19 (June 2016): 70–80.

Mushinski, David and Stephenn Weiler. "A Note on the Geographic Interdependence of Retail Markets." *Journal of Regional Science* 42, no. 1 (2002): 75–86.

Nassim Nicholas, Taleb. *The Black Swan: The Impact of the Highly Improbable*. New York: Random House, 2010.

Pyhrr, Stephen A., Waldo L. Born, Rudy R. Robinson III, and Scott R. Lucas. "Real Property Valuation in a Changing Economic Market Cycle." *The Appraisal Journal* 64, no. 1 (January 1996): 24–26.

Rattermann, Mark R. "Highest and best use problems in market value appraisals." *The Appraisal Journal* 76, no. 1 (2008): 23–25.

San Francisco Enterprise GIS program, http://sfgov3.org/index.aspx?page=3959 also see: https://data.sfgov.org/ (2014).

Smith, Charles A. and William C. Forrest. "Forecasting likely job growth or decline with net deficit techniques." *The Appraisal Journal* 61, no. 2 (April 1993): 254–260.

Thomas, Kuhn. *The Structure of Scientific Revolutions*. Chicago, Illinois: University of Chicago Press, 1996.

Tversky, Amos and Daniel Kahneman. "Rational choice and the framing of decisions." *The Journal of Business* 59, no. 4 (Part 2 1986): 251–278.

Tversky, Amos, Paul Slovic, and Daniel Kahneman. "The causes of preference reversal." *The American Economic Review* 80, no. 1 (March 1990): 204–217.

Waddell, Paul and Gudmundur F. Ulfarsson. "Dynamic simulation of real estate development and land prices within an integrated land use and transportation model system." *TRB Annual Meeting*, 2002.

Wadell, Paul. "Integrated land use and transportation planning and modelling: Addressing challenges in research and practice." *Transport Reviews* 31, no. 2 (March 2011): 209–229.

Wadell, Paul, Brian Berry, and Irving Hoch. "Residential property values in a multinodal urban area: New evidence on the implicit price of location." *Journal of Real Estate Finance and Economics* 7, (1993): 117–141.

Weber, William and Mike Devaney. "Can consumer sentiment surveys forecast housing starts?" *The Appraisal Journal* 64, no. 4 (October 1996): 343–348.

Wendt, Paul F. "Highest and best use – fact or fancy." *The Appraisal Journal* 40, no. 2 (1972): 165–174.

Wilson, Donald C. "Rank correlation analysis of comparable sales from inefficient models." *The Appraisal Journal* 65, no. 3 (July 1997): 247–254.

Wilson, D. "Highest and Best Use Analysis: Appraisal Heuristics versus Economic Theory," *The Appraisal Journal* (January 1995): 11–26.

Zhong, Ming, John Douglas Hunt, and Lu Xuewen. Duration modeling of Calgary household weekday and weekend activities: How different are they? *TAC/ATC 2007–2007 Annual Conference and Exhibition of the Transportation Association of Canada: Transportation – An Economic Enabler*, 2007.

Zong, Ming. "Critical issues in transportation system planning, development, and management," *Proceedings of the 9th International Conference of Chinese Transportation Professionals, ICCTP 2009*, 358 (2009): 1300–1305.

# 4 An introduction to financial modelling

## Introduction

Every day, we use financial models to help us make decisions on how best to spend and invest our money. Often, our decisions involve choices between two or more scenarios. Consider the decision whether to rent or purchase an auto. My plan is to buy or lease a new car every four years. Establishing which is the better deal will require us to consider the number of payments, the amount being paid, and whether there will be a cash payment required at the end of the lease. Also, are there different costs associated with each option? Will the lease cover oil changes and insurance? If I lease a car, will there be additional charges for mileage above a number specified in my contract? If I plan to the buy an auto, what will be the value at the end of the four-year term? Ultimately, making a decision requires a framework within which we can compare different outlays of cash over time. Without a measure of the relative cost of one scenario against the other, how will I know which is the better deal? A financial model can give us insights into the true costs and benefits of one alternative over another.

In real estate development, we are often reviewing multiple opportunities as we look for the best project among a field of possibilities. If we are acquiring a property for an investment fund such as a real estate investment trust (REIT), we will need a basis for choosing between several prospective projects for our portfolio. Often, these projects extend over many years, each project with different costs and revenue streams. Financial models can be useful tools in evaluating the potential success of a real estate development project. They can also consider how changes in economic circumstances may impact the success of a project. Many things can change over the lifetime of a project that can impact its financial success. Rates and terms can change for borrowing. Changes in short-term rates will impact the construction costs of a project, while long-term rates will impact the overall return over the lifetime of the investment. Changes in the cost of labor and materials will impact the cost of the project, but even delays caused by strikes or skill shortages can extend the time to completion and impact the project by increasing borrowing costs. Rents can change with fluctuations in demand and supply of space in a community. For example, in a town or small city where the demand for commercial space is limited, the completion of two commercial projects at the same time has the potential to reduce rents. During the period when excess space is being absorbed, cash flows can be negatively impacted for some time to come. Similarly, if economic times improve, creating a temporary housing shortage, rents and prices can rise. The supply curve is very inelastic in the short term for real estate. It is difficult to increase the number of units overnight. If we are dealing with commercial real estate, once the tenants have signed a lease for ten years, for example, there is more confidence on the rate of returns on this investment. In some leases, you might find an escalator clause that adjusts rents upward to mirror changes in the consumer prices index (CPI). However, this would only make the numbers look better. Changes in the tax laws, which can impact depreciation schedules, tax credits, and the deductibility of expenses, though infrequent, will ultimately impact the long-term investment returns for a project. For a developer selling a project to a financial institution, understanding the buyer's perspective will be important in arriving at the terms of the sale.

Financial models can assist in evaluating multiple scenarios for a project. Best, worst, and expected cases can be tested against possible changes to the underlying assumptions for a projection. Though, in the short term, we expect businesses to proceed as usual, sudden changes in the economy can impact the success or failure of a project in the long term. When you are in an upward expansion cycle, it is difficult to visualize a time of gloom and contraction. Likewise, when the economy is doing poorly, there is general pessimism about the future. The length of cycles in real estate development, though, tracks closely with the business cycles. Knowing exactly when the next peak or trough will occur is difficult to predict.

### *Capitalization rates*

One approach you could use to decide which of several properties to acquire is to compare the capitalization rate (or cap rate) of each building. The cap rate is the net operating income (NOI) revenue *less* expenses, divided by the property value. As a simple measure of profitability, it is often used as one of the standard methods by means of which appraisers determine the value of investment property.

The cap rate is defined as:
Cap rate = NOI/Property valuation

If NOI is $100,000 and the property value is $8 million, then the cap rate can be calculated as:

Cap rate = $200,000/$8,000,000 = 2.5%

Like a bond, we are receiving an annual income from the real estate investment. Like fixed-rate investment, depending on the rating we would also expect investments with a higher risk to offer higher interest rates. For two bonds of similar maturities and risk, we would prefer those with higher rates. Often measured as differential above the risk-free rate, cap rates reflect local differences in risk and return in each particular sector. For example, if commercial office space development is considered riskier than multifamily apartment buildings, this should be reflected in the cap rate. Cap rates can also reflect regional differences. Every locale will differ in size, age, and quality. The demand for each type of space will also vary within a city by neighborhood. Cap rates will reflect these differences. However, acquiring accurate data on cap rates can be difficult. Though government agencies publish data on cap rates, data can be based on limited surveys and can be out of date. Even with these limitations, some knowledge of cap rates could help us decide which of two properties is a better deal.

In the example that follows, two apartments are under consideration by a group of investors. Looking at the valuations based on cap rates, their decision about which property is most valuable is obvious: if there is little room for negotiating on price, then Apartment Building XYZ would be their choice.

Valuation = NOI/Cap rate

However, the choice is never this clear. Making a decision on which property to acquire involves considering not only current NOI, but also future revenue and expenses, which can change. The direction of this change can have a major impact on future investment returns. If, for example, we know revenues are growing faster for Apartment Building ABC relative to expenses when compared against Apartment Building XYZ, then it might

| | *Cap rate* | *NOI* | *Valuation* | *Asking price* |
|---|---|---|---|---|
| Apartment Building ABC | 7.8% | $150,000 | $1,923,077 | $1,995,000 |
| Apartment Building XYZ | 7.8% | $134,000 | $1,829,268 | $1,775,000 |

be better to pay a small premium to capture future gain. Likewise, if the property manager of Apartment Building ABC has deferred maintenance and significant expenditures will need to be made after taking possession, then Apartment Building XYZ becomes a better choice. Clearly, a decision about the future of an asset that will be acquired as a long-term investment must not be based on a single measure. Only after the examination of the physical condition of the building and its financial history and management can we be confident of the proper course of action to take. Making a decision about the projection of future revenue and expenses will require a method of weighing the value of a future cash flow stream in the present. Having some understanding of methods for considering the time value of money in our calculations will be critical to this analysis.

### *Time value of money*

Creating simple financial models requires an appreciation for the time value of money or present value (PV). Built on the premises that money received in the future is not worth the same to me as cash in my pocket today, PV provides a value of money received in the future in present-day dollars. Consider a simple problem of making a loan to a close associate in the amount of $100,000. The plan is for my associate to pay back $100,000 after a year. The loan is secured by a property of much greater value, so the risk is minimum. In making this loan, I lose the opportunity to invest the $100,000 in another project. At a minimum, I should earn the risk-free rate on the loan or what I would receive if I were to buy a 1-year Treasury bond. If, in this example, the risk-free rate is 3%, the $100,000 investment at the end of year would return $103,000. In our example, the $103,000 received when I cash out the certificate of deposit (CD) a year from now is worth $100,000 at present. At a minimum, I would ask the borrower to pay me $103,000 at the end of the year if there were little risk of default. However, if the borrower plans to make a payment of $100,000 in a year's time, I would loan them something less than $100,000.

Using the formula for PV, it is possible to calculate the value of cash received any time in the future in present-day dollars. In this example, the loan I would be willing to make is $97,087. However, if the payment were made three years from now, then the PV of the $100,000 would be only $91,514:

$$PV = R_t/(1+r)^t$$

Where:

$R$ = cash flow received for a time $t$ in the future
$r$ = discount rate, or the percentage rate earned during the period $t$
$t$ = time period

**Example 1**
For a one-year period after repayment of a $100,000 loan, if
$r = 3\%$
$R_t = 100{,}000$
$t = 1$
then:

$$PV = 100{,}000/(1+.03)^1 = \$97{,}087$$

**Example 2**
For a three-year period after repayment of a $100,000 loan, if
$r = 3\%$
$R_t = 100{,}000$
$t = 3$
then:

$$PV = 100{,}000/(1+.03)^3 = \$91{,}514$$

If I plan to make this loan, I should acknowledge that the risk is much higher than merely placing the money in the bank and earning CD rates. In using a higher interest rate or discount factor, the amount I would be willing to lend will be far less than $91,514 if I am to receive $100,000 in three years' time.

Selecting the appropriate discount rate for a project should reflect the risks associated for that particular class of projects. If the borrower has excellent credit, the discount rate may be only marginally higher than the risk-free rate. In contrast, a project with a high degree of risk will have a markedly higher rate.

### *Net present value (NPV) and internal rate of return (IRR)*

In applying the concept of net present value (NPV) to a financial model, the timing of payments and cash received matter. Income received in the future from a real estate project will not be worth as much on an NPV basis as cash in hand. Likewise, the timing of expenses will impact the overall performance of a real estate investment. As is often the case, we would like to know if a series of payments over time generates a specific rate of return. In real estate, unlike a bond or CD, our cash flows can vary from month to month. It might take two years or longer to complete the project, during which time we are paying out cash to our architect, lawyer, accountant, and contractor. Once completed, the property would be rented and would generate a net positive cash flow after expenses are paid. Hopefully, the rents would go up over time – at least to match or exceed the cost of living index. After 10 or 15 years, the property might be sold and a large cash payment be received. The calculation of NPV allows us to answer the question of whether a series of cash flows would generate our expected rate of return. Net present value merely calculates the PV for each year of our project and sums them up over the entire lifetime of the project. If the discount rate is 6%, then, when the NPV is greater or equal to 0, we have met our goals. Less than 0 means we have not achieved our goal of generating a 6% rate of return on our cash flows.

To calculate the NPV for any cash flow stream, we could construct a simple spreadsheet using Microsoft Excel. We will need to become familiar with a few of the financial functions available in Excel. Using Microsoft Excel or any other spreadsheet program, the user has access to financial functions, including net present value (NPV) and present value (PV):

$$\mathrm{NPV} = \sum_{n-0}^{N} \frac{C_n}{(1+r)^n} = 0$$

In our example, we are in building an office building (Figure 1.4). The project takes two years to complete. Two payments are made to the general contract: one for $400,000 and another for $100,000, or $500,000 in total. During the nine years of the project, there are cash outflows for real estate taxes, insurance, maintenance, and other expenses, and cash inflows from rent and other fees. At the end of the project, the property is sold (a large payment in the ninth year includes rents and proceeds from the sale). If we consider the difference between the outflows ($788,000) and inflows (–$535,000) over the lifetime of the project, the net is $253,000, which might suggest we have exceeded our goals. After all, the annualized return for the years beginning in the third year range from 4.4% to 112%. However, calculating the difference between cash in and cash out tells us little about our rate of return during the life of the investment. If, in this example, our objective was for a minimum of 6% overall return on the investment, using the NPV function

Discount Rate = 6.00%

| Year | Cash Outflows | Cash Inflows | Net | PV | Annual return on an investment of $500,000 |
|---|---|---|---|---|---|
| 1 | | -400,000 | -400,000 | -377,358 | 0.00% |
| 2 | | -100,000 | -100,000 | -89,000 | 0.00% |
| 3 | 27,000 | -5,000 | 22,000 | 18,472 | 4.40% |
| 4 | 29,000 | -5,000 | 24,000 | 19,010 | 4.80% |
| 5 | 34,000 | -5,000 | 29,000 | 21,670 | 5.80% |
| 6 | 39,000 | -5,000 | 34,000 | 23,969 | 6.80% |
| 7 | 44,000 | -5,000 | 39,000 | 25,937 | 7.80% |
| 8 | 50,000 | -5,000 | 45,000 | 28,234 | 9.00% |
| 9 | 565,000 | -5,000 | 560,000 | 331,463 | 112.00% |
| **SUM** | **788,000** | **-535,000** | **253,000** | ***2,397*** | ***NPV*** |
| | | | | ***6.079%*** | ***IRR*** |

*Figure 1.4* Calculation of NPV and IRR for a $500,000 investment made over two years. NB Net cash flows are calculated over a nine-year period.

will answer the question, "did we achieve our goal?" When we use the NPV function in Excel, we obtain a value of $2,397 for this series of cash flows. Simply stated, we satisfied the requirement of generating a minimum return of 6%. In this case, it is slightly greater than 6%, since the NPV is greater than 0 ($2,397).

In many cases, we are considering a number of projects with different cash flow projections. We may also need to consider the impact of different financing options, each with a different impact on cash flows. As is often the case, we would like to know the rate of return for our cash flow stream.

Internal rate of return (IRR) can provide us with the discount rate for the cash flow stream. With an NPV greater than 0, we would expect the IRR to be slightly greater than 6%. The formula, which is based on our NPV calculation in our example above, gives a value of 6.079%. Net present value is a useful number in comparing projects with different cash flow streams, unlike our holding period return (HPR) calculation for bonds that provide us with a specific rate of return based on a coupon that does not vary from year to year (see Chapter 2). For our real estate investment, we can generate both positive and negative cash flows that vary from year to year in different amounts. The IRR provides us with some measure of investment performance. One caveat that must be acknowledged in relying on IRR as a measure of return, however, is that IRR assumes that the cash generated from the investment will also provide returns equal to the IRR from the original investment. Often, as is the case in real estate investments, cash placed in a holding account is generating money market rates that are substantially lower than the expected rate of return for the investment.

## Applying theory to practice

In building a financial model for a real estate project, the level of detail can vary significantly depending on the goals and objectives of the exercise. If the developer is concerned only with the period of acquiring the land, planning, and construction, the time frame will be much shorter than for the investor who acquires the property and holds it for 10 or 20 years before selling it to another investor. It may also be important for the developer who is concerned only with a period of a few years to show cash flows on a monthly or quarterly basis in order to address management's accounting concerns, while for our long-term investor annual summaries showing overall returns are sufficient.

### *Sensitivity analysis*

In testing possible outcomes, we can construct best-case, worst-case, and most likely scenarios. In making these projections, it must be recognized that selecting values for our constants can be a challenging exercise. The process of testing how models respond to changes in constants is often referred to as "sensitivity analysis". In many models, making a small change to the underlying assumptions can have a major impact on the long-term projections. This is because most projections are dependent on our calculation of future value (FV):

$$FV = PV(1 + r)^t$$

Where:

$PV$ = present value (amount invested)
$FV$ = future value at some time $t$ in the future

$r$ = discount rate, or the percentage rate earned during the period $t$
$t$ = time period

Because FV is calculating a compound rate of growth, small changes in the rate will have a significant impact on the projected value 10 or 15 years in the future. To illustrate the impact of compounding on future values, a commonly used rule of thumb in finance is the "rule of 72". The rule of 72 provides a quick estimate for the number of years it takes an investment to double in value: to calculate the time it takes for an investment to double in value, divide 72 by the rate of return. For example, an investment that earns a 6% compound rate would double in value every 12 years, while one earning 3% would take 24 years to double in value. When this concept is applied to a pro forma, estimating future values using FV can have a marked consequence on future rents, market value, and expenses. For example, consider a project that was appraised at $1 million with the assumption that it will appreciate 4% per year (Figure 2.4) At the end of 15 years, the value of the building is projected to grow to $1,800,944. However, if we are to miss our growth rate by 1%, then our FV for our market value is $1,557,697 or an approximate 13.5% difference.

Missing targets can also occur if there is a downturn in the economy with growth rates of 0% a year. For example, in our previous case, if there are four years when the rate of return is 0%, our growth in the investment will be reduced by 14.5% at the end of 15 years (Figure 3.4).

| | Case 1 | Case 2 | Difference | |
|---|---|---|---|---|
| PV= | $ 1,000,000 | $ 1,000,000 | Case 1 - Case 2 | |
| Return= | 4.0% | 3.0% | | |
| Year | FV | FV | Difference in $'s | Difference in % |
| 1 | 1,040,000 | 1,030,000 | 10,000 | 1.0% |
| 2 | 1,081,600 | 1,060,900 | 20,700 | 1.9% |
| 3 | 1,124,864 | 1,092,727 | 32,137 | 2.9% |
| 4 | 1,169,859 | 1,125,509 | 44,350 | 3.8% |
| 5 | 1,216,653 | 1,159,274 | 57,379 | 4.7% |
| 6 | 1,265,319 | 1,194,052 | 71,267 | 5.6% |
| 7 | 1,315,932 | 1,229,874 | 86,058 | 6.5% |
| 8 | 1,368,569 | 1,266,770 | 101,799 | 7.4% |
| 9 | 1,423,312 | 1,304,773 | 118,539 | 8.3% |
| 10 | 1,480,244 | 1,343,916 | 136,328 | 9.2% |
| 11 | 1,539,454 | 1,384,234 | 155,220 | 10.1% |
| 12 | 1,601,032 | 1,425,761 | 175,271 | 10.9% |
| 13 | 1,665,074 | 1,468,534 | 196,540 | 11.8% |
| 14 | 1,731,676 | 1,512,590 | 219,087 | 12.7% |
| 15 | 1,800,944 | 1,557,967 | 242,976 | 13.5% |

*Figure 2.4* FV of $1 million growing at a compound rate of 3% vs 4%.

The investor should be asking: how were these projections made? Are these illustrations based on unrealistic growth rates? Are the growth rates assumed to be constant, even though most markets are affected by business cycles that vary in duration and intensity from decade to decade? Overestimating the rates of growth can also compound when two or more factors can have a negative impact on future values in a pro forma. Consider the case in making an overly optimistic projection for the growth rates of rents while, during the same period, the actual expenses incurred for maintenance and improvements were much greater than expected. In this case, we might not have sufficient cash to meet our annual obligations.

In building our model, the level of detail reflects the concerns and knowledge about the project. Depending on the needs of the client, investor or developer, a pro forma will show varying levels of detail. Statements can display one year or several years. It is not unusual to see a projection for 15 years into the future. In constructing these statements, assumptions must be made about interest rates, rent increases, and market value appreciation, which can be adjusted to test possible scenarios. In testing possible outcomes, it is possible to construct the best-case, worst-case, and most likely scenarios for a project. Arriving at reasonable values for the assumptions needed for each scenario can be a challenging exercise. However, even given the challenges of model building, testing the sensitivity of a model under various constraints can be an enlightening experience.

Clearly, we need to be prepared for the worst of times. Nobody was fired for exceeding expectations. For a project that may have a duration of 15 years or longer, it is

| | | Case 1 | | Case 2 | Difference | |
|---|---|---|---|---|---|---|
| PV= | | $ 1,000,000 | | $ 1,000,000 | Case 1 - Case 2 | |
| Year | Annual rate of growth | FV | Annual rate of growth | FV | Difference in $'s | Difference in % |
| 1 | 4% | 1,040,000 | 4% | 1,040,000 | - | 0.0% |
| 2 | 4% | 1,081,600 | 4% | 1,081,600 | - | 0.0% |
| 3 | 4% | 1,124,864 | 4% | 1,124,864 | - | 0.0% |
| 4 | 4% | 1,169,859 | 4% | 1,169,859 | - | 0.0% |
| 5 | 4% | 1,216,653 | 0% | 1,169,859 | 46,794 | 3.8% |
| 6 | 4% | 1,265,319 | 0% | 1,169,859 | 95,460 | 7.5% |
| 7 | 4% | 1,315,932 | 0% | 1,169,859 | 146,073 | 11.1% |
| 8 | 4% | 1,368,569 | 0% | 1,169,859 | 198,710 | 14.5% |
| 9 | 4% | 1,423,312 | 4% | 1,216,653 | 206,659 | 14.5% |
| 10 | 4% | 1,480,244 | 4% | 1,265,319 | 214,925 | 14.5% |
| 11 | 4% | 1,539,454 | 4% | 1,315,932 | 223,522 | 14.5% |
| 12 | 4% | 1,601,032 | 4% | 1,368,569 | 232,463 | 14.5% |
| 13 | 4% | 1,665,074 | 4% | 1,423,312 | 241,762 | 14.5% |
| 14 | 4% | 1,731,676 | 4% | 1,480,244 | 251,432 | 14.5% |
| 15 | 4% | 1,800,944 | 4% | 1,539,454 | 261,489 | 14.5% |

*Figure 3.4* Difference between two cases: (a) continuous growth of 4% and (b) growth of 4% interrupted by four years of 0 growth (years 5–8).

difficult to consider every possible scenario. Using historical averages and current rates may offer the investor some assurances that a reasonable rate of return is achievable. Based on historical data, it is possible to consider how the numbers will reflect changes in inflation or deflation if they reach all-time lows or highs. Particular attention must be paid to factors that can expose the project to financial ruin. Periods of low rents combined with high vacancy factors can present considerable problems in meeting cash flow demands. Making payments on debt obligations, real estate taxes, utilities, and other expenses cannot be overlooked. Some of the potential increase in expense can be borne by the tenant through the use of escalator clauses and participation factors, which can help remove some of the risk that changes in the economy can have on a project.

## Case study: multifamily development – the value of forecasting

To illustrate how financial projections can be used to evaluate a potential project, consider the case of a small group of investors planning on acquiring an apartment building as an investment. The plan is to buy the property under a limited partnership agreement whereby each of the investors would benefit from the past through profit and loss. The apartment building under consideration was built 20 years ago and is modest in size for the area (62,100 sq. ft.), with 60 units on three floors. Each two-bedroom unit is approximately 850 sq. ft. The apartments are in good repair, though they still have their original kitchens and bathrooms. Close to shopping and mass transit, their location is considered highly desirable among young professionals. Newer apartments in the immediate area may have modern kitchens, but are generally 150 sq. ft. smaller for the same rent. Rents are $956 a month, with the tenants paying for utilities. Currently, occupancy is at 85%. A small lobby provides a small space for mailboxes and a waiting area for guests. Two elevators service the building. There are also 60 underground parking spaces, which tenants may rent for an additional monthly fee of $220. The group of investors plans to buy the property plan to hold it for 15 years and sell it for a profit. The asking price for the property is $11,810,100. The plan is to acquire financing. A local bank in the area has offered a 25-year fixed mortgage at 3.3% after reviewing the preliminary numbers on the project. The asking price is 10% lower than the last appraised price because of the construction of several new apartment buildings in the immediate area. Investors are now discounting older buildings, because renters find these apartments less desirable. Each of the five investors will have to come up with their contribution towards the 20% downpayment required by the bank, or $472,400 each for a total of $2,362,000. Each investor is in the 39.6% marginal tax bracket, which gives them a capital gains rate of 19.8%. Several scenarios are being considered as a test of the merits of the investment. The first scenarios assume that business will continue as usual, with little deviation from the recent past. However, there is some concern that the current growth of the local economy could be stalled by a recession. The hope is that even if economic conditions should deteriorate, the project will not require cash infusions to stay solvent.

### *Scenario 1: business as usual*

In this first scenario, the assumption is that there is little change in the business climate from current conditions. Market values advance in lockstep, with annual inflation at 2.5%, along with expenses of real estate taxes, maintenance, management fees, and all operating

| | | |
|---|---|---|
| **Rent, Year 1 (Less vacancy)** | 846,900 | |
| **Debt** | | |
| Short term rates | 4.5% | |
| Interest rate on mortgage (annual) | 3.30% | |
| Amortization Period | 25.00 | |
| Percent Financed | 80% | |
| Amount Borrowed | $9,448,000 | |
| Down payment | $2,362,000 | |
| Depreciation, Years | $428,000 | 27.5 |
| Commission at sale in the 15th year | 5.50% | |
| Appreciation: Market Value Increase per year | 2.50% | |
| Ordinary tax rate | 39.6% | |
| Capital gains rate at | 50.0% | |

| Year | Income | Total Expenses | NOI | NOI - Total Debt service | Market Value | Equity | Debt Balance | After Tax Cash Flow | Cash on Cash | Return on Equity AFTX |
|---|---|---|---|---|---|---|---|---|---|---|
| 1 | $ 868,073 | $ 244,363 | $ 623,710 | $ 62,835 | $ 12,105,250 | $ 2,724,787 | $ 9,198,909 | $ 108,800 | 2.31% | 3.99% |
| 2 | $ 889,774 | $ 250,472 | $ 639,303 | $ 78,428 | $ 12,407,881 | $ 3,107,351 | $ 8,941,598 | $ 114,963 | 2.52% | 3.70% |
| 3 | $ 912,019 | $ 256,733 | $ 655,285 | $ 94,410 | $ 12,718,078 | $ 3,510,617 | $ 8,675,796 | $ 121,254 | 2.69% | 3.45% |
| 4 | $ 934,819 | $ 263,152 | $ 671,667 | $ 110,792 | $ 13,036,030 | $ 3,935,550 | $ 8,401,222 | $ 127,675 | 2.82% | 3.24% |
| 5 | $ 958,190 | $ 269,731 | $ 688,459 | $ 127,584 | $ 13,361,931 | $ 4,383,159 | $ 8,117,587 | $ 134,229 | 2.91% | 3.06% |
| 6 | $ 982,144 | $ 276,474 | $ 705,671 | $ 144,796 | $ 13,695,979 | $ 4,854,497 | $ 7,824,593 | $ 140,919 | 2.98% | 2.90% |
| 7 | $ 1,006,698 | $ 283,386 | $ 723,312 | $ 162,437 | $ 14,038,379 | $ 5,350,663 | $ 7,521,929 | $ 147,745 | 3.04% | 2.76% |
| 8 | $ 1,031,865 | $ 290,470 | $ 741,395 | $ 180,520 | $ 14,389,338 | $ 5,872,806 | $ 7,209,278 | $ 154,712 | 3.07% | 2.63% |
| 9 | $ 1,057,662 | $ 297,732 | $ 759,930 | $ 199,055 | $ 14,749,072 | $ 6,422,122 | $ 6,886,309 | $ 161,822 | 3.10% | 2.52% |
| 10 | $ 1,084,104 | $ 305,175 | $ 778,928 | $ 218,053 | $ 15,117,798 | $ 6,999,862 | $ 6,552,682 | $ 169,076 | 3.12% | 2.42% |
| 11 | $ 1,111,206 | $ 312,805 | $ 798,401 | $ 237,526 | $ 15,495,743 | $ 7,607,333 | $ 6,208,046 | $ 176,478 | 3.12% | 2.32% |
| 12 | $ 1,138,986 | $ 320,625 | $ 818,361 | $ 257,486 | $ 15,883,137 | $ 8,245,896 | $ 5,852,036 | $ 184,030 | 3.12% | 2.23% |
| 13 | $ 1,167,461 | $ 328,640 | $ 838,821 | $ 277,946 | $ 16,280,215 | $ 8,916,974 | $ 5,484,278 | $ 191,735 | 3.12% | 2.15% |
| 14 | $ 1,196,648 | $ 336,856 | $ 859,791 | $ 298,916 | $ 16,687,221 | $ 9,622,051 | $ 5,104,385 | $ 199,595 | 3.11% | 2.07% |
| 15 | $ 1,226,564 | $ 345,278 | $ 881,286 | $ 320,411 | $ 17,104,401 | $ 10,362,677 | $ 4,711,954 | $ 207,614 | 3.09% | 2.00% |

| | |
|---|---|
| IRR Before Tax | 17.51% |
| IRR Aft Tax | 16.48% |

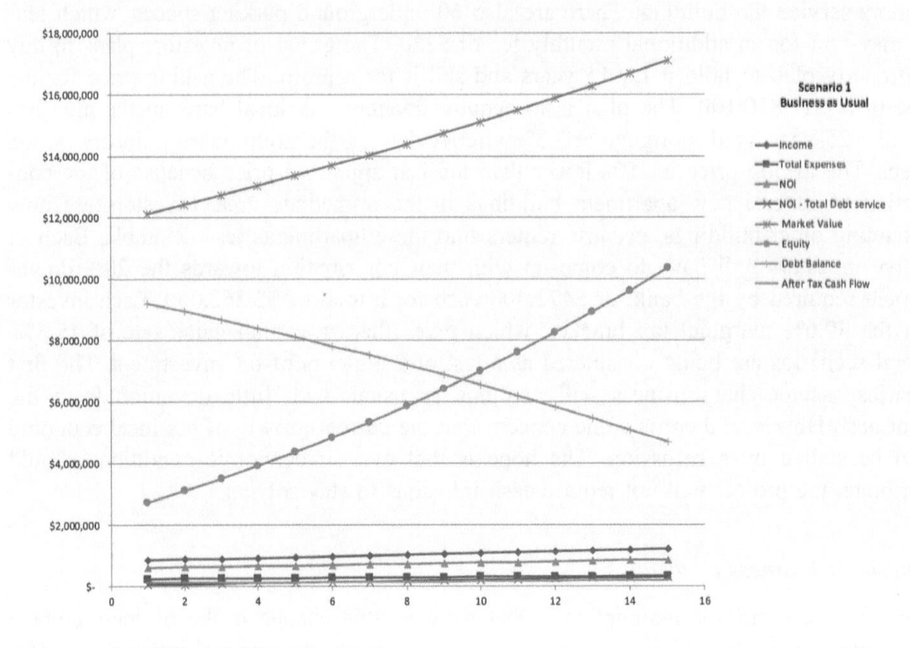


*Figure 4.4* Scenario 1: business as usual.

expenses, which are also expected to increase at 2.5% per year. The occupancy factor, which is currently at 85%, remains unchanged during the entire 15 years (Figure 4.4).

The net result is a fairly good return, with an IRR of 17.51% before tax over 15 years. This is even with a 15% vacancy rate and a modest growth in the asset set to the current growth rate of 2.5% a year. The other positive aspect of this analysis is that there is excess cash each year. The actual return cash/cash is 2.31% in the earlier years rising to 3.09% in the last year. Under this scenario, the rents are sufficient to cover the mortgage and expenses, yet still generate some positive cash flow. This excess cash flow can be used to cover any unexpected expenditure increases, including taxes and insurance, capital investments or those passed through to the investors. The assumption that short-term rates are the same for borrowing and saving does make the projection look slightly more positive, but this may account for only 0.25 differences in the IRR over the 15-year period. Also, a significant portion of the after-tax cash flow is the result of writing off $428,000 a year in depreciation. If the individual investor does not have other taxable income to offset these losses, then the IRR for this individual would also be less.

### *Scenario 2: optimistic projection*

If economic projections call for improved conditions in the future, making a more optimist projection may be warranted. In this first scenario, the assumption is that there is little change in the business climate. But if market values advance at a rate of 1% above the rate of inflation and vacancy falls from 15% to 5%, then the IRR before tax is 21.6%. Having bought the property during a period of rising values, it would appear to be a reasonable assumption. As in Scenario 1, we will assume a 2.5% annual increase in expenses, including real estate taxes, maintenance, management fees, and all other operating expenses (Figure 5.4).

### *Scenario 3: level property values*

Given the positive results of the first two scenarios, looking at a gloomier future is prudent. What if property values do not increase at all over the next 15 years? During this period, real estate taxes, maintenance, management fee, and all operating expenses are expected to go up by 2.5% annually, while occupancy will remain at 85%. Though an unlikely scenario, it is not unusual for property values to advance at a very low annual rate (Figure 6.4).

In this scenario, the tenants are paying down the mortgage each month and, given the leverage of 1:5, we can still see a healthy return on the investment. The IRR of 14.8% before tax is roughly equal to five times the pretax cash flow/equity. With an asset generating a positive cash flow, leverage greatly improves the overall return. In many ways, when the market value has a zero growth rate, it resembles an investment in mortgages during a period in which long-term rates are unchanged. Investors see a return on the investment from the repayment of principal and interest. If the after-tax cash flow is positive, then this will make even higher returns. In established communities where there is a very stable real estate market, appreciation grows at very modest rates. A college town without industry or a community dependent on government expenditures, such as a county seat, state or provincial capital, can have a very stable employment base. In these cases, excess cash flow and the repayment of debt principal will be the largest component of the investment return. It is important, when buying properties that fall into this class, to ensure that they generate positive cash flow throughout their life, since large increases in property value at rates above inflation are unlikely.

| | |
|---|---|
| **Rent, Year 1 (Less vacancy)** | 927,900 |
| **Debt** | |
| Short term rates | 4.5% |
| Interest rate on mortgage (annual) | 3.30% |
| Amortization Period | 25.00 |
| Percent Financed | 80% |
| Amount Borrowed | $9,448,000 |
| Down payment | $2,362,000 |
| Depreciation, Years | $428,000 27.5 |
| Commission at sale in the 15th year | 5.50% |
| Appreciation: Market Value Increase per year | 3.50% |
| Ordinary tax rate | 39.6% |
| Capital gains rate at | 50.0% |

| Year | Income | Total Expenses | NOI | NOI - Total Debt service | Market Value | Debt Balance | After Tax Cash Flow | Cash on Cash | Return on Equity AFTX |
|---|---|---|---|---|---|---|---|---|---|
| 1 | $ 951,098 | $ 244,363 | $ 706,735 | $ 145,860 | $ 12,223,350 | $ 9,198,909 | $ 158,947 | 5.25% | 5.72% |
| 2 | $ 974,875 | $ 250,472 | $ 724,403 | $ 163,528 | $ 12,651,167 | $ 8,941,598 | $ 166,364 | 5.09% | 5.17% |
| 3 | $ 999,247 | $ 256,733 | $ 742,513 | $ 181,638 | $ 13,093,958 | $ 8,675,796 | $ 173,940 | 4.94% | 4.73% |
| 4 | $ 1,024,228 | $ 263,152 | $ 761,076 | $ 200,201 | $ 13,552,247 | $ 8,401,222 | $ 181,678 | 4.80% | 4.36% |
| 5 | $ 1,049,834 | $ 269,731 | $ 780,103 | $ 219,228 | $ 14,026,575 | $ 8,117,587 | $ 189,582 | 4.68% | 4.05% |
| 6 | $ 1,076,080 | $ 276,474 | $ 799,606 | $ 238,731 | $ 14,517,505 | $ 7,824,593 | $ 197,655 | 4.57% | 3.78% |
| 7 | $ 1,102,982 | $ 283,386 | $ 819,596 | $ 258,721 | $ 15,025,618 | $ 7,521,929 | $ 205,901 | 4.46% | 3.55% |
| 8 | $ 1,130,556 | $ 290,470 | $ 840,086 | $ 279,211 | $ 15,551,515 | $ 7,209,278 | $ 214,321 | 4.36% | 3.35% |
| 9 | $ 1,158,820 | $ 297,732 | $ 861,088 | $ 300,213 | $ 16,095,818 | $ 6,886,309 | $ 222,921 | 4.26% | 3.16% |
| 10 | $ 1,187,790 | $ 305,175 | $ 882,615 | $ 321,740 | $ 16,659,171 | $ 6,552,682 | $ 231,703 | 4.17% | 3.00% |
| 11 | $ 1,217,485 | $ 312,805 | $ 904,680 | $ 343,805 | $ 17,242,242 | $ 6,208,046 | $ 240,670 | 4.08% | 2.86% |
| 12 | $ 1,247,922 | $ 320,625 | $ 927,297 | $ 366,422 | $ 17,845,721 | $ 5,852,036 | $ 249,827 | 4.00% | 2.73% |
| 13 | $ 1,279,120 | $ 328,640 | $ 950,480 | $ 389,605 | $ 18,470,321 | $ 5,484,278 | $ 259,177 | 3.92% | 2.61% |
| 14 | $ 1,311,098 | $ 336,856 | $ 974,242 | $ 413,367 | $ 19,116,782 | $ 5,104,385 | $ 268,724 | 3.84% | 2.49% |
| 15 | $ 1,343,876 | $ 345,278 | $ 998,598 | $ 437,723 | $ 19,785,870 | $ 4,711,954 | $ 278,470 | 3.76% | 2.39% |

| | |
|---|---|
| IRR Before Tax | 21.60% |
| IRR Aft Tax | 19.70% |

Income = Gross Income (net vacancy) Note this is an end of year calculation
Total Equity = Down payment + Accumulated Principal + Cash Account
After Tax Cash Flow (Net spendable) = Pre TaxCash Flow - taxes
Cash on Cash = Pre Tax Cash Flow/Equity
Return on Equity after Tax = Cash flow after Tax/Equity

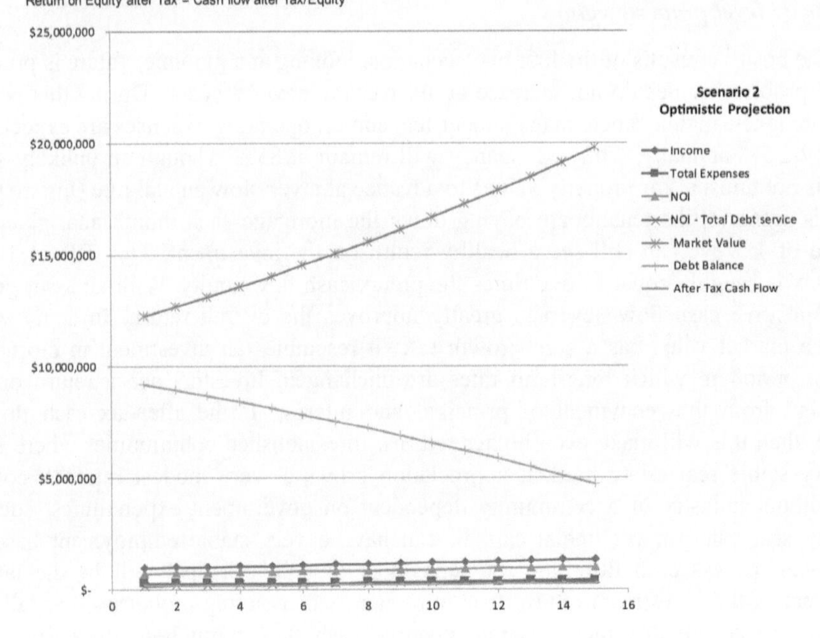


*Figure 5.4* Scenario 2: optimistic projection.

| **Rent, Year 1 (Less vacancy)** | 846,900 | |
|---|---|---|
| **Debt** | | |
| Short term rates | 4.5% | |
| Interest rate on mortgage (annual) | 3.30% | |
| Amortization Period | 25.00 | |
| Percent Financed | 80% | |
| Amount Borrowed | $9,448,000 | |
| Down payment | $2,362,000 | |
| Depreciation, Years | $428,000 | 27.5 |
| Commission at sale in the 15th year | 5.50% | |
| Appreciation: Market Value Increase per year | 0.00% | |
| Ordinary tax rate | 39.6% | |
| Capital gains rate at | 50.0% | |

| Year | Income | Total Expenses | NOI | NOI-Debt | Market Value | Debt Balance | After Tax Cash Flow | Cash on Cash | on Equity after Tax |
|---|---|---|---|---|---|---|---|---|---|
| 1 | $ 868,073 | $ 244,363 | $ 623,710 | $ 62,835 | $ 11,810,000 | $ 9,198,909 | $ 108,800 | 2.31% | 3.99% |
| 2 | $ 889,774 | $ 250,472 | $ 639,303 | $ 78,428 | $ 11,810,000 | $ 8,941,598 | $ 114,963 | 2.52% | 3.70% |
| 3 | $ 912,019 | $ 256,733 | $ 655,285 | $ 94,410 | $ 11,810,000 | $ 8,675,796 | $ 121,254 | 2.69% | 3.45% |
| 4 | $ 934,819 | $ 263,152 | $ 671,667 | $ 110,792 | $ 11,810,000 | $ 8,401,222 | $ 127,675 | 2.82% | 3.24% |
| 5 | $ 958,190 | $ 269,731 | $ 688,459 | $ 127,584 | $ 11,810,000 | $ 8,117,587 | $ 134,229 | 2.91% | 3.06% |
| 6 | $ 982,144 | $ 276,474 | $ 705,671 | $ 144,796 | $ 11,810,000 | $ 7,824,593 | $ 140,919 | 2.98% | 2.90% |
| 7 | $ 1,006,698 | $ 283,386 | $ 723,312 | $ 162,437 | $ 11,810,000 | $ 7,521,929 | $ 147,745 | 3.04% | 2.76% |
| 8 | $ 1,031,865 | $ 290,470 | $ 741,395 | $ 180,520 | $ 11,810,000 | $ 7,209,278 | $ 154,712 | 3.07% | 2.63% |
| 9 | $ 1,057,662 | $ 297,732 | $ 759,930 | $ 199,055 | $ 11,810,000 | $ 6,886,309 | $ 161,822 | 3.10% | 2.52% |
| 10 | $ 1,084,104 | $ 305,175 | $ 778,928 | $ 218,053 | $ 11,810,000 | $ 6,552,682 | $ 169,076 | 3.12% | 2.42% |
| 11 | $ 1,111,206 | $ 312,805 | $ 798,401 | $ 237,526 | $ 11,810,000 | $ 6,208,046 | $ 176,478 | 3.12% | 2.32% |
| 12 | $ 1,138,986 | $ 320,625 | $ 818,361 | $ 257,486 | $ 11,810,000 | $ 5,852,036 | $ 184,030 | 3.12% | 2.23% |
| 13 | $ 1,167,461 | $ 328,640 | $ 838,821 | $ 277,946 | $ 11,810,000 | $ 5,484,278 | $ 191,735 | 3.12% | 2.15% |
| 14 | $ 1,196,648 | $ 336,856 | $ 859,791 | $ 298,916 | $ 11,810,000 | $ 5,104,385 | $ 199,595 | 3.11% | 2.07% |
| 15 | $ 1,226,564 | $ 345,278 | $ 881,286 | $ 320,411 | $ 11,810,000 | $ 4,711,954 | $ 207,614 | 3.09% | 2.00% |

| IRR Before Tax | 14.80% |
|---|---|
| IRR Aft Tax | 13.94% |

Income = Gross Income (net vacancy) Note this is an end of year calculation
Total Equity = Down payment + Accumulated Principal + Cash Account
After Tax Cash Flow (Net spendable) = Pre TaxCash Flow - taxes
Cash on Cash = Pre Tax Cash Flow/Equity
Return on Equity after Tax = Cash flow after Tax/Equity

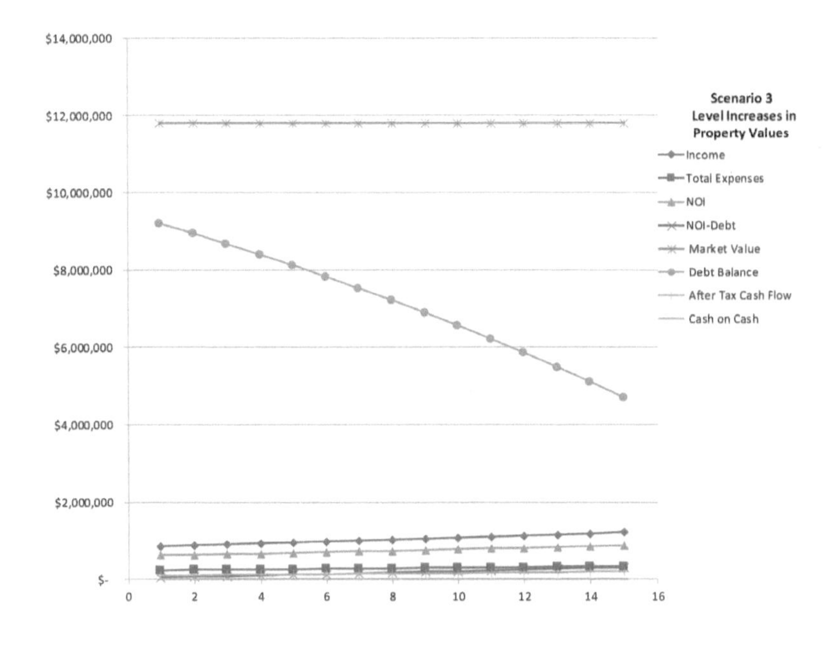


*Figure 6.4* Scenario 3: level property values.

| | |
|---|---|
| **Rent, Year 1 (Less vacancy)** | 743,625 |
| **Debt** | |
| Short term rates | 4.5% |
| Interest rate on mortgage (annual) | 3.30% |
| Amortization Period | 25.00 |
| Percent Financed | 80% |
| Amount Borrowed | $9,448,000 |
| Down payment | $2,362,000 |
| Depreciation, Years | $428,000 27.5 |
| Commission at sale in the 15th year | 5.50% |
| Appreciation: Market Value Increase per year | 2.125% |
| Ordinary tax rate | 39.6% |
| Capital gains rate at | 50.0% |

| Year | Income | Total Expenses | NOI | NOI-Debt | Market Value | Debt Balance | After Tax Cash Flow | Cash on Cash | on Equity after Tax |
|---|---|---|---|---|---|---|---|---|---|
| 1 | $ 762,216 | $ 243,469 | $ 518,747 | $ (42,128) | $ 12,060,963 | $ 9,198,909 | $ 45,403 | -1.58% | 1.71% |
| 2 | $ 781,271 | $ 248,642 | $ 532,629 | $ (28,246) | $ 10,469,669 | $ 8,941,598 | $ 50,532 | -0.95% | 1.70% |
| 3 | $ 800,803 | $ 253,926 | $ 546,877 | $ (13,998) | $ 10,692,150 | $ 8,675,796 | $ 55,775 | -0.42% | 1.69% |
| 4 | $ 820,823 | $ 259,322 | $ 561,501 | $ 626 | $ 10,919,358 | $ 8,401,222 | $ 61,135 | 0.02% | 1.68% |
| 5 | $ 841,343 | $ 264,832 | $ 576,511 | $ 15,636 | $ 11,151,394 | $ 8,117,587 | $ 66,613 | 0.39% | 1.66% |
| 6 | $ 862,377 | $ 270,460 | $ 591,917 | $ 31,042 | $ 11,388,361 | $ 7,824,593 | $ 72,211 | 0.71% | 1.64% |
| 7 | $ 883,936 | $ 276,207 | $ 607,729 | $ 46,854 | $ 11,630,364 | $ 7,521,929 | $ 77,933 | 0.98% | 1.63% |
| 8 | $ 906,035 | $ 282,077 | $ 623,958 | $ 63,083 | $ 11,877,509 | $ 7,209,278 | $ 83,780 | 1.21% | 1.61% |
| 9 | $ 928,686 | $ 288,071 | $ 640,615 | $ 79,740 | $ 12,129,906 | $ 6,886,309 | $ 89,755 | 1.41% | 1.59% |
| 10 | $ 951,903 | $ 294,192 | $ 657,710 | $ 96,835 | $ 12,387,667 | $ 6,552,682 | $ 95,860 | 1.58% | 1.56% |
| 11 | $ 975,700 | $ 300,444 | $ 675,256 | $ 114,381 | $ 12,650,905 | $ 6,208,046 | $ 102,098 | 1.73% | 1.54% |
| 12 | $ 1,000,093 | $ 306,828 | $ 693,264 | $ 132,389 | $ 12,919,737 | $ 5,852,036 | $ 108,471 | 1.86% | 1.52% |
| 13 | $ 1,025,095 | $ 313,349 | $ 711,747 | $ 150,872 | $ 13,194,281 | $ 5,484,278 | $ 114,982 | 1.97% | 1.50% |
| 14 | $ 1,050,723 | $ 320,007 | $ 730,715 | $ 169,840 | $ 13,474,659 | $ 5,104,385 | $ 121,634 | 2.06% | 1.48% |
| 15 | $ 1,076,991 | $ 326,807 | $ 750,183 | $ 189,308 | $ 13,760,996 | $ 4,711,954 | $ 128,428 | 2.14% | 1.45% |

| | |
|---|---|
| IRR Before Tax | 11.83% |
| IRR Aft Tax | 11.76% |

Income = Gross Income (net vacancy) Note this is an end of year calculation
Total Equity = Down payment + Accumulated Principal + Cash Account
After Tax Cash Flow (Net spendable) = Pre TaxCash Flow - taxes
Cash on Cash = Pre Tax Cash Flow/Equity
Return on Equity after Tax = Cash flow after Tax/Equity

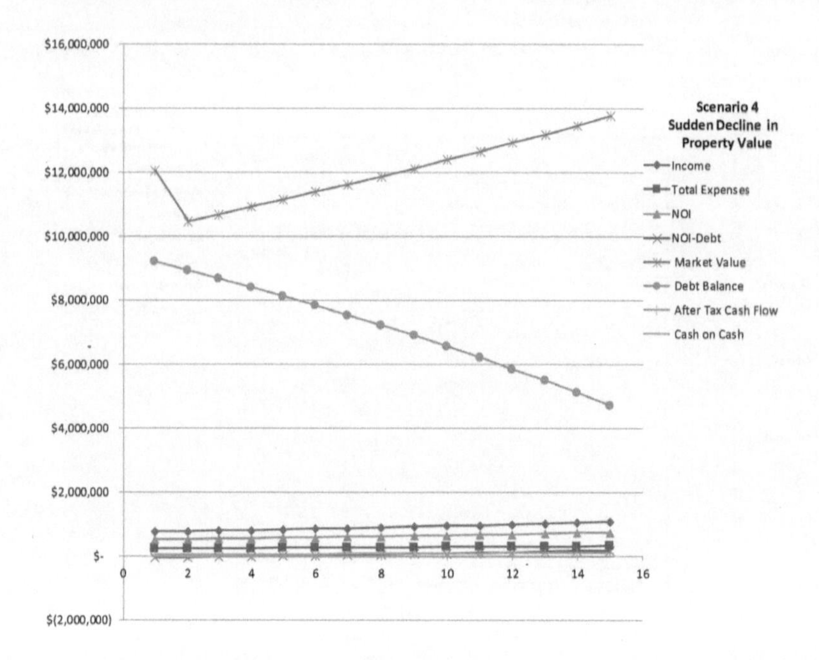


*Figure 7.4* Scenario 4: sudden decline in property values.

***Scenario 4: sudden decline in property values***

With the positive results of the first two scenarios, looking at an even gloomier future would be an interesting exercise. What if a recession takes place a year after taking possession? Property values would fall quickly. Assume that property values fall by 15% in year 2 of the pro forma. Concurrently, rents could fall by 15% to $813 a month in the first year, then rise by the inflation rate of 2.125%. Under this scenario, the rate of growth in property values is also reduced by 15%, from 2.5% to 2.125%. The rate of increase for all expenses, including real estate taxes, maintenance, management fee, and all operating expenses, is now at 2.125% and the occupancy factor, which is 85%, remains unchanged during the entire 15 years (Figure 7.4).

Perhaps surprisingly, a reasonably good rate of return, with an IRR of 11.83% before tax over 15 years, is generated under this scenario. This is even with a 15% vacancy rate and a reduction in rents by 15%, as well as a more modest growth rate set below inflation at 2.13% a year. The unfortunate aspect of this scenario is that pretax cash flow (NOI *less* debt) is negative for three of the years. The one positive aspect of this analysis is that the pass-through of losses can offset some of this pain as long as the individual investor has sufficient income to shelter them. If the investor can take advantage of all the losses generated from this project, their after-tax net spendable after-tax cash flow (net spendable) is positive.

***Scenario 5: sudden decline in property values with no recovery***

Given the positive results of the fourth scenario, constructing an even more disastrous future is advisable. What if rents were to fall 15% to $813 a month and the rate of growth in property values, which is flat at 0%, were not to increase at all for the next 15 years? If there were a serious extended recession, this could easily be the case. In this scenario, the inflation rate for expenses of real estate taxes, maintenance, management fees, and all operating expenses remains at 2.125% and the occupancy factor, which is 85%, remains unchanged during the entire 15 years (Figure 8.4).

In this case, the main benefits of the project are that the tenants are paying down the mortgage each month and, given the leverage of 1:5, we can still see a positive return on the investment at the end of 15 years. The net result still offers a reasonably good return with an IRR that is 8.84% before tax over 15 years. This is even with a 15% vacancy rate and a reduction in rents by 15%, as well as a flat growth rate in property values. The other positive aspect of this analysis is that though cash flow is not excessive, it is still positive each year and should be able to cover the occasional expenditures.

There are two reasons why this scenario still generates a positive IRR over its life. The analysis assumes the pass-through of losses in the first three years of the project. Of course, the individual investor will need income from other investments to shelter these losses. If the investor can take advances of all the losses from this project, their after-tax net spendable after-tax cash flow (net spendable) is positive throughout the 15 years. However, if this is not the case, this project could become a burden on the investors if the losses generated from the project cannot be used to shelter other income. Changes in the tax rates and policy could increase the burden of a negative "NOI *less* debt" of this project on each investor. In both scenarios 4 and 5, the project is short actual cash to cover debt service and expenses.

| | |
|---|---|
| **Rent, Year 1 (Less vacancy)** | 743,625 |
| **Debt** | |
| Short term rates | 4.5% |
| Interest rate on mortgage (annual) | 3.30% |
| Amortization Period | 25.00 |
| Percent Financed | 80% |
| Amount Borrowed | $9,448,000 |
| Down payment | $2,362,000 |
| Depreciation, Years | $428,000 27.5 |
| Commission at sale in the 15th year | 5.50% |
| Appreciation: Market Value Increase per year | 0.000% |
| Ordinary tax rate | 39.6% |
| Capital gains rate at | 50.0% |

| Year | Income | Total Expenses | NOI | NOI-Debt | Market Value | Debt Balance | After Tax Cash Flow | Cash on Cash | on Equity after Tax |
|---|---|---|---|---|---|---|---|---|---|
| 1 | $ 762,216 | $ 243,469 | $ 518,747 | $ (42,128) | $ 11,810,000 | $ 9,198,909 | $ 45,403 | -1.58% | 1.71% |
| 2 | $ 781,271 | $ 248,642 | $ 532,629 | $ (28,246) | $ 10,038,500 | $ 8,941,598 | $ 50,532 | -0.95% | 1.70% |
| 3 | $ 800,803 | $ 253,926 | $ 546,877 | $ (13,998) | $ 10,038,500 | $ 8,675,796 | $ 55,775 | -0.42% | 1.69% |
| 4 | $ 820,823 | $ 259,322 | $ 561,501 | $ 626 | $ 10,038,500 | $ 8,401,222 | $ 61,135 | 0.02% | 1.68% |
| 5 | $ 841,343 | $ 264,832 | $ 576,511 | $ 15,636 | $ 10,038,500 | $ 8,117,587 | $ 66,613 | 0.39% | 1.66% |
| 6 | $ 862,377 | $ 270,460 | $ 591,917 | $ 31,042 | $ 10,038,500 | $ 7,824,593 | $ 72,211 | 0.71% | 1.64% |
| 7 | $ 883,936 | $ 276,207 | $ 607,729 | $ 46,854 | $ 10,038,500 | $ 7,521,929 | $ 77,933 | 0.98% | 1.63% |
| 8 | $ 906,035 | $ 282,077 | $ 623,958 | $ 63,083 | $ 10,038,500 | $ 7,209,278 | $ 83,780 | 1.21% | 1.61% |
| 9 | $ 928,686 | $ 288,071 | $ 640,615 | $ 79,740 | $ 10,038,500 | $ 6,886,309 | $ 89,755 | 1.41% | 1.59% |
| 10 | $ 951,903 | $ 294,192 | $ 657,710 | $ 96,835 | $ 10,038,500 | $ 6,552,682 | $ 95,860 | 1.58% | 1.56% |
| 11 | $ 975,700 | $ 300,444 | $ 675,256 | $ 114,381 | $ 10,038,500 | $ 6,208,046 | $ 102,098 | 1.73% | 1.54% |
| 12 | $ 1,000,093 | $ 306,828 | $ 693,264 | $ 132,389 | $ 10,038,500 | $ 5,852,036 | $ 108,471 | 1.86% | 1.52% |
| 13 | $ 1,025,095 | $ 313,349 | $ 711,747 | $ 150,872 | $ 10,038,500 | $ 5,484,278 | $ 114,982 | 1.97% | 1.50% |
| 14 | $ 1,050,723 | $ 320,007 | $ 730,715 | $ 169,840 | $ 10,038,500 | $ 5,104,385 | $ 121,634 | 2.06% | 1.48% |
| 15 | $ 1,076,991 | $ 326,807 | $ 750,183 | $ 189,308 | $ 10,038,500 | $ 4,711,954 | $ 128,428 | 2.14% | 1.45% |

| | |
|---|---|
| IRR Before Tax | 8.84% |
| IRR Aft Tax | 9.04% |

Income = Gross Income (net vacancy) Note this is an end of year calculation
Total Equity = Down payment + Accumulated Principal + Cash Account
After Tax Cash Flow (Net spendable) = Pre TaxCash Flow - taxes
Cash on Cash = Pre Tax Cash Flow/Equity
Return on Equity after Tax = Cash flow after Tax/Equity

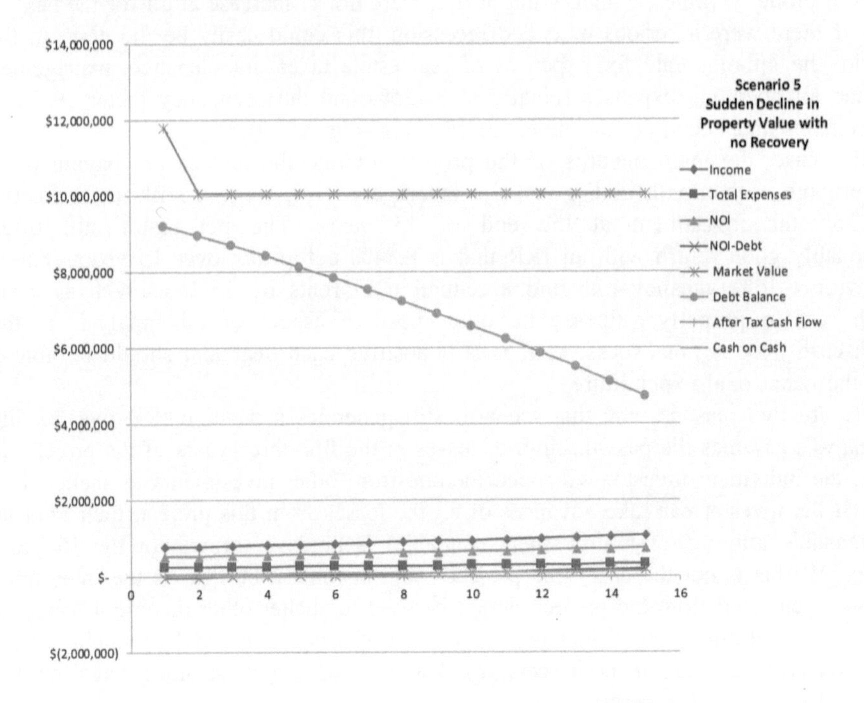


*Figure 8.4* Scenario 5: sudden decline in property values with no recovery.

| | | |
|---|---|---|
| **Rent, Year 1 (Less vacancy)** | 1,081,800 | |
| **Debt** | | |
| Short term rates | 4.5% | |
| Interest rate on mortgage (annual) | 3.30% | |
| Amortization Period | 25.00 | |
| Percent Financed | 80% | |
| Amount Borrowed | $9,448,000 | |
| Down payment | $2,362,000 | |
| Depreciation, Years | $428,000 | 27.5 |
| Commission at sale in the 15th year | 5.50% | |
| Appreciation: Market Value Increase per year | 4.50% | |
| Ordinary tax rate | 39.6% | |
| Capital gains rate at | 50.0% | |

| Year | Income | Total Expenses | NOI | NOI-Debt | Market Value | Debt Balance | After Tax Cash Flow | Cash on Cash | on Equity after Tax |
|---|---|---|---|---|---|---|---|---|---|
| 1 | $ 1,108,845 | $ 470,567 | $ 638,278 | $ 77,403 | $ 12,341,450 | $ 9,198,909 | $ 117,599 | 2.83% | 4.30% |
| 2 | $ 1,136,566 | $ 482,331 | $ 654,235 | $ 93,360 | $ 12,896,815 | $ 8,941,598 | $ 123,982 | 2.99% | 3.97% |
| 3 | $ 1,164,980 | $ 494,390 | $ 670,591 | $ 109,716 | $ 13,477,172 | $ 8,675,796 | $ 130,498 | 3.10% | 3.69% |
| 4 | $ 1,194,105 | $ 506,749 | $ 687,355 | $ 126,480 | $ 14,083,645 | $ 8,401,222 | $ 137,151 | 3.18% | 3.45% |
| 5 | $ 1,223,957 | $ 519,418 | $ 704,539 | $ 143,664 | $ 14,717,409 | $ 8,117,587 | $ 143,942 | 3.24% | 3.24% |
| 6 | $ 1,254,556 | $ 532,404 | $ 722,153 | $ 161,278 | $ 15,379,692 | $ 7,824,593 | $ 150,874 | 3.28% | 3.07% |
| 7 | $ 1,285,920 | $ 545,714 | $ 740,207 | $ 179,331 | $ 16,071,778 | $ 7,521,929 | $ 157,949 | 3.30% | 2.91% |
| 8 | $ 1,318,068 | $ 559,357 | $ 758,712 | $ 197,837 | $ 16,795,008 | $ 7,209,278 | $ 165,171 | 3.32% | 2.77% |
| 9 | $ 1,351,020 | $ 573,341 | $ 777,679 | $ 216,804 | $ 17,550,784 | $ 6,886,309 | $ 172,542 | 3.32% | 2.64% |
| 10 | $ 1,384,795 | $ 587,674 | $ 797,121 | $ 236,246 | $ 18,340,569 | $ 6,552,682 | $ 180,065 | 3.32% | 2.53% |
| 11 | $ 1,419,415 | $ 602,366 | $ 817,049 | $ 256,174 | $ 19,165,894 | $ 6,208,046 | $ 187,741 | 3.31% | 2.42% |
| 12 | $ 1,454,901 | $ 617,425 | $ 837,476 | $ 276,601 | $ 20,028,360 | $ 5,852,036 | $ 195,575 | 3.29% | 2.33% |
| 13 | $ 1,491,273 | $ 632,861 | $ 858,413 | $ 297,538 | $ 20,929,636 | $ 5,484,278 | $ 203,569 | 3.27% | 2.24% |
| 14 | $ 1,528,555 | $ 648,682 | $ 879,873 | $ 318,998 | $ 21,871,470 | $ 5,104,385 | $ 211,725 | 3.25% | 2.16% |
| 15 | $ 1,566,769 | $ 664,899 | $ 901,870 | $ 340,995 | $ 22,855,686 | $ 4,711,954 | $ 220,046 | 3.22% | 2.08% |

| | |
|---|---|
| IRR Before Tax | 20.20% |
| IRR Aft Tax | 18.93% |

Income = Gross Income (net vacancy) Note this is an end of year calculation
Total Equity = Down payment + Accumulated Principal + Cash Account
After Tax Cash Flow (Net spendable) = Pre TaxCash Flow - taxes
Cash on Cash = Pre Tax Cash Flow/Equity
Return on Equity after Tax = Cash flow after Tax/Equity

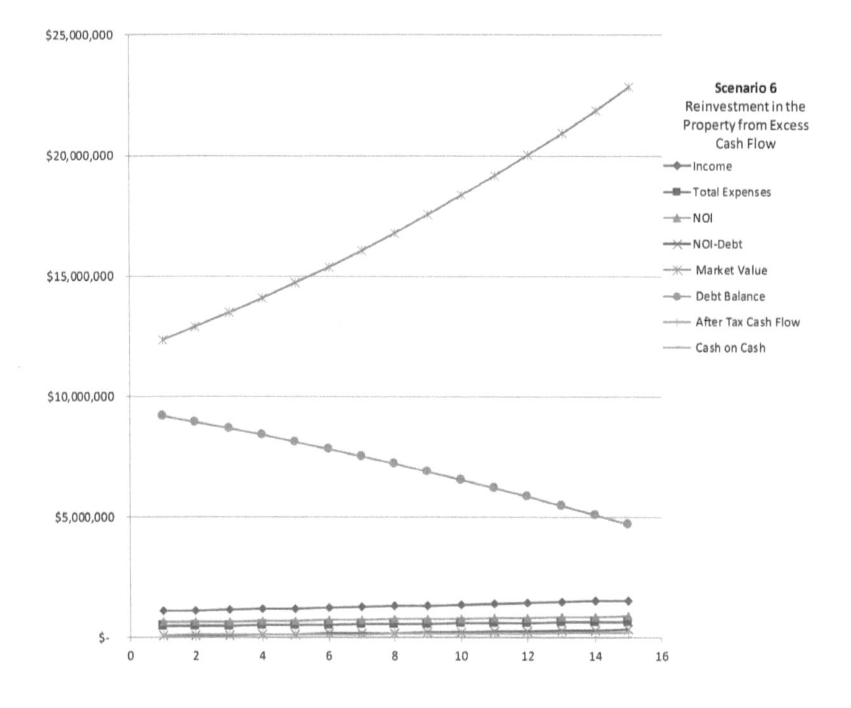


*Figure 9.4* Scenario 6: reinvestment in the property from excess cash flow.

### *Scenario 6: reinvestment in the property from excess cash flow*

Given the results of the first three scenarios, it is useful to look at the potential of further increasing returns beyond their existing levels. Scenario 1, "business as usual", shows an excess in cash flow that might fund some improvements to the building. The goal would be to raise rents and lower vacancy rates, while improving the long-term value of the property. The building is 20 years old and the units look a little dated, which probably accounts for the high vacancy rate of 85%. The property manager suggests that many of the tenants leave after a year for newer properties in the same area, even though the two-bedroom units are, on average, 15–20% larger for the equivalent rents. If the properties are given a facelift, with new kitchens, tiled entrances, bathrooms with jet tubs, and utility rooms, complete with new washer and dryer, these units will be more desirable than those built over the last few years. Other improvements that would be needed include a renovation of the lobby area, new windows, and improvements to the heating, ventilation, and air conditioning (HVAC) system. Some additional plantings and landscaping in the front and along the edge of the property are also in order if the property is to be truly competitive. If all of these improvements are made, it is estimated that rents could be increased by 20% and vacancy rates would fall to 5%. It is hoped that these improvements would be reflected in the IRR for the development. The additional costs, however, are not insignificant. Over a 15-year period, the estimate provided by the architect is that $4.413 million would have to be spent on improvements to the interior and exterior of the building. One issue that will have to be faced is the rising rents: as each tenant leaves, it will be possible to renovate and escalate rents on the newer apartments. However, exterior improvements cannot be done piecemeal. For these expenditures, the entire cost will not be recovered immediately. One solution would be to borrow the needed funds to accelerate the improvements. If these improvements were completed in the first year or two, the projections roughly resemble those shown in Figure 9.4.

As anticipated, spending the excess cash flow on improvements did improve the IRR over the base Scenario 1, "Business as usual". Increases in rent by 20% and an additional 1% increase in market value per year were enough to offset the additional expense. One caveat is that it is always difficult to know what expenditures are going to pay off until they are actually made. It is also important to remember that estimates of proposed improvement costs are never 100% accurate.

### *Scenario 7: negative return*

The first six scenarios all show a positive rate of return, even in the instances of depressed economic times, but negative return is possible if the perfect storm should appear on your horizon. For example, if the property had been purchased just before a decline in property values, lower returns would be expected. To generate a negative rate of return, the investors would have had to overpay for the property. If, for example, a premium amount were paid for the property, the overall return would be impacted (20% over Scenario 1). If the real estate market had been overheating for some time, it is also likely that both interest rates and values were overinflated. As the market drives prices higher, investors will consider properties that would have looked expensive several years earlier. However, with a sense of exuberant optimism moving the market and of the urgency of not being left on the sidelines, investment decisions may assume a market that will continue its upward trajectory for some time into the future. Besides, if the values keep going up, then everyone makes money, right? Unfortunately, if the market

| Rent, Year 1 (Less vacancy) | 743,625 | |
|---|---|---|
| **Debt** | | |
| Short term rates | 5.5% | |
| Interest rate on mortgage (annual) | 4.25% | |
| Amortization Period | 25.00 | |
| Percent Financed | 80% | |
| Amount Borrowed | $13,605,120 | |
| Down payment | $3,401,280 | |
| Depreciation, Years | $428,000 | 27.5 |
| Commission at sale in the 15th year | 5.50% | |
| Appreciation: Market Value Increase per year | 1.250% | |
| Ordinary tax rate | 39.6% | |
| Capital gains rate at | 50.0% | |

| Year | Income | Total Expenses | NOI | NOI-Debt | Market Value | Debt Balance | After Tax Cash Flow | Cash on Cash | on Equity after Tax |
|---|---|---|---|---|---|---|---|---|---|
| 1 | $ 762,216 | $ 251,429 | $ 510,787 | $ (383,267) | $ 14,349,150 | $ 13,289,284 | $ (187,077) | -10.89% | -5.32% |
| 2 | $ 781,271 | $ 256,772 | $ 524,499 | $ (369,555) | $ 11,622,812 | $ 12,960,024 | $ (184,110) | -10.14% | -5.05% |
| 3 | $ 800,803 | $ 262,228 | $ 538,575 | $ (355,479) | $ 11,768,097 | $ 12,616,771 | $ (181,150) | -9.42% | -4.80% |
| 4 | $ 820,823 | $ 267,800 | $ 553,022 | $ (341,032) | $ 11,915,198 | $ 12,258,930 | $ (178,200) | -8.72% | -4.56% |
| 5 | $ 841,343 | $ 273,491 | $ 567,852 | $ (326,202) | $ 12,064,138 | $ 11,885,881 | $ (175,265) | -8.05% | -4.33% |
| 6 | $ 862,377 | $ 279,303 | $ 583,074 | $ (310,980) | $ 12,214,940 | $ 11,496,977 | $ (172,350) | -7.40% | -4.10% |
| 7 | $ 883,936 | $ 285,238 | $ 598,698 | $ (295,356) | $ 12,367,626 | $ 11,091,544 | $ (169,458) | -6.78% | -3.89% |
| 8 | $ 906,035 | $ 291,299 | $ 614,736 | $ (279,318) | $ 12,522,222 | $ 10,668,881 | $ (166,595) | -6.19% | -3.69% |
| 9 | $ 928,686 | $ 297,489 | $ 631,196 | $ (262,858) | $ 12,678,749 | $ 10,228,254 | $ (163,766) | -5.61% | -3.50% |
| 10 | $ 951,903 | $ 303,811 | $ 648,092 | $ (245,962) | $ 12,837,234 | $ 9,768,901 | $ (160,977) | -5.06% | -3.31% |
| 11 | $ 975,700 | $ 310,267 | $ 665,433 | $ (228,621) | $ 12,997,699 | $ 9,290,025 | $ (158,234) | -4.54% | -3.14% |
| 12 | $ 1,000,093 | $ 316,860 | $ 683,233 | $ (210,821) | $ 13,160,170 | $ 8,790,797 | $ (155,542) | -4.03% | -2.98% |
| 13 | $ 1,025,095 | $ 323,594 | $ 701,502 | $ (192,552) | $ 13,324,673 | $ 8,270,352 | $ (152,910) | -3.55% | -2.82% |
| 14 | $ 1,050,723 | $ 330,470 | $ 720,253 | $ (173,801) | $ 13,491,231 | $ 7,727,788 | $ (150,343) | -3.09% | -2.67% |
| 15 | $ 1,076,991 | $ 337,492 | $ 739,498 | $ (154,556) | $ 13,659,871 | $ 7,162,165 | $ (147,850) | -2.65% | -2.54% |

| IRR Before Tax | -15.08% |
|---|---|
| IRR Aft Tax | -23.93% |

Income = Gross Income (net vacancy) Note this is an end of year calculation
Total Equity = Down payment + Accumulated Principal + Cash Account
After Tax Cash Flow (Net spendable) = Pre TaxCash Flow - taxes
Cash on Cash = Pre Tax Cash Flow/Equity
Return on Equity after Tax = Cash flow after Tax/Equity

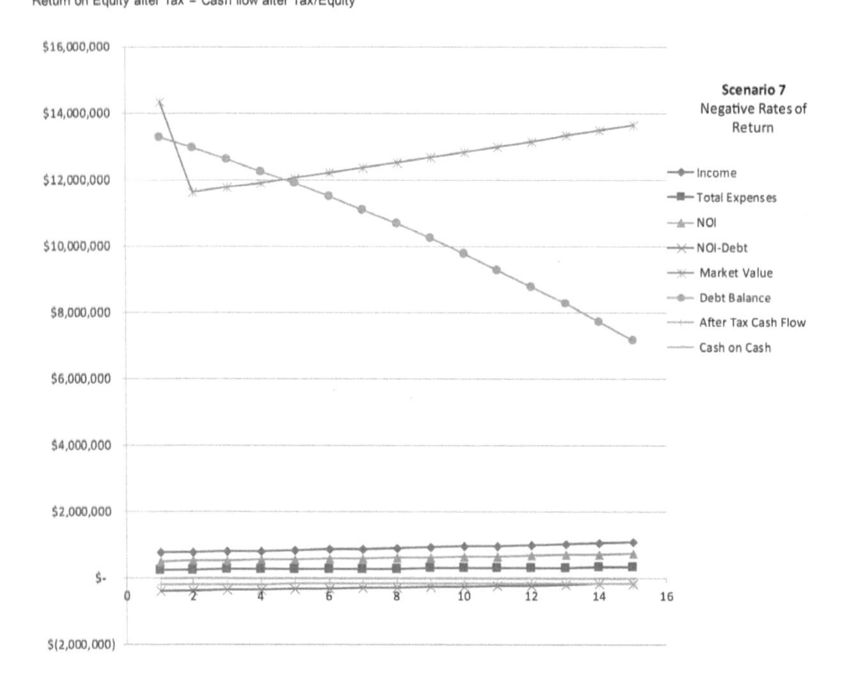


*Figure 10.4* Scenario 7: negative return.

| | |
|---|---|
| **Rent, Year 1 (Less vacancy)** | 743,625 |
| **Debt** | |
| Short term rates | 5.5% |
| Interest rate on mortgage (annual) | 2.00% |
| Amortization Period | 25.00 |
| Percent Financed | 80% |
| Amount Borrowed | $13,605,120 |
| Down payment | $3,401,280 |
| Depreciation, Years | $428,000 27.5 |
| Commission at sale in the 15th year | 5.50% |
| Appreciation: Market Value Increase per year | 1.250% |
| Ordinary tax rate | 39.6% |
| Capital gains rate at | 50.0% |

| Year | Income | Total Expenses | NOI | NOI-Debt | Market Value | Debt Balance | After Tax Cash Flow | Cash on Cash | Return on Equity after Tax |
|---|---|---|---|---|---|---|---|---|---|
| 1 | $ 762,216 | $ 251,429 | $ 510,787 | $ (186,073) | $ 14,349,150 | $ 13,180,362 | $ (111,104) | -5.02% | -3.00% |
| 2 | $ 781,271 | $ 256,772 | $ 524,499 | $ (172,361) | $ 11,622,812 | $ 12,747,109 | $ (106,186) | -4.28% | -2.64% |
| 3 | $ 800,803 | $ 262,228 | $ 538,575 | $ (158,285) | $ 11,768,097 | $ 12,305,191 | $ (101,116) | -3.64% | -2.33% |
| 4 | $ 820,823 | $ 267,800 | $ 553,022 | $ (143,838) | $ 11,915,198 | $ 11,854,435 | $ (95,890) | -3.08% | -2.05% |
| 5 | $ 841,343 | $ 273,491 | $ 567,852 | $ (129,008) | $ 12,064,138 | $ 11,394,663 | $ (90,502) | -2.57% | -1.80% |
| 6 | $ 862,377 | $ 279,303 | $ 583,074 | $ (113,786) | $ 12,214,940 | $ 10,925,696 | $ (84,950) | -2.12% | -1.58% |
| 7 | $ 883,936 | $ 285,238 | $ 598,698 | $ (98,162) | $ 12,367,626 | $ 10,447,350 | $ (79,227) | -1.72% | -1.39% |
| 8 | $ 906,035 | $ 291,299 | $ 614,736 | $ (82,125) | $ 12,522,222 | $ 9,959,437 | $ (73,329) | -1.35% | -1.21% |
| 9 | $ 928,686 | $ 297,489 | $ 631,196 | $ (65,664) | $ 12,678,749 | $ 9,461,765 | $ (67,251) | -1.02% | -1.04% |
| 10 | $ 951,903 | $ 303,811 | $ 648,092 | $ (48,768) | $ 12,837,234 | $ 8,954,140 | $ (60,988) | -0.71% | -0.89% |
| 11 | $ 975,700 | $ 310,267 | $ 665,433 | $ (31,427) | $ 12,997,699 | $ 8,436,363 | $ (54,534) | -0.43% | -0.75% |
| 12 | $ 1,000,093 | $ 316,860 | $ 683,233 | $ (13,628) | $ 13,160,170 | $ 7,908,230 | $ (47,884) | -0.18% | -0.63% |
| 13 | $ 1,025,095 | $ 323,594 | $ 701,502 | $ 4,642 | $ 13,324,673 | $ 7,369,535 | $ (41,032) | 0.06% | -0.51% |
| 14 | $ 1,050,723 | $ 330,470 | $ 720,253 | $ 23,393 | $ 13,491,231 | $ 6,820,065 | $ (33,973) | 0.28% | -0.40% |
| 15 | $ 1,076,991 | $ 337,492 | $ 739,498 | $ 42,638 | $ 13,659,871 | $ 6,259,606 | $ (26,700) | 0.48% | -0.30% |

| | |
|---|---|
| IRR Before Tax | 0.39% |
| IRR Aft Tax | -1.30% |

Income = Gross Income (net vacancy) Note this is an end of year calculation
Total Equity = Down payment + Accumulated Principal + Cash Account
After Tax Cash Flow (Net spendable) = Pre TaxCash Flow - taxes
Cash on Cash = Pre Tax Cash Flow/Equity
Return on Equity after Tax = Cash flow after Tax/Equity

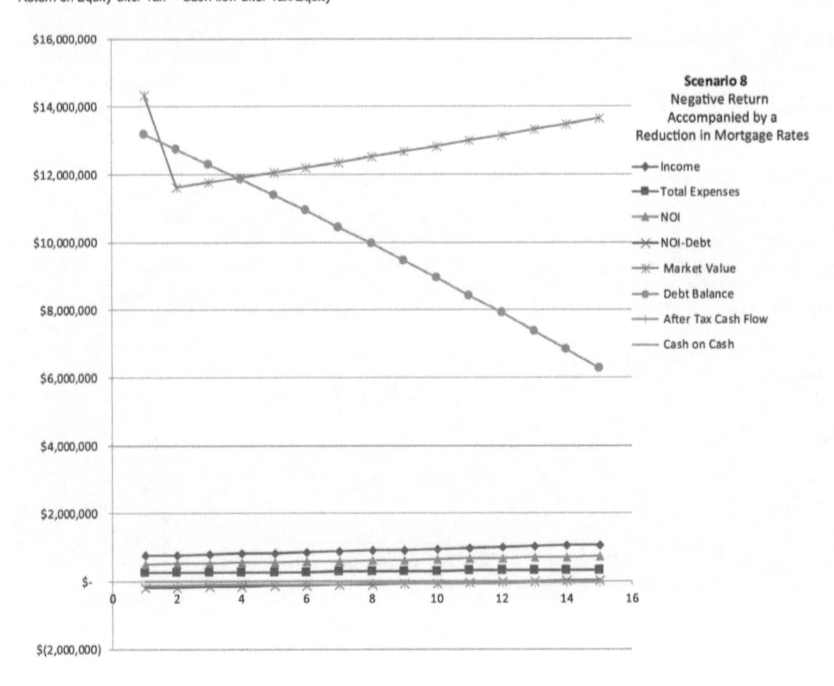


*Figure 11.4* Scenario 7: negative return, accompanied by a reduction of mortgage rates.

should suddenly take a downturn, then what was just purchased may no longer be financially sustainable. Consider the case of Scenario 4, in which rents fall by 15%. We will assume that the reduction in rents keeps the apartment rented at 85% occupancy, though property value falls by 15% and then increases at a rate of only 1.25% per year. In this scenario, we will also assume that the property was purchased at a 20% premium with a mortgage rate 0.75% higher than that in Scenario 4, at 4.25% annually (Figure 10.4).

Under this scenario, cash flows are not able to sustain the project and if the property is sold in the 15th year, the IRR is –15.08% before tax. Given the deteriorating economy, it could be possible to see the federal government support lower interest rates in the hopes of stimulating the economy. Lower interest rates on our current mortgage would reduce our "NOI *less* debt" numbers. If the rate should fall to 3.5% and if we are allowed to renegotiate the mortgage on the balance owed, the IRR is still negative. In order to break even (IRR = 0), banks would have to drop rates to 2%. However, the real issue is that, for most of the lifetime of the project, cash flow on an after-tax or pretax basis is negative. Even if they are able to renegotiate the loan, chances are that investors will want to bail out long before a sale in the 15th year of the project (Figure 11.4).

## Summary and conclusion

As a starting point in any analysis, a most likely scenario, "business as usual" using assumptions based on current values, forms the basis of future projections. Only experience and knowledge can guide the design of alternative scenarios. It is unlikely that anybody would consider buying a property if the numbers were not positive for the "business as usual" case. The best-case scenario, Scenario 2, paints an even more optimistic view of the future with IRRs that may never be achieved, but which may have investors excited about the prospect of higher-than-average returns. Investors looking over the details of the pro forma will also want to consider cash flow before and after tax. In Scenarios 3–6, deteriorating economic conditions result in higher vacancy and lower rents, although these conditions still could produce a respectable IRR over the lifetime of the project. This is because the tax advantages of real estate reduce the impact of negative cash on cash. As long as the investor can take advantage of the losses to shelter other income, the investor can still break even or even show a positive after-tax rate of return if keeping current tenants by lowering rents. It is always advantageous to have a positive cash flow before tax that can be used to improve the property. In doing so, you can improve future property values and grow rents. In Scenario 6, spending some cash could potential improve rents and guarantee a higher selling price at the end of the project's life. Looking at the downside, in Scenario 3, there was sufficient time to recover from the unexpected decline in values of 15%. By the end of the 15-year period, it was possible to sell the property for a value higher than the purchase price even with this initial decline. However, buying a property at a high price with expensive money when values are escalating quickly is probably never a good idea (Scenario 7). In this case, a downturn in the economy will result in a distressed property and investors will have little to look forward to in the future.

Clearly, we need to be prepared for the worst of times. For a project that may have a duration of 15 years, it is difficult to consider every possible scenario. The economics of the project can change overnight. Political events, and changes in trade, fiscal, banking, monetary, and tax policies, can have a direct impact on a local economy. Projections based on historical data are a starting point for any financial assessment. Though it is not possible to consider every combination, historical lows and highs for critical assumptions can assure investors that the project can survive a perfect storm in which economic forces work against the project's success. A basic understanding of the economy can be helpful in testing how economic forces can bring financial ruin to a project. Periods of low rents combined with high vacancy factors can present difficulties in meeting debt obligations, real estate taxes, utilities, maintenance, and other expenses. In particular, interest rates can be a source of great pain during an economic recession.

Interest, though a deductible expense, is always a considerable drain on the bottom line. Having a lower interest rate will always improve the financial return of a pro forma. If a mortgage was negotiated during a period of high interest rates and rising property values, a recession will make it more difficult to maintain a positive cash flow. If rates fall, as they often do when governments are trying to stimulate the economy, then renegotiating a mortgage will be to the investor's advantage. However, properties that are underwater or have had their equity reduced by a decline in real estate prices could be difficult to refinance. Even with a reduction in their debt load, the decrease may still not be sufficient to generate a positive after-tax cash flow and therefore these properties will require infusions of cash to stay afloat. It is difficult to offset the cost of financing a property that was considered reasonable in an inflated market, only now to be overpriced.

Looking for strategies that can protect the downside of any investment is prudent. If possible, sharing the risk of rising costs with the tenants is always a good strategy. Having tenants pay utilities relieves you of the burden of escalating rates. Also, to the degree that any potential increase in expenses can be borne by the tenant, the use of escalator causes and participation factors is also preferable. For non-residential tenants, it is also possible to employ a triple-net lease, which requires the tenants to pay the cost of insurance, taxes, and maintenance.

Financial models are tools that we can use to explore the performance of an investment under various constraints and conditions. Past experience can be helpful in the testing and evaluation of future scenarios. History can help inform the choice of parameters with which to test the financial health of a project under various scenarios. Examination of measures, including cash on cash, net spendable after tax, and IRR can help inform the decision as to how an investment will perform when the market is positive, but will also test the likely survival of the project when economics conditions are dismal. It is impossible to predict the future. Having made contingency plans for various scenarios will at least prepare investors for the action that may be needed to survive difficult economic times.

## Bibliography

Brett, D. and A Schmitz. *Real Estate Market Analysis, Methods and Case Studies*. Washington, DC: Urban Land Institute, 2009.

Brueggman, Willimam and Jeffrey Fisher. *Real Estate Finance and Investments*. Boston, Massachusettes: McGraw-Hill/Irwin, 2001.

Gallinelli, Frank. *Mastering Real Estate Investment, Examples, Metrics and Case Studies*. Southport, Connecticut: Real Data, Inc., 2008.

Havard, Tim. *Financial Feasibility Studies for Property Development, Theory and Practice*. London & New York: Routledge, 2014.

Linneman, Peter. *Real Estate Finance & Investments: Risks and Opportunities*, 2nd ed. Philadelphia, Pennsylvania: Linneman Associates, 2008.

Poorvu, W. J. and J. Cruikshank. *The Real Estate Game, the Intelligent Guide to Decision-Making and Investment*. New York: The Free Press, 1999.

Richard, Reed and Sally Sims. *Property Development*, 6th ed. New York: Routledge, 2015.

Staiger, Roger. *Foundations of Real Estate Financial Modeling*. New York: Routledge, 2015.

# 5 Market analysis

## Market analysis: supply, demand, and geography

Anyone who has purchased a home or rented an apartment implicitly understands how the market forces of supply and demand shape that decision. As a buyer in the marketplace, you have specific needs and preferences. You may want a home with a modern kitchen, three bedrooms, and a yard in which the children and your dog can play. You need to live within a 25-minute drive of work and it would be great if a grocery store were within a 5–10-minute drive. Having good schools in the area is also a major concern. Like all households, there is a limit on how much you can spend on housing. If you are a first-time home buyer, your savings for a down-payment and your household income will determine how much you can borrow. The budget stretches only so far. If you decide to put off your decision to buy this year, you may look to rent something comparable with the hopes of saving more towards your down-payment. If you are lucky enough to have a home and you have equity in your home, you can use it towards the purchase of your next property. This equity may help you step up to a more expensive home. In searching through an online real estate database, you narrow your choice to a few possibilities and arrange to meet with the agent, in the hopes that one of these properties will meet or surpass your expectations.

In this brief discussion of an everyday personal financial matter, many of the issues that must be considered in any market analysis are outlined. In conducting a market analysis for a real estate development project, we consider the collective sum of all decisions made by potential buyers and sellers rather than those of a single person. Ultimately, this research will inform the decision of whether or not to move forward on the project. Knowing what individuals are willing to pay to rent or buy a home is critical when planning to build a single house, an apartment building or a new community. As in solving any problem, it is essential to first define the constraints of the problem. For a market study, one such constraint is geography. Potential buyers and renters are often looking for a property that is confined to only a few neighborhoods in a community. Commercial and retail tenants will also have constraints that limit their search to those sites that meet very specific requirements. They may need to have a site of a specific size, located in the downtown, within a minimum distance of a metro stop and city hall. In any one city, this could constrict the tenant's choices to only a few locations. At any one point in time, the number of potential buyers and tenants looking for a property is fixed. For a specific development, the depth or size of the demand is always a concern for the developer and real estate investor. If the number of buyers are few and the supply is large, there will be little reason for anyone to create new product for this market. However, in markets in which there the demand is growing and outstripping supply, there is an opportunity to create a development project that can satisfy this growing demand. Under these conditions, the risk associated with any project appears to be justified.

Markets are never in perfect balance. For that reason, timing is always a critical factor in market research. If we plan to build condos for the retirement market, then timing our development to meet an increasing number of retirees will be key to our success. Estimating the size of this demand is only one component of our market study. It is also important to know the inventory of existing and future properties that could satisfy this demand. If our competitors are well positioned to satisfy this market, there may not be room for one more additional participant in the future, making it difficult for us to succeed with our proposal. Where your competitors are known for making a quality product in a market with a limited number of potential buyers, entrance can be difficult. Even when there is unsatisfied demand for a particular type of development, a sudden change in the economy will threaten future success. Few predicted the rapid decline in the price of oil in 1979–86, 1996–98, 2008–09, and

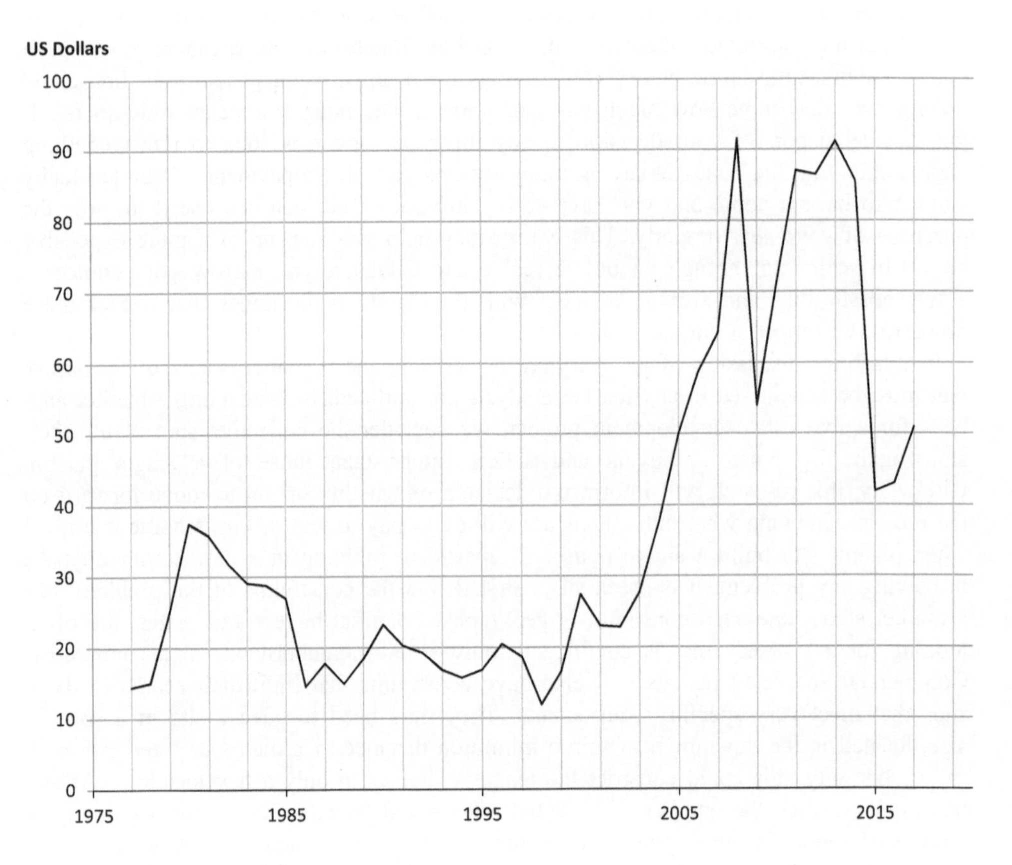


*Figure 1.5* Crude oil prices (WTI), 1976–2017.

Data Source: InflationData.com

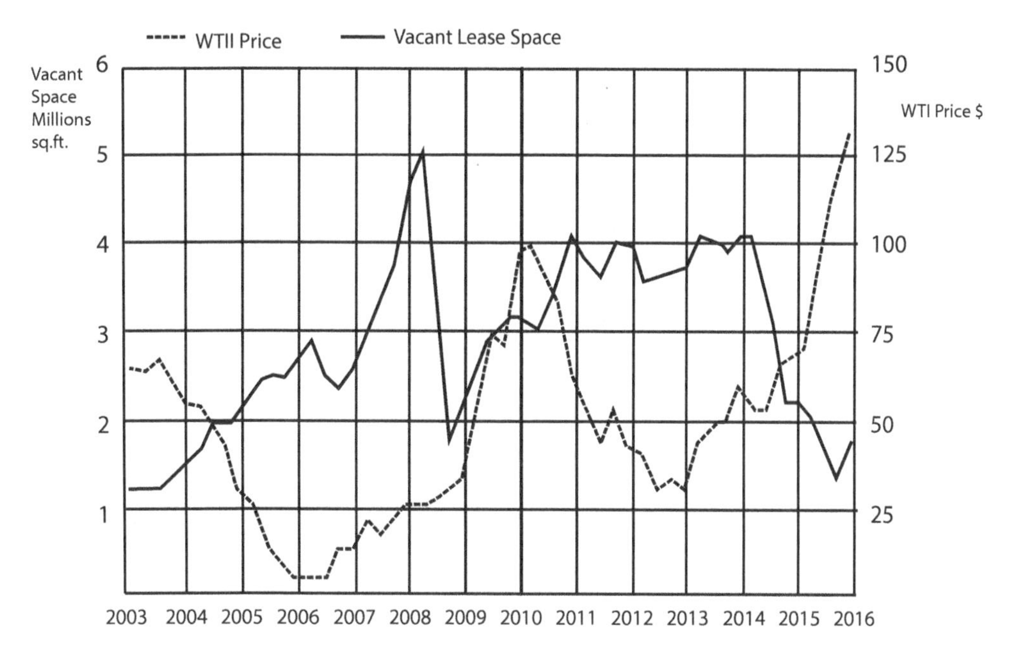


*Figure 2.5* Oil prices (WTI), vacancy in Calgary office space, 2003–16.

Data Source: CBRE 2016.

later in 2014–15, resulting in a surplus in commercial office space in cities including Calgary and Houston (Figures 1.5 and 2.5). Much of market research is built on the premise of predictable growth. As long as our economy proceeds as "business as usual", our projections are reliable.

In summary, this chapter will provide an introduction to the principles and techniques for conducting market studies in real estate. Gauging the forces of supply and demand on the success of a project is not a precise science. When projects have a projected life of 10–15 years, market studies can provide only guidance. Market research is an essential and useful tool that can support decision making by testing the basic premises and assumptions about a project. The first step in conducting a market study is to answer the questions of "what is its purpose?" and "who is the audience?"

## Market studies and the real estate development process

All studies have specific goals largely determined by their client. As the developer, it is important that you establish the terms of reference for any research project. Conducting what might appear to be simple studies can be costly in time and financial resources. Establishing constraints on the research early in a project may help you avoid a poorly defined set of objectives and the expense of additional studies.

### *Concept development*

During the early stages of concept development, an understanding of the market is critical to testing the financial success of the project. For rental property development, one must know rents for comparable developments. Equally important, the direction that rents will take in the future provides data for conducting a sensitivity analysis on the pro forma for the project. Knowing how rents will increase or decrease in the future could determine the financial success or failure of a project. If the project is being built as a speculative project, price data on comparable developments is a critical input. Having some knowledge of the direction of prices is just as important in determining the potential financial success of a project. It takes time to bring any project to market. For small projects, even a year can make all the difference. What may have seemed like a good idea during the planning stages may seem later, near project completion, more like "one of *those* ideas", or you might ask yourself, "what was I thinking?" No two projects are identical. The basic assumptions underlying a pro forma must be reasonable. Where economic data are used to establish the basic assumptions of a pro forma, a margin of error is prudent for factors such as the cost of construction, interest rates for borrowing, and even the purchase price of the lot. Rent and sale price projections should also be conservative. It is unlikely that anybody would pursue a project expecting the worst-case scenario; however, knowing that the project could survive an economic downturn is reassuring.

### *Design and marketing*

Once a project moves into the design phase, more detailed data on sales, rents, and construction costs are often necessary. In making the evaluation of whether to go forward on a project, the details matter. Design decisions made in consultation with the architect must also take into account their costs and what the prospective buyers and renters are willing to pay for. Few people will pay a premium for features that satisfy the designer's aesthetic sensibilities. When a project is being developed for a specific client, design preferences can be discussed and documented. Clients, including large retail outfits, commercial clients, and government agencies, will often bring to this design development process a wealth of data that will help in refining the design. When clients have internal design departments, their experience of working with consumers can help specify many of the finer points of the design. Knowing what buyers value can help us identify the finishes, style, and architectural details that will add value to the design. Having a detailed understanding of consumer preferences and their associated costs can be key to reducing financial risk.

In markets in which there is a low activity level, assembling a data set that can inform every design decision can be difficult, if not almost impossible. In the absence of data on the local market, national and regional trends can help establish consumer preference. Ultimately, the goal is to establish what prospective buyers and renters are willing to pay for various design features. For example, in developing a residential development for a particular neighborhood, four-bedroom houses with a minimum of 2,400 sq. ft. may be the norm. The buyers in this market may also expect oversized large master bedrooms with en suites, complete with dual sinks, jet tubs, showers, and large walk-in closets. If possible, each design decision should be supported by market studies. In analyzing market data, the objective should be a table clearly setting out the costs associated with every feature and the impact on the project price. For each class of buyer, the level of purchasing power must be considered. When designing a home for entry-level buyers, clearly affordability is important. A common mistake made by investors just getting started in real estate development is to design the units for themselves. They will include features that a few would appreciate, but which do nothing for the bottom line. Being sensitive to the wants, needs, and purchasing power of your ultimate client is critical at this stage.

Where data is not sufficient to create a table showing the costs and incremental value to the final price of the unit, it may be helpful to conduct focus groups or conduct surveys targeting potential buyers. Firms that specialize in market research can be used to qualify and narrow design choices. Data gleaned from national design shows and architectural design periodicals can help establish the future direction of consumer preference. To guard against the risk of designing a product that does not satisfy the market, it can be advisable to leave the detail of design until after the sale has been finalized. Building on a speculative basis is always a game of chance. Many large-scale condo and residential development complexes rely on design centers to reduce the risk associated with making the wrong design decisions. Giving the buyer input on the architectural details of their home or unit after they have purchased the property eliminates the risk of making the wrong choices. Many developers adopt a strategy of providing only a basic unit at an initial entry price, knowing the buyer will want additional features, which will escalate the final price. With higher profit margins on the optional features, a design center can be part of a winning strategy for the builder. In a residential development, these features can include appliance packages, custom kitchens, extra bathrooms, finished basements and bonus rooms, higher-quality flooring materials, architectural moldings, and exterior decks and landscaping.

### *Market studies, planners, and government officials*

As a part of the total package needed by planners and government officials, the data assembled as part of the market studies can be used to explain how the project fits into the community. Benefits to the community are numerous for commercial and industrial projects, including increased tax revenues and additional space for businesses expanding their operations. Residential projects can play a critical role in the development of the local economy by providing homes and apartments for an expanding workforce. Communities that have constrained development can experience higher home prices and rents, making it difficult for new residents to find adequate housing, even for those with a secure job. Though affordable housing should be considered an asset for any community, planners may need to assure local residents that its presence will not result in lowering property values.

Market data can also help planners evaluate the impact of proposed developments. When development permits are accompanied by a requested zoning change, measuring the potential impacts of a development on the community becomes essential. As inputs into transportation and public utility studies, data on the number of households, household size, ages, income, and potential work locations will enable planners to evaluate the impact of future development on the city. Demands for health, education, police, fire protection, and other community services will also be affected by any large development. Having the data needed to answer important questions will only speed up the approval process.

Assembling this data set will also be important on projects that can benefit from federal, state, and municipal tax credit and abatement programs. A development that offers some affordable housing can benefit the investors in the form of tax credits from the federal and some state governments. However, to benefit from these programs, it is important to have the necessary documentation on incomes and rents to show that the proposed projects meet the test of affordability. Each community will have unique demands for these projects. In some communities, they also include programs for work–live housing and even artists' workshops.

### *Investors and lenders*

Lenders also need assurance of the financial feasibility of a project. Investors and lending institutions will be particularly concerned about the assumptions regarding the buyers and tenants for your project. For lenders to be assured, the rents and prices charged must be aligned with the market. While there must also be sufficient demand in the community to support the proposal in its early stages, the size of demand must be sufficient to carry the project through each phase in the development plan.

### *Buyers and tenants*

Buying a home is a long-term investment. Purchasing a home only to see values fall will not build confidence in the public for your future projects. Market studies can assure future owners that buying into your project is a good investment. Renters also need assurances that the neighborhood is not in decline. Commercial and retail investors, in particular, need to be assured that their location is going to provide access to their customers and clients. Signing a 10 or 15-year lease represents a significant commitment and, though it is more likely that they have already commissioned market studies of their own, any data that makes your case will help during the negotiations. If a leasing agent is being used to secure tenants, it is more than likely that they have access to proprietary data that has already been shared with prospective tenants. Large leasing firms, including Grubb & Ellis and Colliers International, are already knowledgeable about the benefits of any particular location. However, having a market research study will be useful in any negotiations with your leasing agents.

### *Role of consultants*

For many real estate development firms, conducting a market study will be outside their comfort zone. Even though staff may have some education and experience in this area, their jobs may not be focused on market research. Larger firms may have staff devoted to conducting these studies, but not the skill for intricate projects. Complex projects requiring expertise beyond the staff's knowledge area make hiring a consultant prudent. Appraisers can also help establish the price for comparable properties and parcels. Highly specialized market research firms exist to serve a broad range of clients and may focus on a single or all areas of real estate development: industrial, commercial, residential and retail. Clients contracting with these larger firms have the advantage of benefiting from their experience and, more importantly, their comprehensive databases of properties over many markets in North America. Access to this data will help at each stage of the development process, including the initial stage of considering various locations and design scenarios.

## Market research: basic principles and techniques

Fundamental knowledge of economics, demography, geography, and urban planning are all required in conducting a market research study. Depending on the project, commercial, residential, industrial, recreation or retail, the application of specific tools and techniques may be required. For example, looking at the demand for retail space will require the use of gravity models and transportation models. Knowing which models are most appropriate is useful when drawing up terms of reference for a market study.

### *Market area: drawing the boundaries*

Defining the trade area, or market area, is the first step. Most projects begin with a basic knowledge of the surrounding community, sometimes drawn from personal experience. For many developers, it is likely that their first project will be in their home town. A familiar background can help in defining the boundaries that separate neighborhoods and communities from each other. In his canonical work *Image of the City*, Kevin Lynch developed a mapping approach that is still used by planners today. Based on extensive field studies in Boston, his book, first published in 1960, is still considered a must-read for all planners and urban designers.[1] A careful observer of city life, Lynch demonstrates a process that is commonly used by planners today to describe each neighborhood. In walking or driving through our city, we make mental notes of the "paths, edges, nodes, landmarks and districts". By creating a cognitive map based on these elements, we break down our communities into distinct zones, each with unique qualities. Between one zone and the next, there are subtle differences in building form, age, and use. We may even note a business that has just opened or a restaurant now under new management. We may form an impression of the streetscape, with its trees, benches, sidewalks, pedestrian areas, and street traffic, which we use in our evaluation of its overall character. Constructing this inventory is almost subconscious and forms the basis of our memory of each neighborhood in our community. An initial reconnaissance, using photos, video, and notebook notations, can document the condition of individual buildings and the overall character of a neighborhood.

As the first step in our delineation of trade areas, a drive or walk-through will quickly illustrate how geography can create the boundaries separating trade areas. Anyone who lives in a community with a river or large body of water such as a bay or inlet knows how their daily life is affected by these geographic features. To minimize river crossings, the everyday activities of shopping may be restricted to your side of the river. Also, rivers that have always conveniently served as boundaries for counties, towns, and school districts can define neighborhoods, towns, and cities. Communities that are separated by a river are less likely to share a public library, post office or courthouse. Other natural features, such as mountains, valleys, and wetlands, can also serve as boundaries for communities. In addition to these natural features, built forms, including railroad tracks, highways, and overpasses, will act as boundaries separating communities. Many new communities actually use the road system and natural features to create a sense of identity for their residents. With their own neighborhood shopping centers and schools, residents develop an identification with their community.

A simple mapping exercise can serve as a step in determining the trade area for a project. Initially, a paper map or a geographical information system (GIS) application should be used to locate the proposed project. If the project involves retail or entertainment, the first consideration is "how many patrons do I need to make this a successful location?" Clearly, some businesses, such as a pharmacy or a convenience store, need fewer patrons to prosper, whereas a big box store may need to draw in the entire population of a community to succeed. At the other extreme, neighborhood bars and pubs are businesses enterprises that draw only from their local community. Anyone who has watched an episode of *The Simpsons* knows that Moe's bar has only six loyal customers: Homer, Barney, Lenny, Carl, Sam, and Larry. Though not a picture of reality, Moe's does illustrate that many businesses can exist with the support of a relatively few loyal patrons. Determining the boundaries of our draw requires knowing the distance our potential customers will travel to the proposed project. With this awareness, we can begin to ask questions about the economic and demographic composition of the potential customer base. Understanding the collective buying power of individuals in my trade area is an important factor in evaluating the likelihood that I will have a successful enterprise. However, to complete this evaluation, we must also know the strength of the competition. In doing so, we can determine "the residual demand", or the unsatisfied demand, for the service or development we plan to offer.

### Residual demand

In examining the trade area for your new proposed development, first locate any competing developments. The goal of this exercise will be to calculate residual demand:

$$\text{Residual demand} = \text{Total demand} - \text{Net absorption by competitors}$$

In arriving at the number for residual demand, the calculation should include any new developments planned for the area. In evaluating your particular location, you must consider the demand for the service or space you are offering. Retail services like those offered at pharmacies are needed in every community. If we were the only pharmacy in a community of several thousand, we would certainly be assured of a livelihood. However, for most other businesses, a retail market study is needed to evaluate how the proposed venture will fit into the existing retail landscape. In making this determination, it is critical to know how far people will travel for the particular service you are offering. On one extreme, I may travel more than an hour to reach a big box outlet, but for most other retail I prefer to shop locally. In contrast, the convenience store that offers coffee, milk, eggs, snacks, soft drinks, and a few overpriced items will draw its customers from only a few blocks. Adding one more to the area will dilute patronage for all of the adjacent stores. Depending on the population density, economic purchasing power of your community will determine the total demand for your particular service. If your data show that, on average, an individual will purchase $100 of items from a convenience store annually, then, if you need a minimum annual revenue of $200,000 to create a viable business, you will need to draw in 2,000 individuals from within your trade area. If we find that there are already two stores in the general vicinity and there are a total of 6,000 individuals living in this trade area, then it may be possible that we will divide the customers three ways and still reach our minimum target of $200,000. However, we must realize that, at some point, an oversaturated market will result in a breakeven situation for everyone. An additional retail establishment in the area would only make it more difficult for the three convenience stores to survive. If another store should open during the construction of my location, splitting the total trade of $600,000 four ways, each store would earn on average $150,000 per year – far less than what is needed to make a reasonable profit. Under that scenario, the store with the poorest margins would probably be the first to close its doors.

Calculating the trade area is never that simple. In order to be successful, we may need a certain type of customer: the one who likes to buy snacks on their walk to and from work. In this case, shoppers from the community may matter little in my success. Being located next to the bus stop or a public car parkade may be more important than the surrounding neighborhood. In this case, we may need to employ transportation modeling techniques to predict the traffic past our location. In some locations, it may not even be the residents of my city that drop in to buy coffee and donuts. In fact, it may be the tourists who are my primary customers, which means that it does not really matter that the donuts are stale and that the coffee is not fresh, because they are one-time buyers. Since this case depends on outside residents, merely drawing a circle with a radius of several blocks around my establishment and calculating the number of customers who are inside the trade area will not give me the answer to my problem.

Perhaps a bit of early fieldwork will help you to make the decision on the location for your proposed convenience store, even before you conduct a formal market study. Visiting the local competition in the morning and afternoon, when most of these businesses make a majority of their sales, could be a very useful starting point in this market research project. If the current businesses are marginal, it will be obvious. No lines at the cash register at what should be the busy time of the day should be a sign that the business location is far from ideal. Gauging the business activity of competitors is never easy. Statistics provided by a government census may not offer the fine-grained detail needed to evaluate the demand for a particular service in your small part of the city. The block boundaries of a census and block numbering areas (BNAs) never align perfectly with the trade area of interest. Also, data on the census may be collected every five years and could be out of date for your purposes.

In summary, several issues are highlighted by this armchair exercise. First, defining the physical trade area must acknowledge the importance of natural and built barriers. A river or highway can easily separate one community from another. Trade areas are never simply defined by a circle, but have edges that are often difficult to outline. Each member of a community may, in fact, express differently where the edge of their neighborhood community is. Who lives in the trade area will determine the purchasing power for space and services. Age, income, family size, and myriad other social indicators are strong determinants of spending patterns. However, in defining a trade area, we must consider those who, in their daily lives, might need what you have to offer. When it comes to defining a trade area, customers who drive through the area on a regular basis could be the key to success. Not surprisingly, most fast food restaurants fall into this category.

When we consider rentals and home sales, defining who might be a prospective resident may be determined not by who lives in my community, but by who is planning to move into my community within the next year. Then, there is also the issue of competition: no project exists in a vacuum. We may have competing ventures in our immediate area. Future competition merely exacerbates the problem. Clearly, each of these issues will need to be explored in detail for an accurate assessment of any estimate of residual demand.

## Population projection and cohort models

Understanding population dynamics is important to every real estate decision. The current structure of the population and how it will change in the future will be a critical component of every market study. The ultimate successes of projects that take years to plan and build are dependent on the correct assessment of future demand. If our initial concern is only to identify which cities are growing the fastest in a region, a simple population model may suffice. Presented in Chapter 3, population models can simply be an extrapolation of past trends. Using the following mathematical expression, we can look at growth rates across a broad range of cities:

$$P = P_o e^{rt}$$

Where:

$P$ = population at a future time $t$
$P_o$ = initial population
e = Euler number = 2.71828 …
$t$ = time (in years)

By fitting data to this model, we are basing our estimate of the future population on the past.

Using this estimate, we can calculate future demand for new housing. If we divide our estimate of the increase in population by average household size, we arrive at a gross number of units that will be needed by this population. If we have some information about the existing housing stock and how many units are being planned for the future, we can come up with an estimate for the residual demand. One commonly used indicator is permit data. Using permit data as an estimate of new construction, we can calculate a crude estimate of residual demand:

Residual demand = Total demand – Net absorption by competitors

Total demand = (Population increase/Average household size) – New construction

Based on an estimate for the area per unit needed to satisfy this new construction, we can also arrive at a crude estimate of the amount of land needed in our community to build all these new units.

Though a useful exercise, however, a simple model does not reveal how individual groups within the population will behave. One caveat in using any population projection, based on historic data, is that these models cannot predict when population growth will diverge from past trends because of shifts in the economy. This approach is useful only when a first-order approximation of demand is needed. Also, these simple models cannot tell us how a particular segment of the population will grow. Are young couples moving in? Are old people retiring to more southern climates? What is needed is a model that will predict the demand for a particular type of housing. For example, if the research objective is to determine the number of first-time home buyers, we will need to know more about our target population. Cohort survival models are often used to develop a more detailed picture of population growth. These models all follow the same basic form:

Population increase = Births – Deaths ± Net migration

In applying this model, the population of a city or region is divided into age groups by gender. The census uses a division of five years. In calculating the population increase using the formula, the number of deaths that occur in the population for each cohort or age group is required. This crude death rate is expressed as the number of deaths per 1,000 in the population. In calculating the population increase, we have to account for the number of individuals who will die over the next period. If we are making a five-year projection, we are calculating the number of individuals in each age cohort who will survive another five years. For example, if the number of men living in our city aged between 55 and 60 currently living in our city is 100,000, and if the crude death rate for men aged 55–60 is 13 in 1,000, then:

$$\text{Deaths (over five years for age group 50–59)} = \text{Current population} \times \text{Crude death rate}$$

$$\text{Deaths (over five years for age group 50–59)} = 100{,}000 \times (13/1{,}000) = 1{,}300$$

Given that we know that 13,000 men currently aged 50–59 will die in the next five years, then we also know that 98,700 will survive to live another five years:

$$\text{Survival} = \text{Current population} - \text{Death rate} = 100{,}000 - 1{,}300 = 98{,}700$$

If we express this "survival rate" as a ratio or proportion of the cohort surviving to live another five years:

$$\text{Survival rate} = \text{Number of individuals who will die/Current population} = 98{,}700/100{,}000 = 0.987 = 98.7\%$$

If there were no births in the population over the next year, then the natural increase in the population would be negative. This "natural increase" in population is calculated as the net difference of births and deaths. Of course, in most populations, there will be new births over the same period and, like the crude death rate, this ratio is expressed as births per 1,000 in the population.

In calculating the population increase from new births, this ratio is applied to the number of women in each age cohort. For example, if the number of women aged between 20 and 29 in our population is currently 100,000, and the crude birth rate for women aged 20–29 is 8 in 1,000, then:

$$\text{Births for women aged 20–29} = \text{Current population} \times \text{Crude birth rate}$$

$$\text{Births for women aged 20–29} = 100{,}000 \times (8/1{,}000) = 100{,}000 \times 0.008 = 800$$

Adding up the new births over the next five years will give us the total natural increase we can expect in our city. Similar to survival rates, the birth rates are often expressed as a decimal or percentage. Looking at the equation for population increase, the only number that we need to estimate is the number of individuals who will move away from or into our city. Like our other ratios, it can be expressed as a percentage or decimal. We can use this factor to estimate the growth we will see in our city from net migration over the next five years.

If the number of men aged between 20 and 29 currently living in our city is 100,000, and the net migration rate for men aged 20–29 is 15 in 1,000, then:

| | Cohort population in N = 1 | | | Survival Rates | | | Net Migration | |
|---|---|---|---|---|---|---|---|---|
| Cohort | Age Cohort | Male | Female | Male | Female | Birth Rates | Male | Female |
| 1 | 0-9 | 1,989 | 2,001 | 0.99 | 0.99 | | 0.03 | 0.03 |
| 2 | 10-19 | 2,136 | 2,028 | 0.95 | 0.97 | 0.060 | 0.04 | 0.03 |
| 3 | 20-29 | 2,120 | 2,048 | 0.93 | 0.93 | 1.200 | 0.15 | 0.07 |
| 4 | 30-39 | 2,676 | 2,414 | 0.91 | 0.93 | 0.566 | 0.07 | 0.04 |
| 5 | 40-49 | 2,378 | 2,326 | 0.89 | 0.91 | 0.060 | 0.01 | 0.01 |
| 6 | 50-59 | 1,608 | 1,825 | 0.83 | 0.83 | | | |
| 7 | 60-69 | 1,100 | 1,250 | 0.59 | 0.62 | | | |
| 8 | 70-79 | 534 | 885 | 0.48 | 0.55 | | | |
| 9 | 80-89 | 130 | 530 | 0.28 | 0.3 | | | |
| Total | | 14,671 | 15,307 | | | | | |

| | Cohort population in N = 2 | | | Survival Rates | | | Net Migration | | Births | | |
|---|---|---|---|---|---|---|---|---|---|---|---|
| Cohort | Age Cohort | Male | Female | Male | Female | Birth Rates | Male | Female | births male | Birth Female | Total |
| 1 | 0-9 | 2,002 | 2,083 | 0.99 | 0.99 | | 0.03 | 0.03 | | | |
| 2 | 10-19 | 2,029 | 2,041 | 0.95 | 0.97 | 0.060 | 0.04 | 0.03 | 60 | 62 | 122 |
| 3 | 20-29 | 2,115 | 2,028 | 0.94 | 0.95 | 1.200 | 0.15 | 0.07 | 1,204 | 1,253 | 2,458 |
| 4 | 30-39 | 2,290 | 2,048 | 0.92 | 0.93 | 0.566 | 0.07 | 0.04 | 669 | 697 | 1,366 |
| 5 | 40-49 | 2,622 | 2,342 | 0.89 | 0.91 | 0.060 | 0.01 | 0.01 | 68 | 71 | 140 |
| 6 | 50-59 | 2,140 | 2,140 | 0.83 | 0.85 | | | | | | |
| 7 | 60-69 | 1,335 | 1,515 | 0.59 | 0.62 | | | | | | |
| 8 | 70-79 | 649 | 775 | 0.48 | 0.55 | | | | | | |
| 9 | 80-89 | 256 | 487 | 0.28 | 0.3 | | | | | | |
| Total | | 15,437 | 15,458 | | | | | | 2,002 | 2,083 | 4,085 |

*Figure 3.5* Cohort analysis using Microsoft Excel.

$$\text{Net migration for men aged 20–29} = \text{Current population} \times \text{Net migration rate}$$

$$\text{Net migration for men aged 20–29} = 100{,}000 \times (15/1000) = 100{,}000 \times 0.015 = 1500$$

Calculating the increase and decrease to each age cohort can be done easily using a spreadsheet program such as Microsoft Excel (Figure 3.5) or by using specialized software applications created to solve this problem. By assembling estimates for death, survival, and net migration rates, we can predict the increase in population over the next period for each age cohort. Figure 3.5 shows a sample spreadsheet of a cohort population model. In this model, we can project the population increase for each cohort. In this example, the population increases from 29,978 to 30,782 over the ten-year period (Figure 3.5).

The accuracy of these projections is dependent on our faith in the data we have used in our calculations. Like all models, we assume the past will predict the future. Though estimates of birth rates and death rates do not change suddenly, in most parts of the world birth rates have dropped to what historically are low numbers. In many parts of the world, this downward trend has occurred over decades. In general, in most populations, we do not expect to see large fluctuations in crude death and birth rates.

In the past, population growth was largely a result of natural increase. Communities were isolated and a large influx of migrants was uncommon. Today, populations are highly mobile, and migration represents a significant portion of the population increase in the US and Canada (Figure 4.5). In contrast, countries that have few immigrants combined with low birth rates, such as Japan, are experiencing negative growth rates (Figure 5.5). Changes in immigration

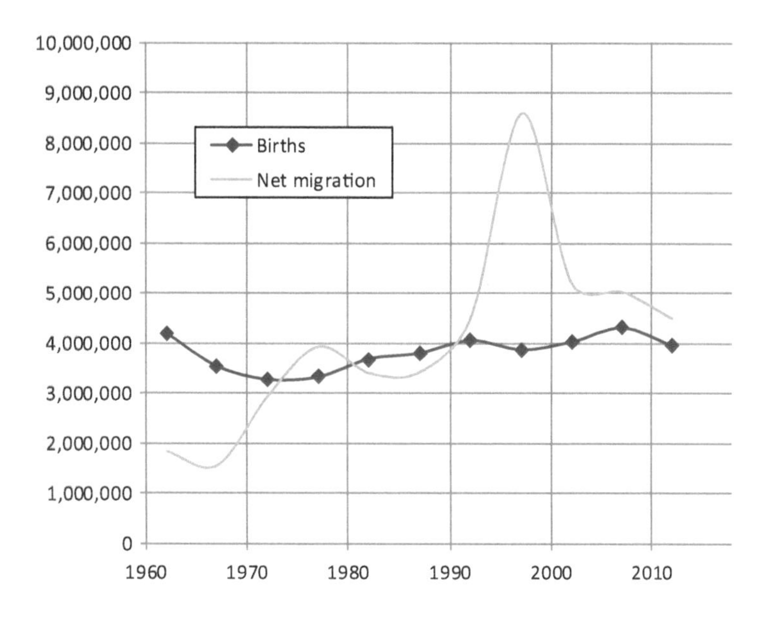


*Figure 4.5* Births and net migration, 1962–2012. US.

Data Source: The World Bank, https://data.worldbank.org/

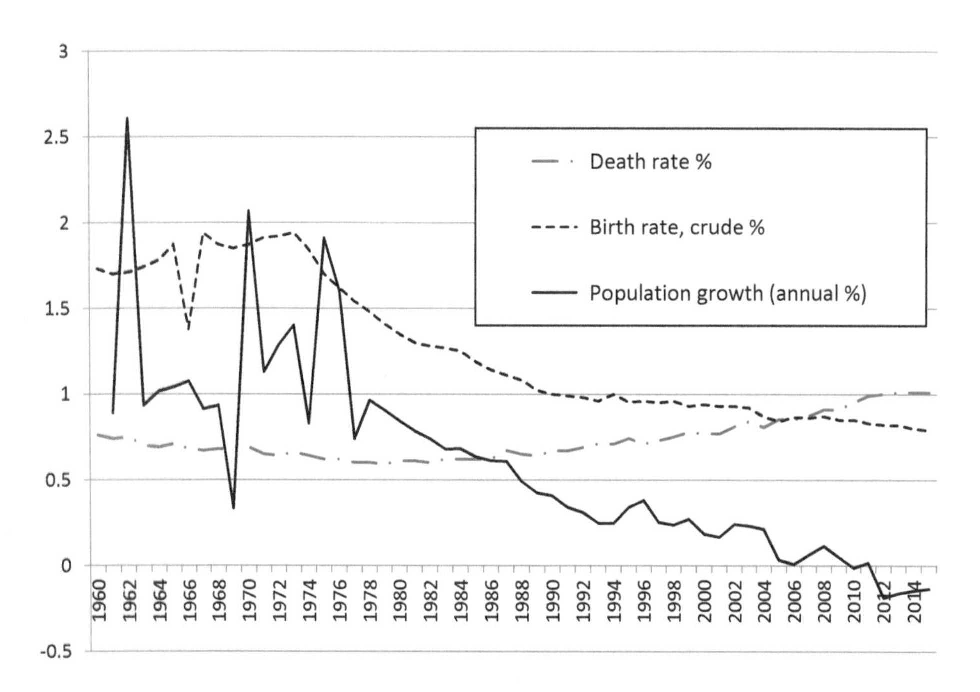


*Figure 5.5* Population growth, birth, and death rates in Japan, 1960–2015.

Data Source: The World Bank, https://data.worldbank.org/

policy will certainly weigh heavily on the growth of the US, Canada, and Europe in the future. Without positive immigration, many industrialized nations will not have a pool of young workers, which is so critical to economic growth and development. The implications for the real estate development market are obvious: new employment opportunities result in a growing demand for residential, commercial, retail, and industrial space.

In estimating future population by cohort, we must consider the impact of net migration for each age group. Clearly, the structure of populations can be greatly affected by who is moving into our community. A large influx of young people will shape the demand for jobs, housing, and services differently than would an influx of new residents in their retirement years. Consider how the demand for restaurants would differ in each of these two scenarios: The Keg vs Golden Corral. We can use employment statistics from the US Bureau of Labor Statistics or Statistics Canada as a starting point. We may also conduct local surveys of employers to learn about their future hiring plans to validate our estimates from published data sources. In general, populations do not change suddenly in size. However, a shock to the local economy can quickly impact the number of people who plan to move into a city. In cities that are dependent on primary industries including mining, and oil and gas, a rapid commodity price change can result almost overnight in reversals in net migration numbers. Since World War II, we have seen several rapid declines in the price for oil. In cities like Houston and Calgary, each time net migration has turned from positive to negative almost immediately. Predicting when a sudden shock to the system will occur is not easy. Few economists predicted the rapid decline in oil in 2014–16. A natural disaster can also have a similar impact. Cities such as New Orleans and Galveston have repeatedly suffered from hurricanes, which have immediate impacts on their populations. Residents of the community are forced to leave their homes, only with hopes to return to the city of their birth to live with friends and family in the future.

Estimates of the population by cohort are provided by both the US Census and Statistics Canada broken down by state, province, and city. Reports on the social, economic, and ethnic breakdown by cohort are also available from these government agencies. In addition to population projection, data on family formation, household size, income, and wealth are useful in building a picture of the population. The advantage of using age cohort modeling techniques is that we will have a better understanding of how demographic and economic factors can provide an estimate of the demand for a particular product. However, in order to use this data, we will need an in-depth knowledge of the buyer. Working with firms that specialize in analyzing demographic data may help in validating an individual real estate development project. For example, knowing that there will be substantial growth in the number of men and women between the ages of 20 and 30 will support a project directed towards first-time buyers. However, knowledge of this single trend is not sufficient to answer all questions about the size or depth of demand for a particular development. In making our evaluation, we will also want to know about the education, employment opportunities, wealth, and earning potential of this younger demographic. In building a profile of our potential customer, ultimately we would like to know more about their spending preferences.

## Demand, affordability, and residential development

In quantifying the demand for a particular project, issues of affordability and preference are critical to any market study. "Affordability" is often defined as a percentage of the household income that can be spent on housing. Though this percentage will vary depending on income and cost of living, for the US Department of Housing and Urban Development (HUD) and Canada Mortgage and Housing Corporation (CMHC), this percentage is considered 30%. When a family needs to spend more than 30% of their disposable income on housing, it becomes difficult to pay for the other necessities of life, including food, clothing, transportation, and medical care. Though banks may often set a limit on the amount they are willing to lend, based on an income statement, clearly many factors enter into this discussion of affordability. A family with children will certainly not be able to spend as much on housing as a couple without children. Other expenses, including outstanding loans, child care, medical expenses, and transportation costs, can weigh heavily on the family budget. What one household can spend on housing will vary depending on the particulars of their situation. In cities such as Vancouver, New York, Toronto, San Francisco, and Los Angeles, rapid price rise has pushed many households completely out of the home purchase market. In Vancouver, in 2016, the percentage of those with a median pretax household income able to afford a home was only 10%. Without government subsidy and support for low- and moderate-income housing programs in these cities, home purchases would be only a dream for most middle-class households.

In addition to the demands of family budgets, the externalities of mortgage lending rates can have a devastating impact on affordability. A 1% increase in interest rates can reduce the purchasing power by 15%. In the 1980s, the high cost of borrowing, with rates as high as 18% in many markets, resulted in a depressed housing market as the upper limit on the amount a household could borrow fell dramatically. One of the implications of changes in affordability for prospective occupants is the impact it has on the demand curve. Because most buyers use debt to purchase their homes, a change in interest rates, lending practices or the tax code will impact directly on the demand curve. With a larger percentage of the mortgage payment going to interest, the maximum price an individual can afford for a house will go down (P2 to P1), limiting choice in the marketplace (Figure 6.5). Simply put, higher

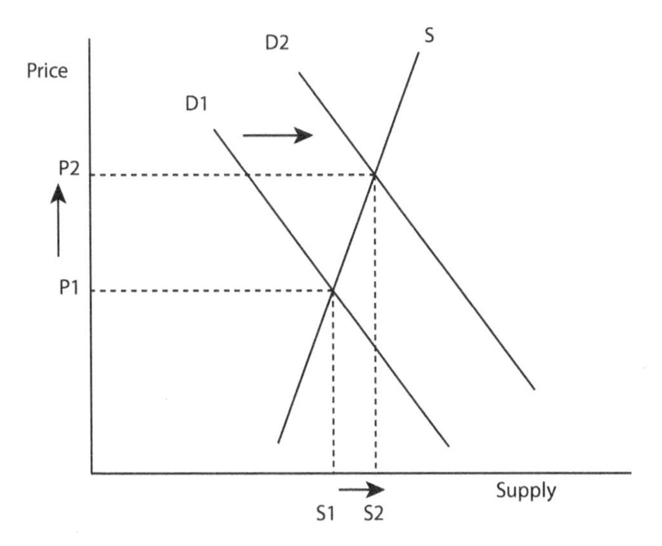


*Figure 6.5* Demand and supply curves for housing.

lending rates reduce the number of housing options. However, when rates are falling, the reverse is true: housing options increase (P1 to P2). Because the housing supply is highly inelastic in the short run, small changes in interest rates can also impact housing prices by increasing or decreasing the number of individuals looking to buy a home (Figure 6.5).

Similarly, banks can change their rules overnight, making it more difficult for households to qualify for a loan. When financial institutions require larger down-payments or apply more restrictive credit requirements, ultimately this reduces the number of households that can purchase a home. Changes in tax policy could also impact the net income for households and what is affordable for many home buyers. In Canada, mortgage interest is not deductible whereas in the US its status appears to be enshrined in the tax code – though since 2018 there has been a limit on how much you can deduct. This almost sacred tax deduction and its elimination would have serious implications for the political party initiating this change.[2]

For the developer, estimating affordability for a particular income group is a moving target. And yet, for a residential project, arriving at this estimate of demand is critical to success. A first step in establishing the size of the demand for a particular project is constructing a profile of the potential buyer or renter. This will also help establish the limits on affordability for this segment of the population. Residential projects can fall into several categories. Each will have buyers with different profiles. Some of the more common type of housing include the following.

**Single-family subdivisions** Since World War II, single-family homes have accounted for the majority of new homes built in Canada and the US. Though lot size will vary from ¼ acre to 1 acre, depending on land costs and zoning, most of the newer homes today will have a minimum of three bedrooms and a two-car garage. In many suburban communities, the lack of public transit makes these households auto-dependent, though these clients are also looking for the convenience of neighborhood shopping and local schools. Priced to accommodate those households with a specific income, developments can comprise a dozen homes to several hundred. Buyers of these homes tend to be families who are looking for the privacy and security of a backyard for their children and pets. Depending on the clientele, these homes can offer many high-end features, including finished basements for play areas, entertainment rooms, family rooms, or bathroom en suites with jet tubs. In warmer climates, in-ground pools are not uncommon in high-end developments. In more costly marketplaces, single-family homes are affordable only to high-income households.

**Townhouses** Built in urban environments as infills or stand-alone projects in more suburban locations, townhouses will attract a range of buyers. Townhouses can be attractive to young couples or those wishing to downsize from larger single-family homes during their retirement years. Good schools nearby may be less important than access to convenient shopping and entertainment. Many townhouse developments will offer one-car garages, though some of the larger units will have two-car garages. Townhouses today designed for the retirement market can feature high-end kitchens and bathrooms with many of the features expected in more expensive single-family homes. One attractive feature of townhouses is a more maintenance-free lifestyle, with the exterior of the home the responsibility of the condo association.

**Condo apartments** In many urban markets, condo apartments offer entry into a housing market in which townhouses and single-family homes are no longer affordable. Young urban professional couples will find condo projects near the downtown core that offer a variety of entertainment and cultural amenities. Marketed to the commuter on public transit or to those who desire an easy walk to work, convenience is a centerpiece of these projects. A car space may even be offered as an extra. Depending on the income levels of the prospective buyers, these units can be small, minimal studios or one- or two-bedroom units designed for more upscale clients. In recent years, couples downsizing from their suburban homes may look to condos as offering a more convenient lifestyle close to cultural amenities. Condo apartments in many cities are also seen as an investment opportunity. If permitted by the condo association, professionally managed units offer rental opportunities for young urban professionals. More recently, many of these condo projects have been marketed as an investment opportunity for buyers from Asia and the Middle East. In cities like New York, Vancouver, and San Francisco, they are seen as lifeboats in the event that the political environment in their home country becomes problematic.

**Affordable housing** In many communities, the need for affordable housing is critical. Without government assistance, it would be impossible to build housing for lower-income households in many North American cities. In the US, affordable housing projects are possible because of the low-income housing tax credit (LIHTC). Under this provision of the IRS code, housing that meets the 30% median income level in a community will qualify for a tax credit generally equal to 30% of the property value. When combined with tax credits for renovating historic properties, these projects can be profitable. Prospective clients include low-to-modest-income households from all age cohorts. In some municipalities, affordable units are only a small proportion of the total mix of a housing market. In cities like Vancouver, requirements for affordable housing are part of the development approval process. In many communities, a large proportion of affordable housing is designed for the elderly, which can make these projects more acceptable to the larger community.

**Second home** Built as vacation properties, second homes can be cottages, houses or condos. Retired Americans and Canadians known as snowbirds, who want to escape the cold in the winter, have bought a large number of second homes in the southwest, Florida, and Mexico. For Canadians, there is a residency requirement of six months a year in Canada if they are to maintain their healthcare coverage.

**Senior housing** With an aging population, a range of housing choices are available to the elderly. Condos and housing developments designed for the active senior (aged 55+) will offer recreational amenities. For older seniors, independent and assisted living housing offers a wide range of social, medical, and recreational services. Depending on the income of their clientele, these senior housing alternatives can offer minimal to luxurious accommodations.

**College student housing** In college towns across North America, the construction of apartment-style units for college students can offer investors a stable source of income. Depending on the marketplace and the incomes of the students' parents, these housing developments can be modest or upscale. Sometimes marketed as an investment opportunity, they can be seen as an alternative to paying dormitory fees. In markets with good upside potential, the parents of college students may find this alternative to dormitory fees attractive on financial terms.

## Case study: a condo apartment development

### *Introduction*

A developer of multifamily homes is considering a location in an advancing neighborhood located within walking distance of the downtown core. A light rail stop makes it possible to get to most downtown locations within minutes. With many new restaurants and boutiques frequented by a growing number of urban professionals who work in finance, engineering design, and IT, the area is an up-and-coming community on the edge of the downtown. Many of the projects in the area that cater to this market are selling quickly. One recently completed project was able to presell 40% of its units within three months. Sales of existing condos are also doing well. If market trends continue, other developments should be similarly successful. First impressions would suggest large studios, and one- and two-bedroom units. A market study will be needed to assess the merits of the location and to assure investors and lending institutions of the financial worthiness of the project.

### *Site and location*

The proposed development would be located on a quiet residential street one block away from the main shopping boulevard. The housing stock dates mostly from the early 1950s. Many of the older one-story bungalows were built as workers' housing when there were considerably more manufacturing jobs in the area. Some of the residents who are in their 80s were residents when many of the houses were first built. Most of the buildings on the main shopping boulevard date from before World War II. About 25% of the buildings are two-story commercial brick structures. Many of these buildings have been renovated under a facade improvement program sponsored by the city. A business improvement district (BID) has been spearheading street improvements since the 1980s. Under BID leadership, shade trees, benches, and several small outdoor plazas have been completed, creating a pleasant area in which to shop or just have coffee. In the 1980s, a grocery store was built across from the light rail stop and two blocks from the proposed condo location. The store is 60,000 sq. ft., and it features a bakery and pharmacy. A bakery and deli offer many options for those who are looking for a quick meal. Other retail located in the area offers a range of services and items, including women's clothing, shoes, sporting goods, and even a small hardware store that seems to have one of everything. There are also numerous options for those wanting specialty food items, including a bakery, an Asian grocery, a greengrocer, and a butcher. Though there are no major fast food options, the neighborhood has a number of restaurants that feature Indian, Moroccan, and Italian cuisine. On the weekends, the restaurants experience a high level of patronage and most require reservations for evening sittings. The area appears to be one of the more active and trendy neighborhoods in the city.

### *Demographics*

Current demographics in this neighborhood are in transition. Older residents (aged 65+) in the neighbored represent 12.9% of the total population of 11,400 (1,471) (Figure 7.5). Many of these retired residents have lived in their homes for over 40 years. The largest proportion of the population is in the age cohort 20–39 (4742, 42%). When compared with the population distribution of the city at large (30%), this neighborhood is distinctly younger. Notably, within this age group (20–39) there are more women living in this community than men (2,468 vs 2,274). Consistent with the population profile of the community, almost three-quarters of all households comprise only two people (74%), split roughly equally between young professionals and retirees. Interestingly, about half of the two-person households are single parents with one child. Recent surveys have shown

| | Neighborhood | | | City | |
|---|---|---|---|---|---|
| Age | Male | Female | Age | Male | Female |
| 85+ | 68 | 80 | 85+ | 6,492 | 7,033 |
| 80-84 | 91 | 103 | 80-84 | 7,574 | 8,656 |
| 75-79 | 160 | 171 | 75-79 | 8,656 | 9,738 |
| 70-74 | 182 | 194 | 70-74 | 17,312 | 17,853 |
| 65-69 | 205 | 217 | 65-69 | 19,476 | 19,692 |
| 60-64 | 251 | 274 | 60-64 | 27,050 | 27,050 |
| 55-59 | 279 | 291 | 55-59 | 34,624 | 35,165 |
| 50-54 | 291 | 302 | 50-54 | 34,083 | 34,299 |
| 45-49 | 410 | 433 | 45-49 | 41,657 | 42,739 |
| 40-44 | 422 | 410 | 40-44 | 41,116 | 41,224 |
| 36-39 | 450 | 456 | 36-39 | 41,982 | 42,198 |
| 30-34 | 570 | 593 | 30-34 | 42,198 | 43,280 |
| 25-29 | 627 | 764 | 25-29 | 41,116 | 41,549 |
| 20-24 | 627 | 656 | 20-24 | 34,840 | 35,652 |
| 15-19 | 120 | 121 | 15-19 | 33,542 | 34,408 |
| 20-24 | 116 | 122 | 20-24 | 30,296 | 30,837 |
| 15-19 | 131 | 137 | 15-19 | 19,476 | 19,584 |
| 10-14 | 194 | 205 | 10-14 | 19,476 | 19,584 |
| 5-9 | 160 | 171 | 5-9 | 15,689 | 15,797 |
| 0-4 | 171 | 177 | 0-4 | 19,422 | 19,584 |
| Total | 5526 | 5874 | Total | 536,077 | 545,923 |

*Figure 7.5* Neighborhood and city population by cohort and sex.

that half of the residents aged 20–39 were born in the city, while the remaining portion has moved from another part of the country in the last year. This trend is consistent with employment data for the city, which indicates that economic growth over the last decade has been attracting young college graduates from other parts of the country. Job growth of 2.2% is roughly equal to the population growth for the city (Figure 8.5).

The population has been growing roughly at 2% (approximately 228 people) per year, which is close to the average in the city. When you factor out single-parent head of households and seniors, it is likely that approximately 100 new households of young professional couples without children or with only one child will be moving into the neighborhood each year. It is also assumed that the demand for these units will keep pace with the annual population growth of 2%. In part, this estimate of demand is an educated guess based on trends over the last ten years. Education and employment data also

| | Neighborhood | | City | |
|---|---|---|---|---|
| | Number | Percent | Number | Percent |
| Family Size: 2 persons | 551 | 74% | 29643 | 12% |
| Family Size: 3 persons | 109 | 15% | 137185 | 54% |
| Family Size: 4 persons | 64 | 9% | 67795 | 27% |
| Family Size: 5 persons+ | 16 | 2% | 20095 | 8% |
| Total | 740 | 100% | 254718 | 100% |

*Figure 8.5* Family size for the neighborhood and the city.

supports the growth of a younger, more educated demographic. In the city, about 45% of all individuals in the age group 20–39 have had some college education, while for this neighborhood it is considerably higher, at 78%. Although incomes for the community are 15% lower than the city on average, this may be partly due to a large number of retirees. Also, young professionals beginning their career will have not yet reached their full earning potential. Estimating incomes for this group is difficult. Census data for the area provides a breakdown percentile, but not by education. Still, at an average, household income of $65,000 should allow the purchase of a condo or home in the $175,000–$200,000 range, with a minimum down-payment.

### *Competition*

Over the last five years, several multifamily buildings have been built to fill the demand for largely one- and two-bedroom units. Part of the boom in multifamily projects has been a response to a zoning change, which now allows buildings of up to four stories on the blocks on either side of the main shopping district. Though the development process can take between 12 and 18 months for approval, the city has yet to deny any developer a permit. A total of 450 units have been built during the last five years. Most of the units are being sold to young professionals. A majority of these units are priced between $150,000 and $275,000. This makes them affordable for incomes above the first quartile; also building units in this price range will be profitable, given current construction costs. About half of these new units are 650 sq. ft. one-bedroom units. The remaining units are roughly split equally among studio, two-bedroom and three bedroom units. The largest units, which constitute about 15% of the market, are 1450 sq. ft. units with three bedrooms and can sell for approximately $400,000. These units, which feature high-end finishes and appliances, appear to be built for working couples over the age of 55. Two-bedroom units range from 755 sq. ft. to 950 sq. ft., while the studios are, on average, 425 sq. ft. Those planning on buying these units as an investment should also do well. Based on current rents in the area, owners who rent their properties are able to cover operating costs of taxes, insurance, mortgage, and condo fees.

Most projects built in the last five years have sold out in less than six months. Only the newest of all the developments have vacant units. Completed just a year ago, the Empire Development Project has a total of 60 condo units. Located several blocks from the proposed development, there are 15 two- and three-bedroom apartments remaining to be sold. It is likely that these units will sell in the current year. Resale in the condo market is also very positive. On average, prices of existing condos have increased 3% per year, which is 1% above the average for the city. At any one time, only 15 units have been on the resale market during the last year. For this study, it is assumed that owners selling their units will relocate outside the neighborhood. Condos remain on the market for between 60 and 90 days, which is about average for the condos in the city.

### *Residual demand estimates*

City permits indicate that two condo projects have been approved. Greenwood has 80 units, while The Elms has 120 condo units. Both have buildings that are mostly one- and two-bedroom units. Construction should be completed in year 2 of your project. Both projects are located several blocks from your planned location. Both projects should come on the market in year 2 and it will probably take two more years to sell all the units in each building. Like most past projects, about 50% of the units presold in their first year even before construction began. Looking ahead, with a demand of approximately 100 units per year, there is a residual demand that will go unsatisfied over the next five years (Figure 9.5).

Best case Scenario

| Year | Resale Market | Empire Development Project | Greenwood | The Elms | Competition Total | Proposed Project | Total Supply | Demand | Residual | Cummulative |
|---|---|---|---|---|---|---|---|---|---|---|
| 1 | 15 | 15 | 40 | 60 | 115 | 0 | 115 | 100 | -15 | -15 |
| 2 | 15 | | 20 | 30 | 50 | 0 | 50 | 102 | 52 | 37 |
| 3 | 15 | | 20 | 30 | 50 | 50 | 100 | 104 | 4 | 41 |
| 4 | 15 | | | | | 25 | 25 | 106 | 81 | 122 |
| 5 | 15 | | | | | 25 | 25 | 108 | 83 | 205 |
| Total | 75 | 15 | 80 | 120 | 215 | 100 | 315 | 520 | 205 | |

Worst Case Scenario

| Year | Resale Market | Empire Development Project | Greenwood | The Elms | Competition Total | Proposed Project | Total Supply | Demand | Residual | Cummulative |
|---|---|---|---|---|---|---|---|---|---|---|
| 1 | 15 | 15 | 40 | 60 | 115 | 0 | 115 | 50 | -65 | -65 |
| 2 | 15 | 0 | 20 | 30 | 50 | 0 | 50 | 50 | 0 | -65 |
| 3 | 15 | 0 | 20 | 30 | 50 | 50 | 100 | 50 | -50 | -115 |
| 4 | 15 | 0 | 0 | 0 | 0 | 25 | 25 | 50 | 25 | -90 |
| 5 | 15 | 0 | 0 | 0 | 0 | 25 | 25 | 50 | 25 | -65 |
| Total | 75 | 15 | 80 | 120 | 215 | 100 | 315 | 250 | -65 | |

Three Year Housing Slump Scenario

| Year | Resale Market | Empire Development Project | Greenwood | The Elms | Competition Total | Proposed Project | Total Supply | Demand | Residual | Cummulative |
|---|---|---|---|---|---|---|---|---|---|---|
| 1 | 15 | 15 | 40 | 60 | 115 | 0 | 115 | 50 | -65 | -65 |
| 2 | 15 | 0 | 20 | 30 | 50 | 0 | 50 | 50 | 0 | -65 |
| 3 | 15 | 0 | 20 | 30 | 50 | 50 | 100 | 50 | -50 | -115 |
| 4 | 15 | 0 | 0 | 0 | 0 | 25 | 25 | 100 | 75 | -40 |
| 5 | 15 | 0 | 0 | 0 | 0 | 25 | 25 | 102 | 77 | 37 |
| Total | 75 | 15 | 80 | 120 | 215 | 100 | 315 | 352 | 37 | |

*Figure 9.5* Market projection of residual demand.

In developing a concept plan for the building, we can assume a total of 100 units. Units would be designed for first-time buyers: singles or couples looking to get into the housing market. They would be predominately one-bedroom units (75%). The remaining units would be studios and two-bedroom units. Changing the mix of unit types could accommodate future demand. If planning for the proposed project begins today, the building could be ready for occupancy in year 4. Hopefully, the approvals will be secured from the city in year 2 of the project. It should be possible to presell 50% of the units in year 3, once all the permits are in order. The balance of the units would hopefully be sold in years 4 and 5. Of course, there could be other developments in the planning stages, which could represent additional competition.

Under the best-case scenario, an additional 41 units would have to be built over the next three years before exceeding the total demand for the community (306 units over three years) (Figure 9.5). However, this assumes that no other project will also be approved. A competing project would result in an oversupply of housing in the community. These estimates also assume that the impact of housing development projects in adjacent neighborhoods has a negligible effect on demand for housing in this neighborhood. However, if a sudden downturn in the economy should occur, demand could easily be cut in half (worst case scenario). If this housing slump begins in year 1 and persists for the next five years, then there could be an excess inventory of units on the market over the next five years, in addition to the 100 units you plan to build. This is perhaps the worst-case scenario (Figure 9.5). It is more likely that a slump in housing could last three years with demand returning to normal. If this downturn began in year 3, then the residual demand would still be sufficient to carry the proposed project (three year housing slump scenario). However, preselling the units when the project came on the market in year 3 could be a challenge. In this event, the market should become very competitive for first-time buyers and it may be necessary to make some changes to the design of the building. Redesigning the building to accommodate the 50+ demographic looking for two- or three-bedroom units could provide a cushion against a disintegrating market during an economic downturn.

## Creating a marketing program

Recognizing the need for a particular housing type in a community is a first step in creating a marketing program. Reaching out to potential buyers of the project will be critical in achieving target sales in years 4 and 5 of a project such as that in the condos example. As part of the marketing program, building trust with those in the target group takes time and effort. Like any marketing program, there will be several stages in moving the potential buyer from first encounter to final sale. Future residents must first feel there is a need, which realization will be followed by some prepurchase activity. This usually involves online research, as well as contact with friends, family, and co-workers. Serious buyers may then engage a realtor who will help them arrange showings and review options in both the new and existing home markets. Finally, during the actual sale negotiations, agreement will be reached on terms, price, and other aspects of the sale. It is also during this final phase that the buyer will choose various design options. This would include upgrades to the appliance package, the flooring, and the wall colors.

In developing a plan for the proposed project, a public relations (PR) firm will be engaged early on in the design and construction process. It is hoped that the city-wide shortage in housing will make this project attractive to young professionals. Local media outlets, including the neighborhood paper, and the community website and newsletter, will be used to get the message out to the public about this project. The PR firm will develop the materials that will be distributed to the media. As part of the marketing plan, a focus group will provide valuable feedback during the design review process. It is hoped that, by engaging the community, the architectural details can be tailored to satisfy the preferences of future buyers. As part of this marketing plan, a web presence will promote interest in the project. It is hoped that, by means of an interactive website, potential buyers will be able to explore the design and leave comments for the design team to consider in arriving at their final design. Visitors to the website will also receive updates on the project. One of the features that will be included in the webpage will be a virtual world that provides anyone visiting the site with an opportunity to explore the different designs. It is hoped that this online presence will offer potential buyers an engaging experience and gather invaluable feedback for the design team. After breaking ground, the usual site board will be posted, with a website URL and QR code. Once the project is in progress, an onsite sales office will be open during the commuter hours on weekdays and weekends. The hope is that, as the project proceeds, it should be possible to use this location for all presales activity.

It is also hoped that, by having a significant online and physical presence, it should be possible to capture s significant proportion of residual demand projected for condos in this community. Because many of the buyers will be relocating from other cities across the country, maintaining good relations with the human resources (HR) directors of major employers will be part of the sales strategy. Emails and in-person meetings scheduled early on in the development process can also help inform the actual design of the building. Knowing more about the potential buyers' incomes and preferences can help in creating a product that will fill future demand.

## Retail

It is not uncommon to include some retail space as part of residential development projects. In many cities, in mixed-use development, ground-floor space is particularly well suited to retail development. If our proposed location had been along the major shopping boulevard, retail would have been a major consideration. Pedestrian traffic along this shopping street would have provided a steady stream of customers going past the location. Determining if the market exists for this space requires knowledge of the rental rates. Like the condo owner who plans to buy your unit as an investment, you need to know their projected rent and whether they can cover their annual costs (insurance, mortgage, taxes, fees, etc.) and still make a profit. Using the pro forma to test the profitability of this scenario requires some knowledge of rents in the area. As a first step, a survey of rents can provide some basic information on the financial feasibility of adding retail or, for that matter, commercial space. Information from listings, in addition to discussions with leasing agents, can be useful in arriving at a table of rents, features, and space for each option currently on the market. However, a survey of rents alone is only part of this analysis. During the lease negotiation, an improvement allowance may have to be given to the tenant to cover the renovations for which the tenant will be responsible if they are to sign a long-term lease. Also, there is the question of how much demand there is for the type of space you are offering. If, because of size, zoning, and availability of parking, the space is suited only to certain types of retail, then it is critical to know the depth of demand for that particular use. We saw elsewhere, in our discussion of a hypothetical analysis of a convenience store location, how the size of the local demand for a particular use can limit the numbers of a particular type of store.

Many mixed-use developments assume that current rents charged by neighboring businesses will carry over to any future development, when in fact there is little need for any more additional space in the community. The neighborhood may have all the convenience stores it will ever need. The failure to rent these retail spaces can also occur when there is an insufficient number of customers to support the proposed business. In our case, it might be possible to find a tenant who will be successful if planning to open a bakery, convenience store or donut shop, but not a women's boutique for designer shoes. Businesses that are not currently represented as part of the mix may be better served by being part of a larger retail district where the draw from the larger community is greater.

## Commercial office space

Like retail, the developers of commercial space require an in-depth market study before deciding on a particular location. As an investment found commonly in real estate investment trusts (REITs), pensions, and insurance company portfolios, these business opportunities must be able to generate a reasonable rate of return. Rents, of course, represent the revenue side of the equation. Estimating what a tenant will pay is a function of a number of factors. Like residential, commercial tenants have different needs and preferences. Commercial tenants come in all shapes and sizes. Some tenants may want to be in a fashionable location, while others will prefer areas of the city in which their employees can find affordable housing. It is common to categorize tenants by three classes: A, B, and C.

The A-class tenant is looking for a building with a good address, quality architectural features, and amenities. These buildings typically are the newer buildings in the city and are often considered architectural gems. Today, they will more likely be buildings certified under the Leadership in Energy and Environmental Design (LEED) scheme, which showcase the latest in green building technology. A good address implies that its location is in an area of the city known for high rents, expensive restaurants, high-end shopping, and, in some cities, a standard of living associated with the ultra-rich. In the past, it was the downtown central business district (CBD) that was the exclusive domain of A-class space; in the 21st century, the trend has been reversed, with now over two-thirds of all A-class space now located in the suburbs.[3] Many of these buildings will house the offices of Fortune 500 companies. These architectural monuments are also the home of major law, accounting, and management consulting firms, where impressing the client is more important than saving a few dollars on rent each month.

In many cities, B-class space was once A class. As a result of wear and tear, and with newer buildings being constructed every year, these older buildings are no longer considered prime locations. Like other buildings in their class, their slow deterioration can create an ambiance that is no longer considered the most exclusive and fashionable address. For many clients, these buildings are considered adequate. They may be suitable as the back-office space for a major corporation. These tenants may also have few visitors who need to be impressed by granite-faced lobbies and conference rooms with panoramic views of the city. In fact, these clients may appreciate the appearance of frugality: few people like to see their consultant drive up to the office in an expensive Mercedes. Likewise, having an office in a B-class space may reassure clients that their fees are not being wasted on expensive surroundings, but instead are being applied to the actual work that needs to get done. Engineering design firms, professionals, tech companies, medical arts buildings, and insurance companies are a few of the tenants that might be found in buildings such as these. Renovated warehouse buildings found in many cities along the east coast and rust belt are often found to be up-and-coming locations for artists, designers, and IT companies. A favorite for start-ups, architects, and advertising firms, they are a unique segment of the commercial real estate market.

Finally, there are C-class spaces. Located in less desirable parts of the city, they can be former B-class buildings that have continued to deteriorate. Alternatively, they may be office buildings that originally were built to minimum standards, sometimes found on the second floors of strip malls in the suburbs and in areas of the CBD that were once considered fashionable or perhaps always had a bad reputation.

In calculating residual demand for commercial space, reliable employment numbers are critical. Employment surveys and census data make it possible to determine hiring trends by industry and professions. Knowing something about the space requirement for various professionals can translate these numbers into the demand for future office space. Take a simple problem of estimating the residual demand for a regional shopping center in the southeast of the US. A major insurance company with 3,000 employees will be relocating here and will require, on average, 100 sq. ft. for each employee. This results in a need for a 300,000 sq. ft. space, perhaps in ten floors of a major office tower or a single building in a suburban office park. Adding this demand to the growth needed by existing businesses, I can arrive at an estimate for total demand. For the sake of this example, if there is 4 million sq. ft. of commercial space in the city and, over the last ten years, it has grown by 3% per year, we can add 120,000 sq. ft. to the 300,000 sq. ft. and arrive at a total demand of 420, 000 sq. ft.:

$$\text{Residual demand} = \text{Total demand} - \text{Net absorption by competitors}$$
$$\text{Residual demand} = 420{,}000 - 300{,}000 = 120{,}000 \text{ sq. ft.}$$

Looking at the supply side of the equation, if we know that the building for the insurance company is under construction by one of our competitors, then there is a residual demand of 120,000 sq. ft. that will need to be found in the city. Satisfying this need may be found in existing office buildings that have vacant space. Some tenant may also have been carrying extra space that they are willing to sublease to other businesses. It is not unusual to sublet space to firms that complement your business. For example, an architect may sublet space to an interior design firm or an advertising firm to a video production company. In our case study, if existing vacancies and sublease space represent 7% of all commercial space, or 280,000 sq. ft., then there should be adequate space to meet the future need for space for the next few years without any new construction. In many cities that have seen a contraction in their workforce, the abundance of space represents a surplus that will exist for many years. Understanding how much vacant space actually exists in a city can be difficult. Often, businesses do not like to advertise how much vacant space they actually have. Calculating future demand must also consider the growing dependency on IT. With the internet, it is now possible to use workers offshore to take on many of the administrative roles of middle managers in downtown offices. Improved management software capable of low-level decision making will also eliminate many of the white-collar jobs found in banking, insurance, and sales. The future cannot be a simple projection of past trends.

### *Clients and landlords*

With the wide variety of building forms and uses, determining the market for each project is a challenging exercise. Though residential, commercial, and retail represent the majority of real estate development projects, other types of development, including hotels, recreational properties, health care, and industrial facilities, will require a market analysis that considers both the supply and the demand for that particular use. In each case, it may be necessary to enlist the expertise of specialists with the experience, tools, and data needed to evaluate the particulars of a project. In many cases, the tenant will have a unique knowledge of the market, as is the case for managers of hotels, retirement communities, fast food chains, and big box stores. Their knowledge of the market will help ensure that the design and location of the project will be successful. Many types of development fall into this category. Clients working with developers will have particular requirements that will need to be met as part of the lease negotiations. One positive feature of projects designed for a specific tenant is that there is some assurance that the financials of the project can be achieved. However, there is no guarantee that involving the tenant in the analysis of the market will ensure success. In 2010, when Target took over the leases of Zellers from HBC, shoppers were looking forward to the opening of more than 133 Target stores across Canada. In 2015, after accumulating losses of over $5 billion, Target Canada sold off its inventory at bargain prices and closed its doors. When Target Canada declared bankruptcy, landlords were left with claims totaling over $365 million. In addition, finding new tenants for spaces that were renovated to meet the specific requirements of Target has not been easy. Building spaces that suit a particular client is not without its risks.[4]

## Summary and conclusion

Market research is critical to the planning of any real estate development project. In conducting a market study for a proposed real estate development project, the goal is to arrive at an estimate for residual demand. There are many constraints that operate in estimating residual demand. Demographic and employment data is used to forecast future need, while knowledge of the existing inventory and future projects is used to calculate an estimate of supply. Though simple in principle, getting the numbers right may be difficult. Awareness of other competing projects may be incomplete. External factors can influence future demand and supply. Interest rate changes, changing technology, new lending practices, changes in the tax law, an economic recession, and even a change of political party in power can place plans in jeopardy. Still, if you are lucky, you might turn a profit. If not, you might be selling at a loss – or end up in bankruptcy. Perhaps there is a message here: during an economic recession, it might be better to buy property at a discount with the objective of getting through the bad times and making a profit in the future.

## Notes

1 Lynch, 1960.
2 Desmond, 2017.
3 Brett and Schmitz, 2009, 159.
4 Wahba, 2015; Austen, 2015; Dahloff, 2015; Strauss, 2015; www.cbc.ca/news/business/target-canada-s-closure-leaves-landlords-with-vacant-real-estate-1.2913138

## Bibliography


Alonso, William. *Location and Land Use, Towards a General Theory of Land Rent*. Cambridge, MA: Harvard University Press, 1964.

Austen, Ian. "Target pushes into Canada: Stumbles." *The New York Times*, February 24, 2014.

Austen, Ian. "Target's hasty exit from Canada leaves anger behind." *The New York Times*, April 22, 2015.

Austen, Ian and Hiroko Tabuchi. "Target's red ink runs out in Canada.." *The New York Times*, January 16, 2015.

Brett, Deborah and Adrienne Schmitz. *Real Estate Market Analysis, Methods and Case Studies*. Washington, DC: Urban Land Institute, 2009.

Cervero, Robert, Tammy Rood, and B. Appleyard. "Tracking accessibility: Employment and housing opportunities in the San Francisco Bay Area." *Environment and Planning A* 31 (1999): 1259–1278.

Cooke, Jason. "Compensated taking: Zoning and politics of building, height regulation in Chicago 1871–1923." *Journal of Planning History* 2, no. 3 (September 2016): 207–226.

Dahloff, Denise. "Why target's Canadian expansion failed." *Harvard Business Review*, January 20, 2015.

Desmond, Matthew. "How home ownership became the engine of American inequality." *The New York Times*, May 9, 2017.

Huff, David L. "A programmed solution for approximating and optimum retail location." *Land Economics* 42 (1966): 293–303.

Huff, David L. "A probabilistic analysis of shopping center trade areas." *Land Economics* 39, no. 1 (February 1963): 81–90.

Lynch, Kevin. *The Image of the City*. Cambridge, MA: The MIT Press, 1960.

Miles, M., L. Netherton, and A. Schmitz. *Real Estate Development: Principles and Process*, 5 ed. Washington, DC: Urban Land Institute, 2015.

Okoruwa, A. Ason, Hugh O. Nourse, and Joseph V. Terza. "Estimating sales for retail centers: An application of Poisson gravity model." *The Journal of Real Estate Research* 9, no. 1 (Winter 1994): 89–97.

Reed, Richard and Sally Sims. *Property Development*, 6th ed. New York: Routledge Press, 2015.

Schmitz, Adrienne, et al. *Multifamily Housing Development Handbook*Washington, DC: Urban Land Institute, 2004.

Strauss, Marina. "How Target botched a $7-billion rollout." *The Globe and Mail*, January 15, 2015.

Wahba, Phil. "Why Target failed in Canada." *Fortune*, January 15, 2015.


## Web resources

Canadian Real Estate Association
www.crea.ca/

National Association of Realtors
www.nar.realtor/

Statistics Canada
www.statcan.gc.ca/eng/start

US Census Bureau
www.census.gov/

US Department of Housing and Urban Development
https://portal.hud.gov/hudportal/HUD

# 6 The planning process

## Introduction

In planning a development project in the US, Canada, and most other parts of the industrialized world, you do not always build what you wish; you build what you are allowed, while hopefully still making a profit. To be successful, it is important to know the rules, the processes, and ultimately the people who will be responsible for deciding the outcome of the project. The process may not be simple nor straightforward. However, knowing the rules and the context on which designs are judged can help avoid many of the conflicts and delays that can arise when securing an approval. In outline, there are six principles that are central to this discussion, as follows.

- Know the laws and regulations for each of the jurisdictions that apply to the project.
- Be aware of the exceptions to rules that will delay a project.
- Understand the history and culture of the community.
- Understand the motivations of those ultimately responsible for decision making.
- Recognize that the bigger the project, the longer it will take to get approved.
- Work with professionals who have an in-depth knowledge of the community and the development process.

### *Getting started*

Before a contract is signed to buy a property, it is important to establish whether the local planning authorities will permit the proposed project. Even some simple changes to an existing property will require approvals of the municipal planning department before permits are granted. For example, you may have acquired a house with the objective of creating a basement suite to rent and earn supplemental income to help defray the cost of a mortgage. Sounds like a good idea. After all, you will be providing housing to a community in need of apartments for low-income seniors, students, and local residents. Everybody wins. Unfortunately, the community may not allow this conversion, because it will increase traffic and make demands on the existing street parking. Getting an exception to this rule from the planning department may not be possible given the objections of residents concerned about declining property values if basement suites are allowed in the neighborhood.

The first step in any development, therefore, is to understand the rules that apply to either a change in use or the physical features of your development. Getting started begins with acquiring some basic knowledge of the city by-laws and zoning. Depending on the locale. there may also be federal, state, and provincial regulations and guidelines that must be considered. Much of this information is available today from municipal websites. However, understanding the implications of laws and regulations that can stretch literally hundreds of pages can be a daunting task. Sometimes, it is not clear if a particular regulation may even apply to your project. Consulting with a professional planner with years of practice will help you avoid major obstacles to a proposed development. However, in all cases, it is important to develop some level of comfort with regulations so that you know what questions to ask.

### *Land use control: a brief history*

Personal property has been regulated for centuries. References back to the Bible and the Code of Hammurabi reflect an awareness that poorly constructed buildings are a hazard to the public and those who occupy them. During the Hellenistic period, cities were planned. Districts were laid out to accommodate commerce, government, and public use. Like cities today, there were areas reserved for industry and commerce, while other areas were reserved for residential use. A walk through the ancient Roman city of Ostia reveals a plan with a retail street of shops, a warehouse, and grain-processing district, multifamily apartments, public baths, temples, and magistrate buildings. Within each residential neighborhood, you may find a temple, public bath, retail, and what even might be considered the first fast food restaurants.

In more recent times, much of the regulations have been associated with reducing the potential destruction of a city from fire. The building code of London changed after the Great Fire of 1666, making it less prone to future conflagrations. Cities that were planned in the 17th and 18th centuries served particular uses, including trade and military defense. During the 19th century, in an attempt to make cities "modern", Paris, London, and Vienna embarked on massive urban renewal programs. The Paris of today, with its boulevards and with wide sidewalks, trees, kiosks, benches, and controlled architectural facades, presents us with a uniformity that often hides a much older city untouched by redevelopment. In response to the demands of industry and commerce, cities in Germany began adopted zoning in the 1870s.[1] In Frankfurt in 1891, a zoning act created broad zones of development. Nuisances were regulated within individual zones of the city. However, unlike zoning today, each zone was allowed a mix of uses with restrictions on activity and massing.[2]

The World's Columbian Exposition of 1893 provided Americans with a vision of the future. Designed to gleam under the new illumination of electric light, the grand scale of Beaux Arts facades viewed along expansive boulevards, with large open public spaces, and reflecting pools and fountains, gave a glimpse into the possibilities of a new urban design sensibility. Under the leadership of Daniel Burnham, a leading architect of the age, plans for Chicago, Washington, DC, and San Francisco were drawn up embracing the ideals of this "City Beautiful Movement". Though none of these plans were implemented in their entirety, they set the tenor for planning in North America, establishing planning as a tool for both economic development and urban beautification. Chicago reflects this new ideal. A fire in 1871 provided the impetus behind the rebuilding of a new city designed by the leading architects of the age. Burnham & Root, Adler & Sullivan, and William Le Baron Jenney were the leading architects of the age who created this new commercial aesthetic, which would become known as the Chicago style.[3]

A growing city that had reached a population of almost 1.7 million by 1890, its central business district (CBD) hosted the retail titans of Sears, Montgomery Ward, and Carson, Pirie, Scott & Co. The new functional aesthetic, characterized by fireproof steel-framed construction and modern conveniences, would now dominate the commercial boulevards of Chicago. Designed by one of the leading architectural firms, Adler & Sullivan, the Carson, Pirie, Scott Store featured a library, restaurant, and atrium where concerts took place. With its large display windows and lavish detailing in the Art Nouveau style, the building was an architectural masterpiece of the commercial age. The Chicago style was characterized by facades with strip windows covered in terracotta tile. Built on a grid plan, Chicago was a city of the modern age, with towering commercial and retail buildings in the CBD linked to residential neighborhoods by a mass transit system of commuter rail, streetcars, and elevated rail lines.[4]

In the last century, the interest in zoning was in part motivated by a desire for healthier cities.[5] With the rapid growth of major metropolises, cities became overcrowded. Overcrowding, poor water, and lack of sanitation were found to be the sources of many illnesses of the age, including tuberculosis, cholera, and a host of other communicable diseases. Access to air and light was considered essential to the elimination of infectious diseases. The germ-destroying power of light was considered a compelling reason for restricting the height of buildings. A number of landmark pieces of legislation were passed to improve the design of buildings and their surrounding urban environment. Though far less ambitious than current laws, these early laws established the basis for regulating urban architecture. The Tenement Acts in New York (1901) and Toronto (1912) barred the construction of apartment buildings with railroad flats or rooms without windows. Though the new dumbbell design was only a small improvement over previous designs, these early regulations placed the groundwork for more substantive regulations. With the passage of the "Building Zone Resolution" in 1916, in New York City, buildings were required to be set back from the street to allow light into the street. The construction of the Equitable Building in New York in 1911 may have been the impetus for regulations that would limit the height and impose setbacks on the design of future skyscrapers, driven by a fear of New York City becoming an assemblage of urban canyons. If you walk down 5th Avenue, along Central Park West, it is easy to see the impact of these regulations on architectural design. Later, these regulations would inspire Hugh Ferriss' studies of skyscraper design. Published in 1922, his dramatic black-and-white drawings of ziggurat-stepped skyscrapers are precursors to the designs of the Chrysler Building and the Empire State Building.

Three broad types of power are delegated to local governments: taxation, eminent domain, and police power. A landmark US Supreme Court decision, *Village of Euclid v. Ambler Reality Co.* (1926), established the basis for zoning in our cities today. In this landmark decision, the Supreme Court gave the Village of Euclid in Ohio the power to enforce a zoning ordinance that limited industrial development in a residential neighborhood and allowed a 1921 zoning law to stand. Under this case, the land had been zoned from industrial to residential so as not to create a nuisance for its residential neighbors. The owner argued that the 14th Amendment, passed after the Civil War in 1868, guaranteed that this action by the city constituted a "taking" without due process. Simply, by rezoning the parcel, the city had reduced the value of the property from $10,000 an acre to $2,500, without due process. The Supreme Court responded in its brief that city governments have the power to act in the public good to regulate land use in order to promote neighborhood stability and economic development. In other words, cities have the authority to impose zoning to achieve their goals for promoting orderly, efficient growth and development of their community, while satisfying the need for amenities and recreational areas. They should also provide for a variety of housing types and minimize potential conflict between residential areas and other uses, including transportation and utilities.[6]

### *Zoning defined*

Zoning can take many forms, depending on the jurisdiction. Smaller communities have little use for complex regulations and may have a zoning ordinance that is only a few pages, while larger cities may have a zoning ordinance that fills multiple volumes, each hundreds of pages in length. In principle, all zoning ordinances share some basic underlying characteristics. Most communities will define classes of use similarly and in terms that most residents will understand: residential, industrial, commercial, recreation (including parks), and government. In creating zoning under their police powers, cities must first establish the definitions of each zoning class. Each class of use defines what activities can take place and, just as importantly, what activities are not permitted. In many respects, the zoning by-laws can define the character of the neighborhood by excluding specific businesses. For example, a particular commercial zone may not allow liquor stores, hair salons, fast food restaurants, and massage parlors. For each class of use, further subdivision will normally distinguish, for example, single-family from higher-density apartments. Commercial zoning can be differentiated between small-scale retail, such as a neighborhood convenience store and pharmacy, or larger-scale development, such as an auto dealership or a big box store. Industrial zones may not allow heavy industry that pollutes the air and water or creates significant noise and light pollution.

In creating a zoning map for a community, planners working within a process requiring political and community consultation must arrive at an outcome that considers the physical characteristics of each neighborhood and the topographical features within their borders. Not all land is capable of supporting development. Where the slope is too steep or a sensitive habitat needs to be protected, areas may need to be set aside as public reserves. Having development along rivers and streams can be particularly problematic in areas of flooding and rising waters, even where there has historically been building for decades. After each flood, the cost of rebuilding is a hardship for the residents and government, who often support the rebuilding efforts through disaster relief funds and low-interest loans. In many areas where there is a likely reoccurrence of flooding, an inability to acquire funding in the future will constrain reconstruction. Even when the terrain is suitable for new development, it is important that utilities and other services are available in the area. Roads, sewers, water, gas, and electric utilities are expensive to provide ahead of demand. With lower-density development, the cost of providing these critical services may be prohibitive. One solution in rural communities is to transfer some of the cost to the individual. Well and septic systems can be used in lieu of municipal systems. However, even rural communities will need schools, hospitals, and fire stations. In low-density communities, having these services available within a reasonable distance of every household can become a real challenge. Consider the placement of a fire station. In order to have a reasonable response time, the station must be placed close to the potential call. Imagine if we placed fire stations so that the farthest outlier would have a response time of an hour or more. This solution would be unacceptable to most residents. One solution is to have smaller stations placed more frequently on the landscape. The problem is that, like many public services, there is a minimum size required if the service is to be effective. Even a fire station staffed by volunteers will need a building and one or two specialized trucks. Finding the funds to support the construction and maintenance and operation of this station can be a challenge in rural areas with a low tax base.

Scale is always an issue in supporting public services. As the density increases, it becomes more difficult to find the land needed to build a school, or a fire or police station. Where lands have been zoned for medium-to-high-density housing, there will also be a need to have good public transportation. This is not to say that public transportation options are not warranted in rural areas. However, the high cost of providing services in a low-density community makes it prohibitive in many. Perhaps, in the future, on-demand driverless vehicles will fill part of this gap. However, at the present time, without public transportation services, residents may not be able to get to work, to shops or to their doctor's appointments unless they own and drive a car. For residents of modest income, the requirement of car ownership is an economic hardship. As a community transitions from lower to higher densities, scaling up services will always be a challenging exercise. The message is simple: merely rezoning areas to accept more residential development will not solve the problems of providing services to these future residents. When communities rezone areas to accept more residential development, it must be done without adding to the burden of increasing costs for new infrastructure that goes beyond any hope of repayment through future tax revenues. In some communities, the burden of establishing new services is placed on the developer. If the costs are too high, no development will occur. However, even when development fees merely cover the cost, it is ultimately the community who must maintain the water and sewer systems out of future tax revenues. Future repairs and maintenance will require hikes in taxes to cover the costs. Debt instruments, including municipal bonds, can be a solution. However, in amortizing these costs, there still must be sufficient tax revenues to pay the principal and interest to the bondholders over the life of this commitment.

Finally, zoning must take into account the historical development of the community. When zoning is overlaid on an established community, creating a rational and workable plan can be a challenge. In older communities, much land will have been built on prior to the creation of a zoning map. For example, many communities find themselves with defunct industrial buildings along their waterfront. In the past, easy access to wharves was important for the location of warehouses and factories. Today, new opportunities in the form of high-density condos with waterfront views would be a logical use for these vacant lands. However, transforming this type of district into a residential community can be challenging. Road and street improvements, pedestrian thoroughfares, and parks will have to be created ahead of any development and the receipt of any new tax revenues. The expense of environmental remediation may be beyond any community's resources. Where government grants exist, there is some relief from bearing the full cost of removing and stabilizing contaminated soils. Creating new communities takes time and financial resources. Rezoning is only a single step in a complex development scenario when communities are considering redevelopment of older and dysfunctional industrial areas.

Included in the definition of use are restrictions that control the physical attributes of a development. This can include setbacks, maximum height, and coverage. It can also include a maximum floor area ratio (FAR), which defines how much development in age and space can be incorporated on the site. For example, a parcel that is 10,000 sq. ft. with an FAR of 2 would result in a building with 20,000 sq. ft. of interior space (Figure 1.6). However, given setbacks and maximum height constraints, it may not be possible to achieve a maximum buildout. If

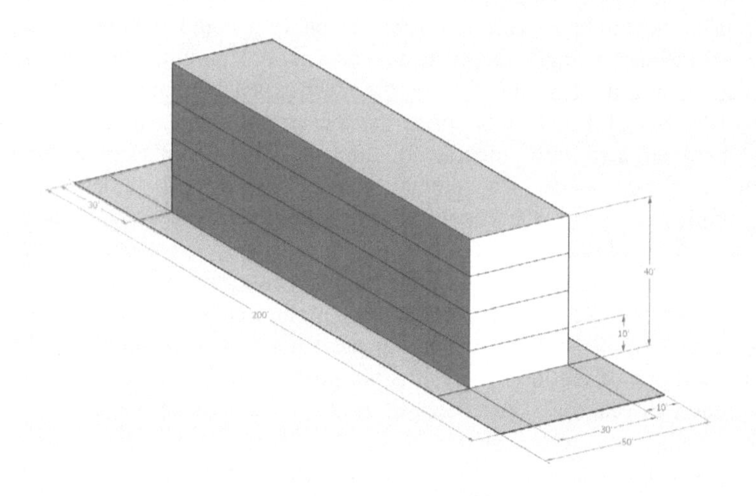


*Figure 1.6* FAR calculation for a four-story building on a 50 × 200 ft. lot with front and rear setbacks of 30 ft. and sideyard setbacks of 10 ft. Maximum height is 40 ft. or four stories.

a property measures 50 ft. × 200 ft. and there are front and rear setback requirements of 30 ft. and side setbacks of 10 ft., then the building footprint (the area you are allowed to build on) is 4,200 sq. ft. If the maximum building height is 40 ft., then you will be allowed to build only a four-story building, which would generate only 16,800 sq. ft. The difference of 3,200 sq. ft. could easily be the difference between profit and loss. For this reason, it is important even in the early stages of land acquisition to have a clear understanding of how zoning will impact the economic feasibility of a project. Other considerations that can constrain development include vehicular access to the site, maximum lot coverage, and the amount of area reserved for landscaping and the number of required parking spaces.

$$\text{Site} = 50 \text{ ft.} \times 200 \text{ ft.} = 10,000 \text{ sq. ft}$$

$$\text{FAR} = 2$$

$$\text{Buildable area} = \text{FAR} \times \text{Site area} = 2 \times 10,000 \text{ sq. ft.} = 20,000 \text{ sq. ft.}$$

$$\text{Max. height} = 40 \text{ ft. (Assume 40 ft.} = \text{four stories)}$$

$$\text{Maximum buildable area that can be accommodated} = (\text{footprint}) \times (\text{stories})$$

$$\text{Footprint} \times \text{stories} = [200 \text{ ft.} - (30 \text{ ft.} + 30 \text{ ft.})] \times [50 \text{ ft.} - (10 \text{ ft.} + 10 \text{ ft.})] \times 4 = 4,200 \text{ sq. ft.} \times 4 = 16,800 \text{ sq. ft.}$$

$$\text{Allowable area} - \text{Possible area} = 20,000 \text{ sq. ft.} - 16,800 \text{ sq. ft.} = 3,200 \text{ sq. ft.}$$

Satisfying the parking requirements can be a non-starter for many projects. Each zoning ordinance will have specific parking requirements for residential, office, retail, and industrial uses. For residential development, the parking space requirement is usually linked to the number of units. Finding space for off-street parking could require the construction of separate or attached garages, or even an underground parking garage. Each has an associated cost. Going underground will be the most expensive solution. Where there are high water tables or solid rock, this option can be prohibitively expensive. In all cases, access to parking can also limit potential solutions. A corner lot may be difficult to build on given the curb cuts needed for both the entrance and exit to underground parking. Most communities require a minimum setback from the intersection to any curb cut. There may also be minimum distances between curb cuts. Given these constraints, there may not be sufficient street frontage to have the curb cuts for both the entrance and exit to underground parking. Rear parking access may require either an alleyway or driveway. In both cases, getting in and out of parking areas may be difficult, depending on the shape and proportions of the lot. Where it is not possible to fit all the required spaces on site, some cities will require a fee in lieu of parking. Though, by paying a fee, a quick fixed has been achieved, from a market perspective there are now fewer spaces on site for your tenants. If the project is for condos, there may not be sufficient parking to offer one space for each unit. One possible solution might be to build fewer, but larger, units. Another possible scenario is to sell the parking spaces separately from the condos. In many urban markets today, residents do not own their own cars. With car-sharing services, many urban dwellers have decided that owning a car is just a nuisance. This is an emerging trend particular to young urban professionals. To capitalize on this trend, it may be possible to allocate a few spaces to car sharing and receive a variance for spaces that cannot be accommodated in the parkade. By having the car space as an optional purchase, it is also possible to lower the entry price for condos, while satisfying the onsite need for those who require personal parking spaces. As in all design scenarios, there is no perfect solution; there are only acceptable ones.

### *Zoning is never perfect*

Zoning must be considered a framework or strategy at best. In some communities, planners may feel that zoning is fixed in stone, given to them on a tablet by Moses. It would be difficult, if not impossible, to create a zoning plan that, from its inception, would address all the economic, social, and aesthetic concerns likely now and in the future. Technological change is inevitable. With economic and social change, land use plans must also change. For example, we now have to provide land for cell towers in our communities – an issue that plans decades ago would not have addressed. The economics of a community may also change, demanding more land be allocated to some uses and less to others. We may need less space for traditional retail along main street and more space for larger big box stores. Or we may want to provide space for living and working, so that young startup businesses can afford to locate in the community. Communities that have parcels that have been vacant for years need to revisit their zoning. Vacant land benefits no one, sitting idle, creating a visual eyesore for the neighbors and a security problem for the police and the community.

In the administration of zoning, there must be a process that allows for continued monitoring and changes to the zoning map. A highly politically charged process, planners may view rezoning as a challenging exercise offering very few rewards. In some communities, the process of updating the zoning map can be an arduous task. Studies, surveys, town hall and public consultation, and ratification can make change a slow process. Everyone in the community who might be affected by the proposed changes will want to voice their opinions. At town hall meetings, planners will need to mediate the inevitable disagreements among community members who feel they are being treated unfairly. Many of these disagreements can stem from issues that are only tangentially related to the rezoning, but instead may actually have originated from past decisions against the individual or one of their family members. The collective historical memory of a community can last decades. However, revisions to the zoning map must be made on a periodic basis in every community if the community is to maintain its social and economic vibrancy.

Aside from rezoning and changes to the by-laws, it is possible to offer property owners some relief on a one-by-one basis. The test in administrating a community zoning must rest on procedures that do not place undue hardship on a single property owner. Relaxations or zoning variances can address both issues of use and development. Zoning boards of appeals also have the opportunity to respond to needed changes through the appeals process. Every community will appoint members to their zoning boards who have knowledge and experience with planning and design. These committees can include former members of the council, planners, architects, engineers, and concerned citizens. In their response to the requests of landowners who need relaxations in the zoning by-laws, zoning boards of appeals make adjustments to the zoning on an ad hoc basis. They can allow for uses not originally part of the plan or relax a rule that will permit a landowner to renovate or build a new project. The relaxation of physical constraints such as a setback or height limitation can benefit the community by promoting improvements to the community's building stock. For example, a landowner may want to make a small addition to the upper floor of their home. Having some additional space would allow them to have two more bedrooms needed for a growing family. However, because of a sloping lot, one corner of the proposed renovations would exceed the maximum height for residential development in the area. The zoning board of appeals in such cases may relax this rule, given the reasonableness of this request. However, before making any decision, the board must first consider any objections from adjacent property owners. Zoning boards of appeals can be swayed by neighbors who feel that the new addition might infringe on their privacy and, ultimately, the use of their property. In general, the appeal board must not place an undue economic hardship on the property owner. Even small requests of this nature are not exempt from politics, which can override such principles. Projects requiring relaxation of numerous requirements can pose a barrier to development. For developments with numerous adjacencies with other property owners, the pressure on boards to do nothing can be enormous.

### *What is wrong with zoning?*

If all else fails, going to court is always a possibility. However, given the cost and time, for most developers the simple solution is to find an alternative location for their projects. Only developers with deep pockets can sustain such appeals. With no guarantee of success, it is better to have projects that conform, with only the most minor requests for relaxations. In cities, where decision making is highly political, developers can win concession because of their political influence. In cities, where councilors are elected by ward, there is a greater likelihood that they will be more involved in the details of any planning application than when councilors are elected at large. Clearly, getting support from the council is important. The council may voice its opinion on a development application and, in some communities, be part of the approval process. In many cities, support for a project may grow when developers make promises to contribute funds for parks, recreational centers, roads, sidewalks, and public art.[7]

One argument against zoning is that communities can use it to restrict particular uses. A plan that does not include land for multifamily units can directly bar those with lower incomes from living in their communities. Arguments can easily be made to support this omission on the basis of an existing growth management plan. With higher densities and larger population, communities may feel that the added burden for transportation, schools, and social services is unreasonable. Communities may also state that high-density development would not be in character with the existing suburban and rural development. The result is that lower-income families are zoned away from the community. Whether some cases are racially or religiously motivated is a question the courts have been asked to examine. In *Golden v. Planning Board of Ramapo* (1972), the courts initially upheld a requirement that future development be tied to an 18-year public improvement project, thus denying the development of multifamily housing. This set the stage for a 1966 court decision when the town board eliminated multifamily housing from the zoning ordinance. Justification for this decision was based on a concern that multifamily housing would change the character of the community and would also place a "burden on public facilities". In 1970, attorney David Coral filed a court brief with the New York Supreme Court on behalf of the developer, Ruth Golden, and the Ramapo Improvement Corporation. The argument was that the improvement plans for such a lengthy period denied the landowner use of the property. The case was eventually appealed to the US Supreme Court and was denied "for want of a substantial federal question". The reaction of the New York Court of Appeals was that the decision not to hear the case would result in denying the poor access to residence outside of cities, with the suburbs remaining more economically prosperous and largely white.[8]

Zoning has also been used to restrict uses within a specific district within a city. Perhaps one of the most prominent uses of zoning was to remove owners of businesses that were viewed as immoral along 42nd Street in New York.[9] In 1994, Mayor Giuliani's administration resorted to changing the zoning to restrict, if not eliminate, all X-rated video stores, movie houses, strip joints, and topless bars from operating on 42nd Street. Once a center of theatre and culture, the 1960s saw the area experience a rapid decline. Many popular theatres once reserved for musicals were now the home for "XXX-rated" movie houses. As a part of a redevelopment scheme that would ultimately see Disney take ownership of two theatres along the "Great White Way", the new by-laws stated that uses that were deemed undesirable could not operate within 500 ft. of schools, religious centers, or residences. The effect was to clear out those businesses that were not compatible with the future plans for the area. This question of First Amendment rights conflicting with zoning legislation was resolved by the Supreme Court in 1976. In *Young, Mayor of Detroit, et al. v. American Mini Theatres, Inc.* (1976), the Supreme Court ruled that communities can use zoning to prohibit negative impacts on a community. Of course, as is often the case in New York, this meant that unwanted business activities were eliminated from 42nd Street only to relocate to Brooklyn.[10]

### *Architectural guidelines and historic districts*

In many communities, development must also conform to architectural guidelines. Guidelines can specify materials for facades, size, and placement of windows and doors, paint colors for trim and facade, and types of sign. In some cases, guidelines can specify the type of windows and doors, roof design, and architectural detailing, including shutters, cornices, railings, balustrades, grills, and chimneys. Location and design of fences, mailboxes, lighting, fire escapes, walkways, landscaping, and free-standing structures can also be detailed in architectural guidelines. A committee's appointed members generally oversee the guidelines and will make suggestions. Often, they can offer help on the design of the facade and signs. Appointed members usually have backgrounds in architecture, construction, and design. If the building is located in a business improvement district (BID) or business retail zone (BRZ), design and improvement grants may be available to help with renovations and with acquiring signs that are within the guidelines. Though many property owners may view guidelines as an imposition, it has been shown that, over time, BIDs contribute to higher property values.

Historic districts and buildings with landmark status can place additional constraints on property development. Similar to architectural guidelines, historic guidelines must be followed when preparing plans for changes to the exterior of a building. Although guidelines are subject to interpretation, they can impose constraints on any renovation plan. Materials and designs not compatible with the age of the building generally will not be allowed in a historic district. For example, vinyl and metal siding cannot be used to replace existing wood siding on 19th-century homes. However, it may be possible to use a fiberglass cast of details, including cornices and balustrades, to replace features that are partially destroyed or missing. Guidelines, in addition to addressing issues of design and construction, will also focus on maintenance, including the appointment of masonry, the proper method of cleaning brick, and the painting of wood surfaces. Ultimately, the design review board will need to be consulted before a developer makes any improvements or repairs. Where a historic building is designated as a landmark building, its status may also require special attention when renovating interior space. For buildings that have a significant history, any improvements must be carefully planned and executed. The advice in such situations is to hire an architect with the expertise and experience of working on historic properties. It is probably just as important for the consultant to have worked with the committee on past projects. This working relationship may provide some assurance that your plans will not be delayed by more reviews than necessary.

During the process of renovating and redeveloping historic properties, issues that were once hidden by decades of use can be uncovered. Lead paint or asbestos can be revealed during the renovation process, adding extra expense and time, while doing nothing for the bottom line. When demolishing older industrial buildings, the expense of clean-up can be prohibitively expensive. In cases in which the development hinges on a future transformation of the area, these costs can sink a project, placing it in the red. Conducting an environmental survey should always be part of the due diligence when acquiring a property. For industrial sites, insurance maps from the 19th century should provide some clue into previous use. A past foundry, or paint or gas works, should give warning that additional environmental testing will be needed to assess the cost of remediation. Though it may be possible for a community to secure funding in the form of grants from the US Environmental Protection Agendy (EPA) for remediation, these issues will need to be resolved prior to the actual purchase of the site.

The added expense involved in renovating and maintaining historic properties can be offset by tax credits and abatement. The US Tax Act of 1986 offers some significant benefits, offsetting some of the added costs of renovating using traditional materials and techniques. Under this Act and until 2017, non-residential buildings placed in service before 1936 received a tax credit of 10% on any improvements, while those buildings put into service before 1986 and deemed historic received a 20% credit. Since 2018, developers have had to claim the credit at a rate of 4% per year over a five-year period. The result is that, when the time value of money is considered, the total sum of these credits is less than 20% of the cost of improvements. Developers knowingly acquire historic properties only if they know in advance that they will be able to make a profit even after all credits are taken into account. It is the person that inherits an historic property who will be surprised by the costs associated with maintaining and renovating a building with landmark status.

In Canada, provincial programs often provide some grants for the repair of historic properties. In addition, individual cities in the US and Canada offer a range of subsidies, including grants, low-interest loans, tax abatements, and the transfer of development rights to other properties in the form of zoning bonuses, which act as economic incentives to help in the preservation of historic properties.[11] The municipal planner and historic resource officer should always be consulted before initiating a project. Ultimately, it is best to hire a consultant with the appropriate expertise to help develop the concept plans. Architects who market themselves as experts in historic architecture renovation can provide advice in this area. Remember, however, that estimating the cost of renovating an historic property is never an exact science. In the course of making changes and upgrades to a property, it is not unusual to discover problems after the project begins. With changes in plans, costs can easily escalate beyond the initial estimates after the work begins.

### *Americans with Disabilities Act of 1990*

Building owners need to be aware that federal legislation can impose regulations on the design and renovation of a building. The Americans with Disabilities Act (ADA) first passed in 1990 guarantees access to those challenged by disabilities. Standards for accessible design were passed in 2010, establishing design criteria for both commercial and public buildings.[12] All registered architects will be aware of the ADA and should be able to advise a developer in the early stages of concept design on how best to satisfy these requirements. In many cases, the costs can be easily managed if these requirements are considered during the early stages of the design process. Exceptions are made in cases in which the requirements are considered impractical.[13] Exceptions are also made in cases involving historic buildings where modifications may destroy the historic value of the building. When buildings are being renovated for new uses, the requirement for elevators can easily make the project prohibitively expensive. This is especially the case when the spaces above the ground floor are modest in size and will never generate sufficient rents to cover the additional cost of an elevator.

Though some European countries have enacted similar laws to the ADA, discussions held over the last 30 years in Canada have not resulted in a equivalent measure. Canadians are afforded protection against discriminatory practices under the Charter of Rights and Freedoms 1982 and the Canadian Human Rights Act 1985. Specific legislation, with established guidelines and requirements, does exist for public transit, employment, education, and social programs. Canada Mortgage and Housing Corporation (CMHC) also provides financial resources for making dwellings more accessible. Additional requirements through the municipal building code also work to make buildings more manageable for those with disabilities. For accessibility issues falling under the heading of "universal design", a registered architect will be able to advise in the planning stages of any development project.[14]

### *Subdivisions*

Often, development involves the subdivision of large tracts of land. This will be the case when farmland is acquired for suburban housing developments. The process can be expensive and time-consuming unless you are planning to partition out only a single parcel for a family member. In this case, some jurisdictions provide a simple process that can be done with much less time or expense. In most cases, however, the process can require years of negotiations before completion. The council will judge the merit of the proposal at various stages in the approval process. At each stage, there will be requests for more detailed information. In the early stages, a drawing of the proposed parcelization will be required, followed later by more detailed studies to address drainage, utilities, road design, and even emergency services accessibility. On larger projects, there will also be a requirement for parks and open spaces, and possibly schools and recreational centers. When these subdivisions approach the scale of new communities, there may also be space in the plan for schools, retail and commercial services.

At later stages in the approval process, additional studies by transportation planners, landscape architects, civil engineers, and real estate analysts may be required to contribute to the overall plan. Each consultant may be asked to supply a detailed design addressing the concerns of the city planner, the council, and the community. Depending on the requirements of the community, it may also be necessary to complete the equivalent of an environmental impact assessment (EIA). Depending on the scale and location of the project, knowledgeable experts in fields of archaeology, botany, and wildlife management may have to be consulted. This would be the case in the development of recreational properties located on or adjacent to ecologically sensitive land. In addition to municipal approval, depending on the location of the property, state, provincial, and federal regulations may apply. Properties along coastlines and navigable waterways may need to adhere to additional federal, state or provincial regulations. One common complaint by developers of properties that fall under multiple jurisdictions is the time and expense required to satisfy each set of legal requirements.

One aspect of the subdivision process that cannot be overlooked is public input. In addition to adjacent neighbors, planning departments will solicit public opinion on major developments. Where subdivisions are located in established communities, opposition from residents can arise when the proposal is for higher-density housing. As a community matures, property values can appreciate, making the land costs for new developments more expensive. One option is to increase the density (units per acre) and thus divide land costs over more residential units. Planning departments are usually supportive of higher-density housing as a strategy for creating affordable housing for its residents. Where there is an existing infrastructure, the increase in tax base can support the repair and upgrading of aging infrastructure. However, local residents who are more concerned about their property values than the need to raise taxes for the support of schools and other services can be opposed to subdivisions. Members of the council not wanting to alienate their constituency are less likely to vote in favor of these proposals if there is significant public opposition. This is particularly true where members of the council are elected by ward. One argument often made against higher-density subdivisions is that the development is not appropriate and compatible with existing development in the neighborhood. This can be a particular problem when new development contains a significant number of multifamily units. One solution is to place single-family development around the perimeter of the new development. In this case, there will be less opposition to larger-scale development when residential development is located around the edge that is similar to what already exists in the community. Taking on large-scale development is only for those with deep pockets, who can sustain years of planning and negotiations. Individuals with family farms who think they will develop a subdivision to help reach their retirement goals will quickly realize that the process may be outside their reach. In these cases, partnering with a seasoned developer may be a solution.

### *Planning process: getting to a successful conclusion*

Knowledge of a community's zoning is the first step in the planning process. In developing a plan, it is important to have both a concept plan for the site and an understanding of the approval process. Planning approval requires communication and interaction with planners, engineers, and the council, as well as design review boards. Being successful requires knowing what is needed at each stage of the approval process. However, it just as important to understand the personalities, background, and motivations of the people with whom you are doing business. Planners, like other professionals, are members of a culture with social norms.[15] With an accredited educational program and national associations, planners subscribe to a set of ethical values dedicated to public service. Though it is difficult to characterize an entire profession on the basis of its websites, the American Planning Association (APA) and the Canadian Institute of Planners (CIP) offer some insight into their professional cultures. Planners, as a group, are more dedicated to issues of "social justice", and preserving "the integrity and heritage of the natural and built environment", than most developers. As part of their code of ethics, planners:

> … seek social justice by working to expand choice and opportunity for all persons, recognizing a special responsibility to plan for the needs of the disadvantaged and to promote racial and economic integration. We (AICP) shall urge the alteration of policies, institutions, and decisions that oppose such needs.
>
> (AICP Code of Ethics and Professional Conduct[16])

To gain an understanding of the role of planners in the development process, John Forester's seminal work *Planning in the Face of Power* presents an informative perspective on the planning process.[17] As public servants, planners serve communities by supplying technical knowledge, developing skills within a community, encouraging independent community organizations, educating the public on critical issues, ascertaining the concerns of the community, and anticipating political pressures. During the course of their professional activity, planners may also control the flow of information as a means to accomplish specific ends. For a variety of reasons, the control of information can be used to influence a decision on a specific development project. By merely controlling the timing of critical information, outcomes can be influenced. In some cases, exaggerations, misinformation, uncertainness, and the use of jargon can all be used to bias solutions and direct decision making. This can be intentional, but in many cases it can be a response to conflicts that arises merely out of self-preservation. Responsible for speaking for the public, planners may be placed at odds with the developers, who are often presented as individuals or corporations motivated only by profit.

Working with planners requires some sensitivity to their position within the organization. Political pressures can place them at odds with the council and the mayor. Planners may be more concerned with issues of social equity, preserving historic properties, promoting public transportation alternatives, and providing community amenities. In contrast, members of the council elected by their constituents are more sensitive to issues of employment, taxation, and delivery of municipal services, and may not always be receptive to programs without immediate economic or political benefit. When engaging with members of the planning profession, it is also important to recognize that there is no single personality profile. However, the ability to recognize the different types of planner may help you in establishing a good working relationship. For projects where the approval process can last years, establishing a good working relationship is critical to the success of your project. Forester describes each personality type.

- Technicians take a bureaucratic approach. In working with a technocratic planner, the specific requirements of a plan, setbacks, and site coverage will be their primary concern. For the technocratic planner, satisfying the requirements of the by-law is fundamental. Rules and requirements are all-important.
- In contrast, incrementalists are pragmatists. For them, knowing how the system works is critical. Getting to a politically acceptable decision may be more important than satisfying every requirement on the checklist.
- For liberal advocates, information serves a cause, while the progressive uses political power to shape outcomes. Liberal advocates address concerns that the community may see as needing attention.
- The urban designer, meanwhile, focuses on the architectural qualities of the project. For the urban designer, how a project will look will be more important than the social or economic underpinnings for the proposals.[18]

A favorable outcome may require a different response to each set of concerns. Where proposals must be reviewed by several planners, each with a different set of responsibilities, it is important that presentations address each of their concerns.

In smaller communities, where planners are not shielded from the realities of local politics, it may be easier to know how proposals are going to be received than in larger cities, where planners are more insulated from the political life. For communities that are in need of economic stimulus, a winning strategy will be to focus on jobs and growing the tax base. Development projects that can show a sustained boost to the economy will be welcomed even if there are planning deficiencies. Commercial and industrial projects will be of particular interest to local governments because they afford a significant boost in tax revenues. Most communities have differential tax rates, with commercial and industrial development paying significantly higher rates than residential development. Without the tax revenue paid by businesses, residential property tax rates would have to be proportionally higher. Not surprisingly, communities that lack the land dedicated to industrial and commercial uses will have significantly higher tax rates on residential properties. Proposals that will place a burden on existing taxpayers will be least favored because, come election time, everybody remembers their last tax increase.

One benefit from working in smaller urban centers is that they are less bureaucratic. In many communities, there may be only one person who looks after all planning approvals. Access to planners and the council is generally more informal. Learning how a proposal is going to be received before it is submitted can be done on a more informal basis. In communities that cannot afford a full-time planner, a consulting firm may be on contract to assist the council and the mayor on planning matters. Clearly, larger cities are more bureaucratic and hierarchical. Planning approval processes can require the review of a proposal by a number of design professionals, planners, and urban designers, including transportation planners and civil engineers. There may also be input necessary from the fire chief, police department, and emergency services as part of the approval process. As the project progresses through the approval process, be prepared for numerous meetings with planners and public officials before getting your final approval.

### *Hiring a consultant*

For many projects, a professional planner will be hired to look after the approval process. Representing the interests of the developer, hiring a consultant that formerly worked as a city planner for the community can have significant benefits. Planners who once worked in the city should have the respect of those who will approve your proposal. They are also are familiar with the technical requirements and can advise you on the particulars of the review process. Having a seasoned professional represent your interest can be critical in addressing the particular concerns of planning departments. You never want to hear "that is not how we do planning in our community" during the negotiation process. In working effectively with a consultant, it is important to know what will be needed at each stage in the approval process. As the project manager, you will need to make sure that the appropriate reports and drawings are prepared and completed for each stage in the development process. It is possible to use a consultant to manage all aspects of the a development. To guarantee that the process progresses smoothly, a detailed list of responsibilities and associated fees should be reviewed at the onset of the project. Even then, fees can escalate when the planning department requests additional studies. These studies can be extensive and numerous. An established planning firm can advise on and arrange for the studies requested during the approval process. In looking for planning services, it may be useful to note that large engineering firms often have planning departments that can also provide services. For larger projects, using a large multiservice consulting company may provide some advantages. Having a planning, surveying, transportation, and environmental departments under one roof may facilitate coordination and cost savings. It is not uncommon for planning services in these larger firms to be offered at a discount with the hope that if the project were to materialize, they will have a substantial consulting contract for surveying, engineering, and design.

### *Enlisting the media*

For larger projects, enlisting the media will be an important part of a successful communications strategy. In the past, print media was the primary outlet for getting the message out on a proposed project. Articles in the local press that announce to the community that a new project is in the planning stages could be useful in shaping public opinion. An article about the benefits of the project to the community could help set the stage for the design approval process. Even today, there are many regional and local newspapers that are willing to publish articles on real estate development. Often staffed by one or two people, they are in constant need of articles to fill their pages and may be very willing to bring your project to the attention of the public. By providing some background information, a press release, and some images, you will make their job a little easier.

Getting the message out before a community association tells your story to the press is critical today in a world in which public opinion is shaped by social media. First, it is unlikely that any community association or public interest group will be able to present the details of your project. If interviewed by the press early in the development process, there is always the potential that community spokespersons will dwell only on the negative attributes of the project. It is easy for a project to go astray if the focus of the press is on how the new project "is going to make current traffic worse" or "will be out of place with the architectural character of the community" or "will provide housing only for upper-income households". Even if this is not true, once these ideas are fixed in the minds of the public and, more importantly, the city council, it may be very difficult to move the project forward. Therefore, it is critical that, at each stage in the planning approval process, the message that your project is going to respect community concerns be continually reinforced. In cases in which there are potential negative impacts, such as having a tall building that may tower over existing development, there is a need to show how the positives outweigh the negatives. Job creation, economic stimulus, the creation of public amenities such as parks and schools, and the generation of sales tax revenues are all positive benefits that can be stressed. However, in each case, the message must be tailored to the audience. Though the council may value the higher tax revenues generated from a Walmart, the majority of small businesses on main street will be in opposition, even though the community may like having greater shopping choices and avoiding longer trips to the regional mall. Likewise, a community may not like higher-density housing on the basis of its architectural character, but may approve of having a retirement complex where aging parents and grandparents can live.

For large-scale projects that will eventually capture the attention of the media, it may be useful to have a public relations (PR) firm on retainer from the beginning of the project. Its connections to the local media, press, radio, and TV can be particularly useful in getting ahead of the public on anything that might be viewed as controversial. As experts in media, the firm can be responsible for looking after the preparation of press releases, arranging for meetings with media outlets, and staging press conferences for large projects. On these larger projects, they can also act as resources for the preparation of video and other media needed for TV and web broadcasting. The use of computer animation and drone footage can also help in presenting the details of a project. It is surprising how a video appearing in social media can promote the project far more than a one-page press release in the local paper. Having a PR firm in tow can also protect your project from the unexpected. At some stage, if an unplanned negative event should occur, your media consultant should be able to manage the fallout. The public will need to be reassured quickly that any issue that could have a negative impact on the project has been addressed responsibly and promptly.

A PR firm working with your marketing consultants can also be useful in soliciting feedback from prospective buyers and patrons for your proposed project. For example, learning about the needs and preferences of potential buyers for your condos can help you avoid making costly design errors. Having focus groups with target buyers can provide important feedback in the early stages of design. Markets can change. Knowing where to position a product can help you avoid unnecessary expenditures on amenities that are considered unnecessary. This can be particularly true in the residential housing market, in which the preference for wall colors, surface treatments, appliances, and lighting can change with the season. Knowing more about market preference can also help in supporting proposals before planning commissions, councils, and community groups during the approval process.

After the building is completed and ready for occupancy, a PR firm can help bring the project to the attention of your target market. As part of a long-term strategy, where projects are built in phases, having a PR firm will help your project in the minds of the public and maintain your reputation. This can be particularly important if you plan to build future projects that will require public support.

## Case study: inner-city development project

As an example of how the planning process might progress from inception to final approval, the following case study – a compilation of several actual projects – will review the likely issues confronting developers during the approval process, including zoning, design, financial feasibility, the planning process, and community involvement.

### *Background*

This case involves the request for approval of the development of a multifamily condo project in a city of 1 million people. Like many cities in North America, the area surrounding the inner city is under development pressure, because lots with older homes built since 1945 are sitting on land that has increased in value over the last few decades. Many of these older homes are in disrepair and their lots could support more intense development. A rezoning of the area occurred over ten years ago, with the goal of more affordable housing within walking distance of the CBD. Under the new zoning ordinance, it is now possible to build a four-story apartment building on a site that is large enough to support some underground parking.

Only a few property owners in the community have sold to developers, who have built, for the most part, duplexes and townhouses on the sites of these older homes. For those residents who plan to live in the community in the foreseeable future, the prospect of living next to a four-story apartment building is hardly welcomed. Developers, if they are to be successful in this environment, will need to satisfy the requirements of the city and the concerns of local residents. The city council is facing re-election next year and the last thing their members want is to earn the wrath of their constituents. With the neighborhood in transition, getting through the development process may be difficult, given the opposition by existing property owners who fear that the proposed development will lower property values and destroy their sense of privacy. Nobody wants an apartment building overlooking their backyard.

### *Site*

The proposed site for your development was assembled from three lots. Each lot measures 50 ft. ×140 ft. The topography is relatively flat, except for a small drop along the corner of the site. One of the three properties was your family home, while the other two you acquired over the last ten years. You purchased one of these two homes directly from your neighbor, while the third property was acquired when it came on the market in the last year. All three single-family homes were built in the late 1950s. One of the properties is located on a corner lot, which will make it possible to offer more natural lighting to the end units. The area on the first floor of each of these bungalows is approximately 1,200 sq. ft., with unfinished basements. They are all currently being rented (Figure 2.6). Without any mortgages, the rentals are paying the expenses, while generating a reasonable positive cash flow. Though all are in good repair, over the next five years some capital improvement will be needed. Most likely, a roof will be replaced and a heating furnace upgraded. One of the renters, a retired city worker, looks after the maintenance on all three homes. You could conceivably hold all three properties for some time while pocketing a reasonable dividend on your investment.

*Figure 2.6* Case study: proposed site for development.

***Development options***

The hope is that, on the three parcels, you will be allowed to build a condo project with 40–45 units. The need for housing among young professionals commuting to the downtown has increased over the last decade. This potential project is well situated. Within walking and cycling distance of many downtown office buildings, prospective buyers would include first-time buyers and investors looking to buy rental units. Market analysis suggests that there is a demand for one- and two-bedroom condo units, ranging from 550 sq. ft. to 700 sq. ft. A few three-bedroom units in the 900–1,200 sq. ft. range also have been selling well over the last year – mostly to young families with children. One of the challenges will be providing adequate onsite parking. Under the current zoning, one space is required for each unit. A loading area is also required for buildings with more than 35 units (Figure 3.6).

Another possibility is to place townhouses on the site. Townhouses in the 1,100–1,300 sq. ft. range have been built in the community for some time. Most are two stories in height, with parking below ground. Larger townhouses in the 1,300–2,000 sq. ft. range are being purchased mostly by empty nesters approaching retirement. These buyers are looking for low-maintenance housing close to the city, fitted out with all the latest appliances and conveniences. It should be possible to put up to ten townhouses on the site depending on their size (Figure 4.6).

A third option being considered is the multifamily flat. Professionals in their 30s who like to live near downtown are the market for these units. Built for individuals with more established incomes, the finishes and features on these types of unit are upscale. Designs for these units can either be up and down or side by side. There are advantages to both styles. The single-floor flat units offer more flexibility in layout, while the side-by-side units provide more privacy for the occupants. It is also possible to build two units per floor. Size could range between 700 and 900 sq. ft. In this case, some units will not have any yard access. If you can offer yards and easy access to garages, they become highly desirable in the current market (Figure 5.6).

*Figure 3.6* Case study: condo development.

In the case of flats, constructing flats on each of the 50 ft. lots offers flexibility. If a design is not successful and does not sell well, you are able to recover by changing the design on the two remaining lots. Besides, smaller-scale projects seem to receive fewer objections from the community.

To begin the process, your architect is engaged to explore three design options allowed under the current zoning: a condo project (Figure 3.6), townhouses (Figure 4.6), and multifamily flats (Figure 5.6). Each will be evaluated for its risk and reward. It is your expectation that one project that takes advantage of the three lots assembled into a single parcel will be more profitable than building a separate project on each of the lots.

*Figure 4.6* Case study: townhouse development.

*Figure 5.6* Case study: multifamily flats.

*Scenario evaluation: design review*

Having received the architect's initial designs showing the massing on the site and various floor plans, you evaluate the cost and selling price of each of the three options (Figure 6.6).

What quickly becomes apparent is that, as expected, the condo project will generate the most total profit. Even with higher construction costs per square foot for the condos, the total profit for the project is significantly higher at $1,785,000 vs $886,500 for flats and $375,000 for townhouses. The higher construction costs per square foot for the condo project can be assigned largely to the need for underground parking and the need for elevators. Also, the city requires a fee of $30,000 for each parking space that cannot be accommodated on site. For the condo project, an additional $540,000 will have to be accounted for in the total cost of the project for these fees. For the flats, this fee for a deficit of onsite parking is $180,000. Looking at the risks the condo project will face, the cost of construction and fees is $9,465,000, which is considerably more than what is required for either of the other two options. For the townhouses, construction costs and fees are less, at $2,525,000. If you were to build flats, it would be possible to build on one site at a time. In this case, the cost of construction and fees is considerably less, at only $1,204,500 for each building (Figure 6.6).

One of the concerns is that though the condos are more profitable, they are only so if all 42 units are approved. If the planning department allows you to build only on three floors (33 units) and not four, the profit falls dramatically from $1,785,000 to $1,087,500 (Figure 7.6). Also, given that the cost of building underground parking can vary in this area depending on the geology, the cost numbers used in the pro forma are only crude estimates and the dollar amount could go much higher. Building a single structure with flats would be less risky.

After reviewing the three options, you decide to meet with the planning department and present a concept for the condo project. If the approval comes through, one alternative would be to sell the project to a developer with more experience and make a profit without taking the risks associated with actually building the condos.

| | Condo | Flats | Townhouses |
|---|---|---|---|
| | | | |
| Sq. footage/unit + common area | 700 | 925 | 1,250 |
| Sq. footage/unit | 609 | 805 | 1,250 |
| Units | 42 | 18 | 10 |
| Total sq. footage above ground | 29,400 | 16,650 | 12,500 |
| Cost/sq. foot above ground | 275 | 180 | 180 |
| Cost above ground | 8,085,000 | 2,997,000 | 2,250,000 |
| Free standing garages | | 270,000 | 150,000 |
| Basement (parking for condos) | 840,000 | 166,500 | 125,000 |
| Fees to the city for parking deficit | 540,000 | 180,000 | - |
| Total construction cost & fees | 9,465,000 | 3,613,500 | 2,525,000 |
| Land cost | 1,350,000 | 1,350,000 | 1,350,000 |
| Total cost: land, construction & fees | 10,815,000 | 4,963,500 | 3,875,000 |
| Market price/unit | 300,000 | 325,000 | 425,000 |
| Cost per unit | 257,500 | 275,750 | 387,500 |
| Profit per unit | 42,500 | 49,250 | 37,500 |
| Total profit | 1,785,000 | 886,500 | 375,000 |

| Parking | | | |
|---|---|---|---|
| Underground spaces | 24 | | 10 |
| Garage spaces laneway | | 12 | |
| Spaces needed | 42 | 18 | 10 |
| Parking space deficit | 18 | 6 | 0 |

*Figure 6.6* Three development options: condos (42 units), townhouses (10 units), and apartment flats (18 units)

### *Step 1: first meeting with planners*

You arrange a meeting to discuss your concept with the planner assigned to review development applications for the community. You know that this planner tends to be a bit reserved and plays it by the book. Your hope is for an informal opportunity at which you can learn more about other projects in the works. Your hope is to merely test the waters to see which direction would be welcomed by the city. You bring a simple sketch of the site to the meeting. It is important at this stage to learn more about the concerns of the approving authorities. Also, by presenting a less detailed design at this stage, you hope to show a willingness to work with the planning department on developing a concept acceptable to both city hall and the community.

Unfortunately, the planner shows up to the meeting late. His last appointment ran over and now you have only 20 minutes in which to present your ideas. Because this is your first project, you have not established a working relationship with the planner and you are not sure where the meeting will lead. You are already wondering when you should schedule your next meeting. You feel at a slight disadvantage: the planner is a 20-year veteran and this is your first project. The planner's first question is the location of the project. You present a plan of the site, with the street addresses. The planner first expresses surprise that these three houses are owned by a single individual. Actually, the properties are currently titled under different family members. The second question asked is about the zoning and the number of units planned for the site. The zoning, which is R3 (medium-density residential), would allow in this community

| | Condo | Flats | Townhouses |
|---|---|---|---|
| | | | |
| **Sq. footage/unit + common area** | 700 | 925 | 1,250 |
| **Sq. footage/unit** | 609 | 805 | 1,250 |
| **Units** | 33 | 18 | 10 |
| **Total sq. footage above ground** | 23,100 | 16,650 | 12,500 |
| **Cost/sq. foot above ground** | 275 | 180 | 180 |
| **Cost above ground** | 6,352,500 | 2,997,000 | 2,250,000 |
| **Free standing garages** | | 270,000 | 150,000 |
| **Basement (parking for condos)** | 840,000 | 166,500 | 125,000 |
| **Fees to the city for parking deficit** | 270,000 | 180,000 | - |
| **Total construction cost & fees** | 7,462,500 | 3,613,500 | 2,525,000 |
| **Land cost** | 1,350,000 | 1,350,000 | 1,350,000 |
| **Total cost: land, construction & fees** | 8,812,500 | 4,963,500 | 3,875,000 |
| **Market price/unit** | 300,000 | 325,000 | 425,000 |
| **Cost per unit** | 267,045 | 275,750 | 387,500 |
| **Profit per unit** | 32,955 | 49,250 | 37,500 |
| **Total profit** | 1,087,500 | 886,500 | 375,000 |

| **Parking** | | | |
|---|---|---|---|
| **Underground spaces** | 24 | | 10 |
| **Garage spaces laneway** | | 12 | |
| **Spaces needed** | 33 | 18 | 10 |
| **Parking space deficit** | 9 | 6 | 0 |

*Figure 7.6* Three development options: condos (33 units), townhouses (10 units), and apartment flats (18 units).

a multifamily building up to 44 ft. in height and the construction of 40–45 units on the site. Looking at the sketch, which shows only the outline of the building footprint, the planner's initial concerns are about satisfying parking, maximum coverage, and the front setback of 25 ft. and side setback of 8 ft. What is of primary concern is the height limitation of 44 ft. As the planner notes, all the projects that have been approved in the last two years were three stories and had a maximum height at the roof peak of much less than the allowed 44 ft. Your hope was for four stories and a flat roof at a maximum height of 44 ft. Already, you are beginning to see your profit margins shrinking. The planner has to leave for another meeting and asks to see a more detailed set of drawings at the next meeting.

### *Step 2: second meeting, bringing the architect into the process*

The architect completes a more refined set of drawings for the project before your next meeting with the planner. You now have an elevation and detailed plan, showing the entrances, curb cuts, and landscaping. It is decided that the architect, who knows the planner from other projects, will come to the next meeting. This should assure the planner that we have followed all the zoning requirements established for medium-density housing. For the next meeting, we were able to get the first appointment of the day. The meeting begins with the usual introductions. Having the architect present at the meeting has created a more congenial atmosphere. After some brief exchanges, the architect presents the details of the project. The planner is still concerned about the number of units and the height of the buildings. Though a building with a height of 44 ft. is allowed under the current zoning, he knows that the neighbors behind the property and to the east of the site will probably raise some objection. There is also some concern expressed about the traffic generated from 42 units on this residential street. The planner suggests that, at this stage, we should begin the process of filing for development approval and that we consider today's discussion in revising the plans.

### *Step 3: review of the development application*

After some discussion with the architect, you decide to go ahead to the next stage. You are aware that, at some point in the process, there will be requests for additional studies, but the anticipated expenditures seem justified. At this stage, there is no reason to think that the project will not be approved, but there are no guarantees. Working with the architect, you file the drawings and necessary forms online and pay the standard fees. After about a month, you receive an email from the planner's office requesting another meeting. Unfortunately, your architect is out of town and the meeting is rescheduled for three weeks later.

At this meeting, the planner is accompanied by the transportation planner for the city and the director of planning. Given the sensitive nature of new development in this community, the planning director felt it necessary to attend. The meeting begins with your architect's presentation. The detailed drawings, which are output from a computer-aided design (CAD) program, show the project in plan, elevation, and perspective. The architect is able to explain how the project complies with the zoning. Because the project is located at the corner of the block, the impact on neighbors' yards is less than if the building were placed between older homes. A shade and sun study reveals that only in winter will the shadows be an issue for neighbors, and this issue is mitigated by the fact that, during the winter, people tend not to use their yards because of the snow. In addition, tree plantings along the perimeter of the site should help to create a visual barrier between the proposed projects and the neighbors. Parking is placed underground and access into the parkade is from the less busy of the two streets.

Concerns are raised again by the transportation planner about the entrance into the property. The concern is traffic: cars entering the building may create traffic issues at the intersection. There is also the concern that the curb cuts are too close to the intersection. While the 20 ft. minimum is met, the transportation planner advises that it would be best to have both the entrance and exits along the front of the building. There is also the issue of grade access to the garage, which appears a bit steep. The director of planning also sees a four-story building as incompatible with the existing architecture of one-story 1950s bungalow-style homes. Clearly, the neighborhood is in transition, so the real question should be how the neighborhood will look in 30 years' time, after these older homes have been demolished for newer development – but that is not their present concern. The planner assigned to the case also comments that neighbors in the community are already aware of a proposed development for the site and have expressed concerns about any building of more than two stories.

After about an hour, the meeting comes to a sudden close: there is another meeting scheduled for the room and it is decided that we consider the comments made today. The plan is to meet again within the next two months, after which it would be timely to gather some input from the community. After the meeting, the architect suggests that perhaps consideration should be given to reducing the height of the building along the side adjacent to the homes to the east of the property. Your concern is that eliminating these two units will reduce your profit margins to a number more comparable to those of the townhouses or flats, which would eliminate the advantages of building a condo project.

***Step 4: review of the revised proposal***

You have instructed the architect to make some changes in the design before the next meeting. The entrance and exit to the underground parkade are now located along the front of the building. The roof design has been modified to lower the profile of the building along the back and next to the neighbor's property to the east. You hope these changes will satisfy the planners at this stage. The meeting this time is held with only the planner in charge of the file. You would have liked the architect to be present, but, given the hourly rate for every meeting, you decided that this time you will go solo. The meeting lasts only half an hour. The planner goes through a checklist and, though you have met all of the zoning requirements, there is still concern about the height of the buildings. The planner suggests that you attend a meeting scheduled next month with the community planning association. At that time, it will be possible to get some input on from the neighbors and members of the community.

***Step 5: meeting the community***

The evening meeting is held at the local community association building adjacent to the neighborhood park. You have been there for BBQs and community events, including the annual yard sale. Many of the people in the room you know from your youth. After all, you and your family grew up in this neighborhood, and many of the residents who are now in their 60s and 70s remember your mom and dad. Some of the neighbors remark that they are surprised that you are now in the development business. Not sure how to interpret this comment, you are nonetheless hopeful, having grown up in the community, that the audience will be receptive to the proposal. You have brought some drawings mounted on boards, which you were able to put up on the walls before the group has assembled. You have also brought some light refreshments, which is common at these events. It is 7.30pm and the chair of the community association introduces you to the group. You begin your presentations of the proposed multifamily development, which would offer those looking for a place to live an affordable housing option. You believe it will attract first-time home buyers. You suggest that many of the community's own children, who are now working in the city and who are having difficulties finding places to live, would be interested in these condos as either renters or owners. You hope this presentation positions your project in a positive light. The architect then proceeds to go through the features of the project. She is particularly sensitive to the neighbor's concerns and assures them. Besides meeting the zoning, the design now incorporates a roofline that is sensitive to the adjacent properties. We have also provided additional tree plantings, which should soften the view from neighbor's properties.

Towards the end of the presentation, there is a question about views from the backyards of properties just across the laneway separating the proposed development from the houses to the rear. One of your neighbors, who has been living in the community for over 35 years, does not want to look at condos every morning when he is gardening in his backyard. Even with a fence and trees located at the rear of the property, the slight incline in topography will result in a facade of more than four stories on the edge of his property. He is particularly concerned that his "privacy and use of his property" will be disturbed. Apparently, he has been to a number of these meetings and now has the planners' jargon memorized. All focus turns to the height of the building. However, this specific issue is not addressed in the drawings and the architect tries to explain the issue away, but is somewhat unsuccessful. Comments from many of the residents are that the building is "just too big" and they were hoping for a building that was only two, or possibly three, stories in height. There is also now a question of traffic. Many of the residents have younger children and are concerned about the potential hazard of cars entering and exiting the parkade. The planner, who has been listening to the comments and discussion for the last hour, suggests that we consider these concerns and return again with a traffic study and revised architectural drawings. At this point, you have no alternative but to thank the group and prepare for the next meeting. You discuss with the planner when they, you, and your architect can meet as a group again and the planner suggests in two months, after the holidays. Later, you discuss with the architect if you should bring to the next meeting views from the neighbor's backyard, showing the facade of the building. You decide not to give them any more ammunition at this stage. It is looking more likely that this project may not get approved in time to break ground in the spring. Luckily, the houses have renters and a delay of a year does not seem unreasonable.

### *Step 6: second revised plan*

After the last meeting, you discussed with the architect the need for a transportation planner. She knows someone who will not charge a lot for a simple study, which is all that is needed at this stage. Concerning the height issue, the architect suggests that you bring in the top floor by another 8 ft. from the edge of the building adjacent to the neighbor's sideyard. You can make the units a little smaller on the top floor or you might consider several large studio apartments. This will reduce your profit only slightly, making all of this effort pay off. Revised drawings are prepared for planning. The transportation planner's report states that the traffic along the block will increase during the commuting hours by 15%, which translates into ten more cars per hour. Not sure how the community will respond, but given the money spent on drawings and consultants, your only choice is to hope for a favorable outcome.

### *Step 7: meeting the community to review the revised plans*

At the next meeting held with the community, association attendance is almost triple that of the first meeting. Apparently, word got out that a development project was to be approved and this sparked the interest of everyone who lives in the community. In addition to the architect and the planner assigned to the file, a member of the local press is also in attendance. Given that most of the audience did not hear the previous presentation, it was decided to present the project in detail. Design changes are noted, so that the community knows that some concessions have already been made. Particular attention is given to the changes in the roof line on the east side. Now effectively a three-story building on this side, you hope that the response to this design will be more favorable. Sun and shade studies reveal that only in late afternoon on winter days would there be any significant shade on adjacent properties. In addition, the transportation consultant's report shows only a slight increase in traffic during the commuter hours, which is of some interest. Towards the end of the presentation, the issue of scale comes up again.

The neighbor living behind the proposed development project reiterates his comment from the previous meeting that the scale of the proposed development is still too massive, even with the modifications to the roof line. It is now revealed that one of the neighbor's sons who is an architecture student at the city university was able to get access to the drawings given to the planner. He has used the architect's drawings as the basis for creating his own computer model of the project, which makes it possible to show the project from each of his neighbor's backyards. You had planned to show these views if required, but now it has become the central issue at tonight's discussions. The attention of the group turns to the series of computer printouts placed on the table. One of the views shows the condo property looming over the neighbor's backyard fence. The change in elevation at the corner of the property places his property slightly lower than the others. The focus on each view generated from the backyards of each resident has now effectively ended all discussion for the evening. While reviewing the views from her backyard, one of the neighbors suggests that a two-story building would be acceptable, but four stories are out of the question. Now it would appear that even a compromise of three floors may be unlikely.

The local reporter seems to have found the story interesting and would like to set up a time to meet with you to discuss the proposal. He then proceeds to get copies of your drawings and those produced by the neighbor's son. It will come out in the press next week in a piece on development in this community. Your understanding is that the piece will focus on how this project fits the goals of the city for new housing in the neighborhoods adjacent to the downtown, but your concern is that the drawings done by the architecture student will be all that anyone remembers. Just before leaving, the planner assigned to the file suggests that you meet within the next week to discuss the future of the project. Your feeling is that now, after spending over half a year on this project, you are not making much progress. Your architect agrees that a meeting with the planner some time in the next week or two would be worthwhile to see where this proposal might go.

### *Step 8: project revaluation*

A meeting held a week later in the planner's office provides an opportunity to review the progress made on the application. He reminds you and the architect that, although you satisfy the zoning requirements, it is unlikely that any permit will be approved until after the election. The article that appeared in the local paper showcased the project, with the architect's drawings juxtaposed against the computer renderings produced by the neighbor's son. The city does not want to appear insensitive to community concerns in light of the newspaper piece, which presented a narrative of a developer working with the city, trying to force higher-density development on the community. He suggests that you just wait it out till the new council is elected. The planner also indicates that if you build a two- or three-story building on a single lot, he would be willing to approve that project. It is an option worthy of consideration, but you state that it would require further study before a decision can be made. Your plan is to contact the planner again in a week or two.

### *Step 9: starting over?*

Later that week, attention focuses on evaluating the option of building a two- or three-story building. Looking at the pro forma, if you were to build a six-unit building with a garage for four cars in the back, it should be possible to sell the properties and make a very reasonable profit of $295,500 on each of the three sites. The architect assures you that, with other projects going up in the community, this approval should be forthcoming. The community would not even need to be consulted on this project, given its height of three, and not four, stories. The plan then would be to create six units above ground and possibly two basement suites, if allowed. You would still need to negotiate on the parking, with space for only four cars along the laneway. However, with the two basement suites designed for students and lower-income individuals, it is less likely they would own a car and have a need for a garage. At worst, you might have to pay a fee in lieu of parking to the city. It might even be possible to break ground in the early fall and have the building completed for the peak selling period in the spring. The loan officer at the bank also mentioned that it could provide a loan using the two other properties as collateral at a much better rate than if you had applied for a construction loan for the condo project. The other advantage is that, by building on a single lot now, the project, once completed, could act as a buffer with the adjacent property owner, removing one serious objection against future proposals for a higher-density project on the two remaining lots. After some thought, the decision is made to contact the planner next week to see which of these two proposals would be acceptable.

### *Lessons learned*

In many cities, the approval process can take years. In this abbreviated case study, almost a year has passed before we considered alternative scenarios. At that point, the developer has spent fees on consulting and filing requirements in the tens of thousands of dollars. Though it might be possible to secure an approval for the original concept after the elections, there are no guarantees. Clearly, the building design met the zoning requirements, but without a clear message from approving authorities, the time required could exceed the patience of the developer. If the developer had only an option on the land, time could run out before approvals can be secured. Communities know that, to stop a project, it is sometimes only a matter of running the clock out on the developer. Another message is that it is difficult to prepare for the unexpected. The neighbor who likes to garden, the architecture student son of a neighbor, and the attendance of the press were all surprises. In retrospect, it might have been possible to have prepared a presentation that would have satisfied the community, but that is unlikely. To the developer's credit, advice was sought from an architect who had a previous working relationship with the city planner. Also, being from the community may have made for a slightly more receptive audience during the design review process. In conclusion, this case study demonstrates that reaching a solution that maximizes profit is probably never possible. However, in the end, getting to an acceptable solution may be all that is required.

## Notes

1 Moga, 2016.
2 Hirt, 2013.
3 Benevolo, 1989.
4 Cooke, 2016; Burg, 1976; Condit, 1973.
5 Kwartler, 1989.
6 American Law Institute Model Land Development Code 1976, in Ellickson and Tarlock, 1981, 387.
7 Moore, 2016.

8 Meck and Retzlaff, 2008.
9 Papayanis, 2000.
10 Ibid.
11 Collins, 1980.
12 www.ada.gov/regs2010/2010ADAStandards/2010ADAstandards.htm
13 See www.ada.gov/regs2010/2010ADAStandards/2010ADAstandards.htm#c1: "historic property in a manner that will not threaten or destroy the historic significance of the building or facility, alternative methods of access shall be provided pursuant to the requirements of § 35.150."
14 http://69.89.31.83/~disabio5/wp-content/uploads/2011/07/CDA-reformat.pdf
15 Wilensky, 1964.
16 APA home page, www.planning.org/; CIP home page, www.cip-icu.ca/
17 Forester, 1989; Forester, 2000.
18 Cuthbert, 2007.

## Bibliography

Attoe, Wayne and Donn Logan. *American Urban Architecture, Catalysts in the Design of Cities.* Berkeley, California: University of California Press, 1989.

Babcock, Richard F. *The Zoning Game, Municipal Practices and Polices.* Madison, Wisconsin: The University of Wisconsin Press, 1977.

Benevolo, Leonardo. *History of Modern Architecture, the Tradition of Modern Architecture.* Vol. 1, Cambridge, Massachusetts: The MIT Press, 1989.

Burg, David F. *Chicago's White City of 1893.* Lexington Kentucky: The University Press of Kentucky, 1976.

Collins, R. C. "Changing views on historical conservation in cities." *The Annals of the American Academy of Political and Social Science* 451, no. 1 (1980): 86–97.

Condit, Carl. *The Chicago School of Architecture, A History of Commercial and Public Buildings in the Chicago Area.* Chicago, Illinois: The University of Chicago Press, 1973.

Cooke, Jason. "Compensated taking: Zoning and politics of building, height regulation in Chicago 1871–1923." *Journal of Planning History* 2, no. 3 (September 2016): 207–226.

Cuthbert, Alexander R. "Urban design: Requiem for an era – Review and critique of the last 50 years." *Urban Design International* 12, no. 4 (2007): 177–223.

Ellickson, Robert C. and A. Dan Tarlock. *Land-Use Controls, Cases and Materials.* Boston and Toronto: Little, Brown and Company, 1981.

Forester, John. *Planning in the Face of Power.* Berkeley, California: University of California Press, 1989.

Forester, John. *The Deliberative Practitioner, Encouraging Participatory Planning Processes.* Cambridge, Massachusetts: The MIT Press, 2000.

Haar, Charles M. and Jerold S. Kayden eds. *Zoning and the American Dream.* Chicago, Illinois: Planners Press, 1989.

Hirt, Sona. "Home sweet home: American residential zoning in comparative perspective." *Journal of Planning Education and Research* 33, no. 3 (2013): 292–303.

Grant, Jill ed. *A Reader in Canadian Planning.* Nelson, Toronto, Ontario: Thomson, 2008.

Kaiser, Edward J., David R. Godshalk, and F. Stuart Chapin. *Urban Land Use Planning.* Urbana and Chicago, Illinois: University of Illinois Press, 4th ed. 1995.

Kwartler, Michael. *Legislating Aesthetics: The Role of Zoning in Designing Cities in Zoning and the American Dream.* Chicago, Illinois: Planners Press, 1989. 187–220.

Levy, John M. *Contemporary Urban Planning.* Upper Saddle River, NJ: Prentice Hall, 2000.

Meck, Stuart and Rebecca Retzlaff. "The emergence of growth management planning the United States: The case of *Golden v. Planning Board of Town of Ramapo* and its aftermath." *Journal of Planning History* 7, no. 2 (May 2008): 113–157.

Moore, Aaron A. "Decentralized decision-making and urban planning: A case study of density for benefit agreements in Toronto and Vancouver: Decentralized decision-making and urban planning." *Canadian Public Administration* 59, no. 3 (September 2016): 425–447.

Papayanis, Marilyn Adler. "Sex and revanchist city: Zoning out pornography in New York." *Environment and Planning D: Society and Space* 18, no. 3 (2000): 341–353.
Steven, Moga. "The zoning map and American city form." *Journal of Planning Education and Research* 37, no. 3 (June 2016): 271–285.
ULI, Urban Land Institute. *Multifamily Housing Development Handbook*. Washington, DC: ULI, 1999.
Wilensky, L. Harold. "The professionalization of everyone?." *American Journal of Sociology* 70, no. 2 (September 1964): 137–158.

## Web resources

Information and technical assistance on the Americans with Disabilities Act of 1990
www.ada.gov/regs2010/2010ADAStandards/2010ADAstandards.htm#c1

# 7 Real estate and the law

## Introduction

An understanding of the law is essential to all those engaged in real estate development and finance. All aspects of development, from acquisition and construction to leasing and management, have a legal component. At each stage of the development process, you will need legal expertise to accomplish your goals. Knowledge of the law is essential in protecting the interests of all parties: investors, builders, architects, developers, lenders, and tenants. Protection of your interests will require legal counsel from experts in real estate, urban, and construction law. Like any other profession, your attorney will need the expert knowledge and experience with which to address your specific issue. As in medicine, no single practitioner is an expert in all areas of practice. A general practitioner may be able to address simple cases of buying a home or reviewing a lease, but a specialist who is an expert in real estate and urban law will be required when dealing with the complexities of major real estate developments. Knowing when you need an attorney is essential to the success of any project.

Real estate law practice incorporates both the law and its administration. Like many business and financial endeavors, success in real estate development requires close attention to the management of numerous parties and institutions. Administering the law can involve the courts and municipal, provincial, state, and federal agencies. Within a single municipality, numerous departments may require legal review of documents during the approval process. Under certain circumstances, projects may also need approval of state, provincial, and federal agencies. As part of a team involving urban planners, architects, and engineers, legal counsel can help advance your proposal through an administrative process that may take years to complete. When disputes occur, actions can proceed from lower to higher courts. However, legal solutions to business conflicts can be costly and sometimes unproductive. In these cases, it may be only the attorneys who profit from your venture. Good legal counsel is critical to all business ventures, but it is a good business relationship that will guarantee success. Having a good working relationship built on trust will do more to advance your project than will threats of legal actions.

The law touches on many aspects of real estate development. During the process of acquisition, the law defines the relationship and responsibilities of brokers and agents engaged to purchase a property. Most important, the law defines what you are actually buying when you acquire real estate. Your rights and responsibilities as an owner will also be governed by the law. Ownership in itself may have implications in both life and estate matters. How property is titled will determine its treatment upon death. Though most housing is owned by individual home owners, most commercial and large-scale projects can be owned by partnerships, corporations, and real estate investment trusts (REITs), each with different tax treatments.

If the project is a new real estate development, contracts with architects and contractors will establish the terms of work, cost, schedules, and penalties. Once the project is complete, tenants will sign leases with terms that specify rents, duration, and other requirements. The law specifies the rights and responsibilities of both parties. When disputes arise, the courts must resolve which party violated the term of the lease and make judgments on matters of restitution and eviction. Most real estate requires financing. The law defines the relationship between the lending institution and buyer. Familiarity with the law will prove invaluable at every stage: acquisition, financing, and development of real estate.

## Foundations and principles of private property

Our knowledge of private property dates back to the beginnings of written history. The Code of Hammurabi of 1750BC dictates how land may be sold, leased or used to pay debt and benefit those in possession of land.[1] During the Middle Ages, land tenure served to structure relationships between the monarch and their lords and tenants.[2] In Anglo-American history, private property is at the foundation in the development of our governmental institutions. To this extent, one of the missions of government is to protect the institution of private property. John Locke (1632–1704) stated: "The reason why men enter into society is the preservation of their property." In the founding of the US, defenders of private property often cited Thomas Jefferson's belief that the yeoman farmer would be the cornerstone of a country that promotes individual freedom and prosperity.[3]

In the US, private property is protected under the Fifth and 14th Amendments of the Constitution. The Fifth Amendment states that no person shall be "deprived of life, liberty, or property, without due process of law", "nor shall private property be taken for public use, without just compensation". The adoption of the 14th Amendment in 1868 expanded the scope of the Fifth Amendment to the state governments, reaffirming that:

> No state shall make or enforce any law which shall abridge the privileges or immunities of citizens of the United States; nor shall any state deprive any person of life, liberty, or property, without due process of law; nor deny to any person within its jurisdiction the equal protection of the laws.

The importance of the 14th Amendment is that states and local governments cannot deny the individual their rights safeguarded under the Constitution.[4]

Although the 14th Amendment has been the legal foundation of many civil rights cases since its passage, it has also served the interest of expanding the rights of corporations in the US.[5] Although these rights were extended, corporations are, as Chief Justice Marshall stated in the landmark case *Trustees of Dartmouth College v. Woodward* (1819), "artificial being(s) invisible, intangible and existing only in the contemplation of the law" and merely "creatures of the law". Under the Supreme Court's decision in *Santa Clara v. Southern Pacific Railroad Co.* (1886), individual rights afforded by the Constitution and the 14thAmendment were extended to corporations. The more recent Supreme Court decision *Citizens United v. Federal Election Commission* (2010), corporations, labor unions, and associations, as "associations of citizens", have been given the

| Canada | United States |
|---|---|
| No expressed constitutional protection in Canadian Charter of Rights and Freedom | 5th Amendment, American Bill of Rights – No taking with just compensation |
| | Denial of rights to use all of one's property will constitute a taking |
| Right to affect use in public interest (as long as not for a public purpose) | Right to affect use in the public interest – if affects 75 to 87.5% of usablity, then compensation |
| Availablity of private rights: trespass, nusiance, negligence | 14th Amendment – no state shall "deprive anyone of property without due process" |
| Rights viz. municipal bodies: 1. ultra vires; 2. procedural protection; 4. substantive protection; 5. bias; 6. discriminiation; 7. under inclusiveness; 8. unreasonable bylaws. | |

*Figure 1.7* Protection of civil rights within Canada and the US.

First Amendment right of unrestricted political contribution, dramatically changing the landscape for campaign funding in the US.[6]

Unlike the US, Canada's Constitution does not provide the same protection of private property, with the Constitution Act of 1867 giving the provinces the authority to deal with civil matters and property. The Canadian Bill of Rights enacted in 1960 recognizes the right of the individual "enjoyment of property and the right not to be deprived thereof except by due process of law". The Canadian Bill of Rights is a federal statute and not a constitutional document, so it applies only to the federal government (Figure 1.7).[7]

## The law and real estates: statutes, common law, and regulations

Real estate development and finances embrace both public and private law. Public law concerns itself with the relationship between individuals and the government. Public law regulates professionals providing services and expertise to developers, including real estate brokers, architects, engineers, attorneys, and accountants. Public law is also concerned with enforcement of regulations by both levels of government. Much of real estate development is concerned with satisfying regulatory requirements. Sources of public law include the constitution, statutes, and municipal by-laws. Private law, on the other hand, comprises laws of property, contract, and tort.[8] Real estate development, by its transactional nature, will touch upon private law.

In the US and Canada, elements of common law developed after 1066 in England can be found in our laws today. Under the common law principle of *stare decisis*, the courts would look to decisions in the past for guidance. Reliance on a principle that would "respect the precedent established by prior decisions" supports a view of fairness and consistency present in the Anglo-American legal system. By the Middle Ages, the Court of Chancery arose out of the inadequacies of common law. In England, matters of equity would include issues pertaining to property, such as the issuance of injunctions to bar activities that could do harm. Under the Court of Chancery, there was no right to a trial by jury. The Court of Chancery was dissolved in England only in the 19th century.[9] Statutes passed by municipal, state, provincial, and federal legislative bodies speak to the manner in which property is bought, sold, and developed. Real estate brokers also must comply with laws that bar discrimination. Courts of equity deal with bankruptcy and lawsuits in which damages are being sought.[10]

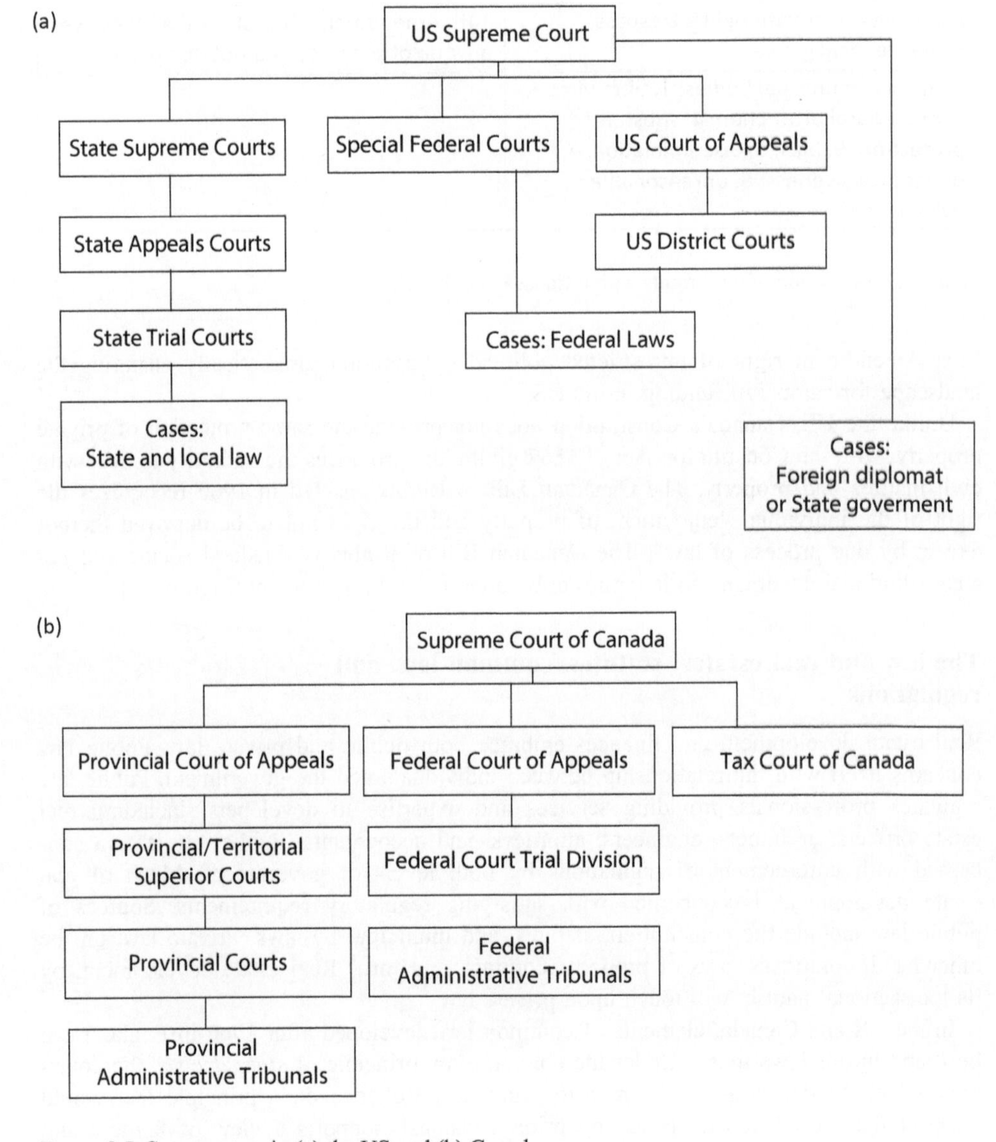


*Figure 2.7* Court system in (a) the US and (b) Canada.

The courts in the US and Canada are the primary mechanism for resolving disputes (Figure 2.7). Understanding legal procedure will be important in disputes concerning real estate. Unresolved issues concerning property and contracts may eventually end up in the courtroom. Evictions and foreclosures might be part of a property manager's normal business, while breach of contract, design errors, substandard construction, and non-payment issues will be that of the developer. Poor construction and non-payment of condo fees are the domain of condominium boards. In all cases, litigation can be a costly and time-consuming affair. Cases including eviction and foreclosure will be a matter for lower state and provincial courts. However, cases involving substantial financial claims or stemming from conflicts with the administration of regulations may end up in federal court. It has been only on rare occasion that cases relating to real estate have been decided by the Supreme Court. In addition, specialized courts may adjudicate on matters of property, tax and tenants' rights.

Finding solutions outside the courtroom may be less costly in time and legal fees. Arbitration can offer a more cost-effective alternative. However, arbitration is not a panacea. The cost of arbiters can be as high as that of attorneys. In addition, most parties will hire attorneys to advise them on the arbitration process. Arbitration may offer little recourse when settlements appear unreasonable. Finally, if the arbitration process should go awry, the opposition will have a detailed knowledge of your circumstances, which may put you in an indefensible position should you go to court. In all cases, legal counsel should be able to advise on the best course of action. Avoiding legal challenges and battles by maintaining good communication, best management practices, and well-written contracts is always a better course of action.

## Collective ownership of property

In the history of human settlement, religious and utopian communities have been based on the collective ownership of property. In North America, the Hutterites, Shakers, New Harmonists, and native communities owned land collectively. For these hunting and agricultural communities, success relied heavily on tribal, family, and religious order. These communities, often based on family relationships, succeeded together as a group and reinvested in their land and buildings over time. For the Hutterites, who live in the west, and the Amish, who live in the east and midwest, this successful strategy has been marked by a continued growth in population and land holdings over the last century-and-a-half.

Private land must also be distinguished from public lands. Public lands are held and administered by the government for the benefit of all its citizens. As an attack on public lands, the "tragedy of the commons" is often cited as a justification. This is a reference to the common green set aside for grazing of animals in England and colonial America, available to everyone in the community, and the assertion that overgrazing is inevitable as each individual maximizes their benefit from the commonly owned land without paying rent. Without wise husbandry of the resource, it is said, the value of the commons is destroyed. However, traditional societies prove that this is unlikely to occur.[11]

The law differentiates real property from that of other types of individually owned property. Unlike personal property and other forms of property, real property refers to land and all attached structures and improvements. One obvious distinction between real and personal property is that personal property is movable and easily transported.[12] Historically, personal property law regulates how goods may be transferred in trade by merchants. Unlike personal property, real property includes issues of inheritance as central in its definition. Also, unlike personal property, right of use may be constrained by law.

## Right of use

Private property ownership denotes a specific "bundle" of rights. Those rights include the rights to use, sell, destroy, enjoy, and give away the property as you may wish. Real property has some important differences that distinguish it from other types of personal property. First, real property is fixed in space, with defined boundaries.[13] An owner may want to live in the property, rent it to another individual or use it as collateral for a loan. Also, the right of exclusive possession allows you to enjoy its use, but also to exclude others. However, there may be times when you must give access to others. Established by use or by need, a fire marshal, police officer, firefighter or Occupational Safety and Health Administration (OSHA) inspector will have access to your property without permission.[14]

Easements (non-possessory rights) may give neighbors rights or interest in your property. For example, a farmer who has used a road to get access to one of their fields may have an established an easement on your property. An adjacent landowner may have acquired an easement barring you from building a structure on your property that will obstruct their view. Easements can also be created to provide access to utility companies for the purpose of maintaining telephone, cable, gas, and electric services. The land that benefits from the easement is known as the dominant tenement, while the other land is the servient tenement. Easements are encumbrances on the deed and can be created through a formal grant, which notes the consideration given, location, use, terms, liability, and any maintenance or repairs required by the grantee and grantor. Easements can also be created when properties are subdivided. If the buyer needs to go through your property to gain access to a road, then there is an implied easement. Often, these easements are unwritten. Easements can also be prescriptive when a person uses a property without permission for some period of time. Because these easements are considered adverse use (use without permissions), this access can expire if there is lapse in use. Easements that are encumbrances on the property will be recorded on the deed of the property.[15]

Covenants restricting use can appear on the deed, which may also limit a property owner's use of the land. Commonly used by developers, these "restrictive covenants" can place a constraint on a landowner by restricting the size of a house that can be built on a property or limit the number of outbuildings. Similar to a zoning restriction, property setbacks or maximum height restrictions can be set through the use of covenants. These types of "restrictive covenant" can even be used to establish architectural review and control over the homes built in a neighborhood. Affirmative covenants also provide access and maintenance of walls, fences, docks, and parks that benefit multiple landowners. Affirmative covenants are not recognized in some provinces and states because of the difficulties that arise in their enforcement. Restrictive covenants also face legal challenges when they conflict with public policy or an individual's rights. In the first half of the 20th century, restrictive covenants were used by developers and communities to bar specific racial, ethnic, and religious groups from buying property in specific neighborhoods. With the intervention of the US Supreme Court in 1948 and later the passage of the Housing Rights Act of 1968, the use of restrictive covenants for this purpose was prohibited.[16]

Real property is also subject to taxation. Non-payment of taxes can result in a forced sale to meet this obligation. Nuisance laws may restrict your "enjoyment' of the property if you plan to have loud parties late at night. Zoning may bar certain uses. In urban communities, it is common for zoning by-laws to bar farmyard animals and junked cars on your property. Municipal law may also restrict your ability to subdivide your property for sale or transfer to another individual. If your property is located in an historic district, you may also be required to follow specific guidelines in the repair and maintenance of your property, including the color of paint, the type of siding, and the mortar used in reappointing the masonry. Federal and state statutes will also bar landowners from polluting the air and rivers, and from restricting navigable waterways.

### *What is a fixture?*

In taking title to a piece of real estate, you would clearly expect possession of the surface rights and any structure attached as a right of ownership. However, defining what "fixtures" are will determine what the purchase actually is. A fixture is an item of personal property that has been permanently attached to the structure. You would not expect someone to remove a free-standing garage sitting on its foundation after the completion of a sale. Nor would you expect the sellers to remove chandeliers, built-in dishwashers or even kitchen cabinets prior to your possession date. However, there may be items that should not be considered part of the sale, including items of trade, such as shelving and tools.[17] In the US courts, tests of annexation, adaptation, and intention provide guidance on what is considered personal property.

- Though subject to interpretation, annexation refers to objects that are attached to the house, such as a screen for a window or door, but can also include "constructive annexation", for example window screens stored in the basement that have been used by the home owner. In an industrial facility, equipment or machinery fixed to the floor will also be considered a fixture under constructive annexation.
- The adaptation test will cover features required for the enjoyment of the property, such as a furnace.
- The intention test focuses on the intent associated with the removal of an object from the premises. For example, if a home owner had always intended to take the front hall mirror or chandelier, which was a family heirloom, when moving to their next home, then the seller may have the right to remove the item from the house.

Clearly, going to court to decide on whether an object is personal property can be avoided by listing these items in the sales agreement.[18]

### *Water rights*

Real property implies the right to use the land for the landowner's benefit. Traditionally, farming was the major activity upon the land, so it is not surprising that many of the laws contained in the ancient Babylonian Code of Hammurabi pertain to the rights of those who planted and harvested crops or to those who rented land for framing.[19] The ownership of land implies surface rights, as well as control over the subsurface and the air above. For farmers, the control over water resources represents a critical component for a successful harvest. Interestingly, the Code established water rights and the responsibilities of landowners to use water responsibly: "If any one opens his ditches to water his crop, but is careless, and the water floods the field of his neighbor, then he shall pay his neighbor corn for his loss."

In common law rule, subterranean water belongs to the landowner. In addition, the landowner may use also use water from springs and sources not part of river systems (percolating waters) to their benefit no matter what the consequences to their neighbors. In the US and Canada, water rights concerning the flow of water along rivers and streams (riparian rights) are dictated by state, provincial, and federal legislation. The US states east of the Mississippi subscribe to a doctrine that entitles all adjacent landowners to share flowing water equally, while in the western states, water rights are given to the first user ("first in time, first in right"). Currently, many western states have adopted a principle of "beneficial use", whereby the landowner must show that water taken from the source does not injure others and will be used for some purpose. In addition, they cannot build a permanent structure for means of diverting water.[20]

In Canada, water rights are dictated by the provinces. In the western provinces of Alberta, Saskatchewan, British Columbia, and Manitoba, and the eastern province of Nova Scotia, for the most part the "first in time, first in right" doctrine applies. The principle of "beneficial use" also applies. As in the US, this dictates that water must be used for socially accepted purposes. Water rights belong to the property and not to the individual. In the north, including the Yukon, Northwest Territories, and Nunavut, water use is controlled by the public authority (water boards). Rights may be reallocated and transferred to meet the interest of the community. In Quebec, the civil code dictates how surface and ground water will be used. In Quebec, the ministries of municipal affairs, fisheries, natural resources, and agriculture have a say on how water is used. In Ontario, "riparian rights doctrine" entitles landowners adjacent to a water source to benefit, with the province requiring special use permits when subsurface water exceeds 50,000 liters per day for industrial purposes. Native water rights became constitutionally protected in Canada in 1982.[21]

To the extent that land borders navigable waterways, federal legislation applies to the use of water. With the passage of the Clean Water Act, owners are not permitted to affect the water quality of rivers running on or adjacent to their property.[22]

### *Minerals rights*

Under common law, property owners have the right to benefit from mineral deposits from the surface to the center of the earth (the *ad infernos* doctrine).[23] In Canada and the US, over time, the rights to underground deposits were separated from surface rights. In Canada and many western states in the US, subsurface rights could be leased to a separate party. In Canada, the Crown maintained rights to all mineral deposits.[24] With exploration of oil and gas in North America, the separation of mineral and oil gas deposits can result in multiple parties needing access to the land. Since conflicts can occur in the use of the land, some constraints may apply on the techniques of recovery so as not to impact use by all.[25] Clearly, in states where mineral, oil, and gas extraction is prevalent, surface land uses will be complicated by subsurface property rights.

### *Air rights*

Air rights follow the common law principle of *ad coelom* ("to the heavens"). Air rights means that the invasion of someone's air space is considered trespassing. Until the 20th century, the ability to trespass into someone's air space was not an issue. However, with the arrival of flight in the 20th century, some additional regulations and restrictions have been placed on the property owner. To the extent that the civil aeronautics authorities have placed restrictions on interference with commercial flight, landowners can expect aircraft to fly no closer to the ground than 500 ft. To the extent that flight can be considered a nuisance, a landowner may be able to sue in cases in which noise is considered a "taking" under the 14th Amendment. In an unusual case, *United States v. Causby* (1946), the owner of a chicken farm lived adjacent to a military airstrip. With his property on the flight path, planes would pass less than 70 ft. overhead. The noise levels were so intense that the chickens were frightened to death. Causby sued on the grounds that he had not been compensated for this "taking".

Today, drones have complicated the matter of trespassing into the air space of a property owner. Drones regularly fly below 500 ft. above ground level (AGL). In the US, the Federal Aviation Administration (FAA), as of June 21, 2016, has restricted drones' operation to 400 ft. AGL. Drones may not operate at night nor near moving aircraft.[26] Transport Canada recommends that drones not fly higher than 90 meters (300 ft.) above the ground, and not approach buildings, animals, and structures within 150 meters (500 ft.).[27] Several states and municipalities have also passed legislation regarding the use of drones in populated areas. As drones are increasingly employed in both recreation and commercial applications, their legal use may need additional clarification.[28]

To the extent that owners have rights to air space, many cities regulate the height of structures and buildings. In Europe, until the 20th century, new buildings could not exceed the height of prominent buildings, including cathedral towers. With few buildings reaching heights of five or six stories, building heights were rarely an issue. One notable exception was the Eiffel Tower. Constructed as a centerpiece for the Universal Exposition of 1889, the reaction from artists, writers, and the public was not overwhelmingly positive.[29] With the construction of taller and taller skyscrapers in the US, height and setback restrictions have had to be addressed in the zoning by-laws to keep cities from having dark urban canyons.[30] Today, cities restrict skyscraper development to specific areas within the metropolitan area. Setbacks and maximum height restrictions on all structures and buildings are generally designed to maximize light and air penetrating to the street levels. Where allowed, air rights may be sold and transferred to other properties and under certain circumstances.[31] Cities have also allowed for the transfer of air rights as part of an economic incentive package that assists the owners of historic properties, while promoting economic development. Under these provisions, individuals benefit from owning an historic property by having the opportunity to sell or use their air rights to construct a taller building on another property. The additional revenue can be considered compensation for the profit that would have accrued if the owner had demolished the building and replaced it with a taller one.[32]

## Forms of ownership

How real estate is owned in the US and Canada can have implications for taxation, inheritance, and use. Land ownership is born out of a historic precedent that can be traced back to the time of feudalism. Under feudalism, all rights to land rested with the Crown. Lands could be parceled out to lords, who were under obligation to the monarch. The lords had the right to use the land, but only with certain obligations to the Crown. The sublords or tenants on the land would provide rent and military service in return for use of the land. The actual transfer of land was by a ceremony of livery (delivery). These sublords could also make similar arrangements with other tenants for purposes of collecting rent. The word "estate" today has multiple meanings. For the feudal lord, "estate" was synonymous with class: noble, clergy, commoner.[33] Today, estate generally refers to all property that a person owns at the time of their death. Ownership can be absolute or be limited to an interest. Division of ownership falls into two categories: leasehold and freehold. Leasehold limits the period of time of possession, while freehold generally lasts for a specific period of time.[34]

### *Fee simple or fee simple absolute*

In both Canada and the US, fee simple is a common form of ownership. Under the category freehold, fee simple or fee simple absolute, the owner has the largest possible estate interest in a property. In addition to possession, owners may sell, gift or give away the property under fee simple. Also, under a will, the owner may direct their executor to take specific actions upon the owner's death. Under fee simple, a portion of the estate may be given to another, subject to zoning and other legal restrictions.[35]

### *Life estate*

Though less common than fee simple, it is possible under freehold to grant only a life interest in a property. Under a life estate, the owner's interest in the property ends upon their death, with interest reverting to a third party. This remainder interest (fee tail) would allow a family home to be used by its current owner or an aging grandparent until their death. Then, upon their death, the property would be transferred to a grandson or granddaughter (the remainderman).[36]

### *Co-ownership*

Co-ownership can be used to solve a number of problems when interests must be shared or divided between two or more parties. Under common law principles, co-ownership can take several forms. Interests may be shared equally or proportionally. Rights may also be divided so that one person may have rights to the surface, while another has air rights and a third has rights to any minerals. Common forms of co-ownership include joint tenancy, tenancy and tenancy in common, community property, tenancy by the entirety, and tenancy in partnership.[37]

### *Joint tenancy with right of survivorship*

Most home owners in the US and Canada are familiar with joint tenancy. Under joint tenancy, the right of survivorship guarantees that, upon the death of either owner, their interest passes directly to the survivor. With its origins dating back to feudalism, the history of joint tenancy solved a number of problems for the landed gentry in England. Lands were the source of wealth, status, and power into the 17th and 18th centuries; joint tenancy in common kept land holdings from being subdivided at death. With land being the primary source of economic and political power, land holdings remained undivided under joint tenancy with right of survivorship. If four siblings were to own

| No of parties | Undivided Interest | Event |
|---|---|---|
| A, B, C, D | 25%, 25%, 25%, 25% | |
| A, B, C | 33%, 33%, 33% | Party D dies |
| A, B | 50%, 50% | Party C dies, leaving A & B as owners |
| A | 100% | Party B dies, leaving A as sole owner |

*Figure 3.7* Joint tenancy with right to survivorship.

a property, under joint tenancy, with the passing of each, the undivided interest in the property would be shared by fewer and fewer family members (Figure 3.7).[38]

Joint tenancy can be created by will or title. Under common law, joint tenancy requires four unities.

- **Unity of possession** Each owner has equal right to possess the property.
- **Unity of title** Each owner will have acquired interest in the property with the same document or instrument for establishing title.
- **Unity of interest** Each owner will have an equal interest, for example one owner cannot have a life interest while the other has a fee simple interest.
- **Unity of time** Each of the owners will have taken interest in the property at the same time.[39]

During life, under joint tenancy, each co-owner has an equal right to the possession of the property in its entirety. However, under the law, occupancy by one of the owners is considered occupancy by all. For example, if this property were a family farm and only one of the siblings lived on the property and ran its operation, any income generated would not have to be shared with the other children unless there were general agreement. Joint tenancy does not require the sharing of this income, which may be the source of many family disputes where one sibling is in charge. However, that said, it is possible to terminate joint tenancy under specific actions. Severance of joint tenancy can occur in the event of divorce. Divorce decrees may sever joint tenancy. Similarly, court-ordered partitions will terminate joint tenancy. Mortgages taken out by only one of the owners can also result in its termination.[40]

### *Tenancy in common*

Similar to joint tenancy, tenancy in common shares unity of title, time, interest, and possession. Under tenancy in common, ownership is not limited to its creation by a deed or title. A single owner under tenancy in common may sell, lease or mortgage their portion of the property. Joint tenancy can occur when the remaining owner dies without a will. Under the laws of intestacy, in certain states and provinces the courts may decide how property is divided among the heirs if the intention of the deceased is unclear. For example, where one partner dies intestate (without a will), the state may divide the property equally between the surviving parent and the children, creating a tenancy in common ownership among the children and surviving parent. If, for example, there were three children, each would receive a one-sixth interest (a third of a half-interest) in the property, with the remaining half-interest going to the surviving spouse. Under tenancy in common, all parties would have an undivided interest in the property. Now, all three children and the surviving parent have the right to enjoy the entire estate. When a parent dies without a will, the surviving parent may be surprised to learn the consequences of not having an estate plan. Under tenancy in common, each child now has the right to sell, lease or mortgage their interest to someone outside of the family, creating even more potential strife.[41]

### *Community property*

In predominantly western states of the US, it is possible for property to be titled as community property. Based on French and Spanish legal traditions, community property assumes that, when married couples acquire property, each has a half-interest in the property and when that property is acquired by one person's labor or skill, that person becomes a co-owner with the other. If one spouse owns interest in real estate, then in a community property state, upon that party's death, 50% of the interest will belong to the remaining spouse. Community property can be in direct conflict with other laws of the state in that ownership of property held by one party is the sole property of that individual. Community property can offer some advantages in estate planning. Upon death, 50% of community property can qualify for a step up in basis.[42]

### *Tenancy by entirety*

Tenancy by entirety can be considered a form of joint tenancy. Designed for married couples, it adds unity of person to the four unities of possession, title, interest, and time. Under tenancy of entirety, both spouses are treated as a single person. Upon death of either party, the surviving spouse has sole ownership. In addition, when both parties are alive, sale of the property requires both signatures to complete a sale or mortgage the property. With a divorce, tenancy by entirety will become tenancy in common.[43]

### *Condominium*

The purchase of a single unit in an apartment-style building today is often done using condominium ownership. Under a condominium agreement, the individual has fee simple ownership to their unit, while having an undivided interest in the land and common areas of the development. These could include the pool, athletics club, meeting rooms, and lobby and circulation space. When creating a condominium, the developer must file two documents. The master deed, condominium document or enabling declaration establishes the creation of a board of directors who will manage the affairs of the condominium. The directors are responsible for the day-to-day management and maintenance of the building. The condominium document also provides a general and legal description of all the units and common areas. Finally, the condominium document notes the undivided interest in common areas. It should also state the market price. In addition to the condominium document, the condominium by-laws details how board members will be elected. Rights and responsibilities of unit owners, regulations, and fees and charges are also detailed in the by-laws. Rules and regulations complement the by-laws by providing additional detail on the how owners may use their unit. Though these rules cannot deny the owner of a right to enjoy their condo, they can place restrictions on the owner's lifestyle. In the rules and regulations, owners may not have the privilege of renting out their unit and may be limited to having guests on only a specific number of days a year. They may also be restricted from operating a business from their residence. Though difficult to enforce, these restrictions can be a source of conflict among residents.

Anyone contemplating the purchase of a condominium should review all documents prior to sale. Also, buyers should review the financial statements of the condominium, noting expenditures on maintenance. When buying a condominium, you should know whether there are sufficient funds in reserve for the continued maintenance and replacement of major components, including heating and cooling systems. Prospective buyer should also review the condo's insurance contract. Insurance should provide coverage for damages to common areas, as well as general liability. Condominium ownerships are now used not only for apartment buildings and recreational properties, but also for commercial, industrial space, parking structures, marinas, and campgrounds.[44]

### *Cooperatives*

Cooperatives, which are often used to acquire apartment-style buildings, are often corporations whereby the tenants acquire shares in the company. Shares in the company convey rights to occupy an apartment. Apartments are rented to the shareholders. The collected rent is used to cover general expenses. In establishing a cooperative, the sale of stock can be used as a way of raising capital for the construction or purchase of the building and land.[45] Cooperatives have notably been used to build affordable housing in the US, Canada, and Europe. In the 19th century, cooperative ownership was used as a way of creating better workers' housing. In England, the Rochdale Society of Equitable Pioneers was established in 1844, which, among other things, aimed to create housing for workers. In the US, cooperative ownership was not only used to establish utopian communities, but also luxury housing. In a period before the existence of condominium ownership, cooperative ownership made it possible for upper-income households to own an apartment in The Randolph on West 18th Street in 1887. In the 1930s, cooperative ownership was used to create affordable housing for union members. Most notably, a complex of 1,500 units was built by the union Amalgamated Clothing Workers of America (ACWA) in 1926. The New Deal in the 1930s sponsored cooperatives built under the Roosevelt Administration, which included the Greenbelt communities (1,500 units) and Armistead Gardens (1,600 units), both in Maryland. Under various US Department of Housing and Urban Development (HUD) programs, including the National Housing Acts of 1954, 1959, and 1974, cooperatives have been established to provide affordable housing throughout the US.[46]

## Ownership and the investment community

Joint tenancies with right to survivorship, tenancy in common, and community property are forms of ownership generally used by individuals to transfer property from one family member to another. Issues of tax, liability, and ownership will need to be considered in selecting the most appropriate form of ownership for a particular business enterprise. In the US and Canada, the appropriate form of ownership for an investment must also consider a host of federal, state, and provincial regulations. Though tax considerations are always critical to this decision, liability exposure must also be weighed in making this judgment.[47]

### *Sole proprietorships*

Sole proprietorship is the simplest and perhaps most common form of ownership for a business. As a sole proprietor, the owner maintains complete control over all affairs. They may operate under their own name or have a registered trade name. For those first starting out in a business, sole proprietorship may be preferred when income is modest and sporadic. A first-time consultant, freelance writer, artist, caterer, landscaper or contractor might find the absence of any formal incorporation very attractive. There are no registration requirements for a sole proprietorship. At tax time, all income and losses pass through to the personal return. However, the law might require other filings, including those relating to employee income and withholdings, sales tax, and health and safety regulations. Sole proprietorship faces the serious drawback of unlimited liability. If a sole proprietor is unable to pay creditors, personal assets could be seized, including the home, bank accounts, and investments. Also, when securing financing, the bank will probably want sufficient collateral before granting the loan. Finally, if the owner should die or become disabled, the business will most likely expire as well.[48]

### *Partnerships*

Like sole proprietorships, partnerships suffer from unlimited liability. And, like a proprietorship, income is taxed at the individual's personal bracket. Partnerships can help in solving the need for capital by drawing on the resources of two or more individuals. It may also be useful to have another partner to help manage day-to-day affairs. A partnership agreement can be an advantage in business planning in the event that one partner dies or needs to remove themselves from the arrangement. However, like their response to a proprietor's application, banks may still shy away from making loans to a partnership unless personal assets are provided as collateral. In the US, The Uniform Partnership Act (UPA) provides uniformity among most states on partnership matters. Under the UPA, partners do not have equal rights to all property nor can property be assigned to a single partner. Property in the partnership cannot be used to settle personal debt. Upon a partner's death, property should pass to surviving partners.[49] In Canada, provincial laws should be reviewed in order to understand partnerships.

The partnership, as a means of managing and investing in real estate, can have several benefits. First and foremost, all income and losses "pass through" to the individual. However, there are several shortcomings. In a partnership, the liability for both partners is unlimited. All partners have an equal right to control of the operations. However, if one partner enters into a contract that proves disastrous, all the partners are liable for the losses.[50]

### *Ventures*

Joint ventures are similar to partnerships in that two or more individuals are engaged in business. However, it is more of an arrangement than a business, designed to allow two enterprises to work together for a limited period of time. Many joint ventures are conducted without formal documents and are generally used to advance a mutually beneficial business enterprise. Unlike partnerships, joint ventures will have an end date. Again, you should consult state and provincial law (Canada) before entering into a joint venture.[51]

### *Corporations*

With the expansion and growth of trade and industry in Europe, new forms of ownership were developed that were better suited to business enterprise. In the 17th century, the corporation, with powers that rival those of nations, appeared with the growth of international trade in England and the Dutch Republic. The Dutch East India Company – *Vereenigde Oost-Indische Compagnie (*VOC) – established in 1602 was granted monopoly over the spice trade with the East. The authority of this multinational company included the right to wage war, negotiate treaties, and imprison and even execute those in violation of the law. One of the most successful corporations in history, it continued until the company was nationalized in 1798 after the invasion of France, which brought an end to the Anglo-Dutch wars. Like corporations today, "perpetual life" insures that the future of a corporation is not in jeopardy when an owner dies. Under the VOC Charter of 1602, the board of directors and stockholders managed operation with a sensitivity to national goals.

Corporations, unlike partnerships or sole proprietorships, benefit from limited liability for their investors, with some exceptions. Corporations also benefit from perpetual life: the death of their founder in itself will not terminate the life of a corporation. Shareholders may also buy, sell, gift or bequeath their shares in the corporation, making it possible to transfer interest to another party. Incorporated under state or provincial charters, the articles of incorporation establish a board of directors, which will elect the executive officers. Corporations are required to hold annual meetings, and to file financial reports and tax returns with the province and/or state. There are certain circumstances in which corporations are required to file returns in both the US and Canada. Corporations benefit from limited liability, which allows investors not to be at risk beyond their investment. However, when corporations do not follow standard procedures, "piercing the corporate veil" may result in claims against the board and stockholders. For example, board members and executive officers of corporations who knowingly violate environmental regulations will not be exempted from prosecution under the Comprehensive Environmental Response, Compensation, and Liability Act of 1980 (CERCLA 1980).

One drawback of a corporation is that shareholders are also subject to double taxation. Income is taxed at the corporate level and again when received as dividends by the individual shareholders. Double taxation is generally cited as a major drawback of corporations. However, the ability to raise funds from a potentially large group of investors makes corporations an attractive business form. "Going public" gives corporations access to capital by selling shares on the stock market. In the US and Canada, listing a company on an exchange is an involved process requiring the filing of a prospectus to the both the US Securities and Exchange Commission (SEC) and state/provincial regulators. In Canada, each province and territory has a security commission that regulates investments. Submitting an initial public offering (IPO) will require specialized legal and accounting expertise if all the requirements are to be met before going public.

Small businesses and startups who have incorporated may still find it difficult to raise funds. Selling shares in a closely held corporation is difficult. Having a share in a small company without having any real say in managing operations probably offers little assurance to the individual investor. Without any real assurance of dividends, there is little incentive in owning shares in a closely held corporation. Unless there is specified price for repurchase at some future date, the investor may never see any return on their investment. One use of the corporation is to split income between family members. By having a corporation, profits can be distributed to those family members who have lower tax brackets. Some care must be taken in adopting this tax planning strategy. Children who have not reached the age of majority will be taxed at their parent's marginal bracket. However, in these cases, a trust may be a suitable vehicle for holding stock to be held for a child's benefit. Another weakness of startups and closely held corporations is that, without assets, banks may require personal guarantees and collateral for loans to companies with a brief track record.[52]

### *Limited liability partnerships*

In Canada and the US, the limited liability partnership (LLP) is a business form well suited to real estate development. Like the corporation, it can raise money from investors (limited partners). As limited partners, they enjoy limited liability unless there is misconduct by the general partners. The general partners look after the management of the enterprise, and share in the profits and losses of the partnership. Unlike a corporation, LLPs are not subject to double taxation. Income and losses in LLPs will pass directly to the limited partners and will be taxed on each investor's individual tax return. In the US, the Revised Uniform Partnership Act provides the basis for creating LLPs in most states. Limited liability partnerships are commonly used in Canada and US to hold real estate investment.[53] Generating income offset by depreciation, these investments can be particularly attractive to those in upper income brackets. Past abuses of tax shelters have been barred by the Internal Revenue Service (IRS) and Canada Revenue Agency (CRA). In the past, the use of LLPs to generate significant losses to shelter personal income from other sources, but this has been curtailed by the "passive loss" rules (TRA86 in section 469 of the Internal Revenue Code). With this change in the law, losses generated from a real estate LLP cannot be used to offset income from other sources. Also, the requirements to file an alternate minimum tax (AMT) limits an individual paying no tax through the use of deductions that produce significant losses and tax credits.[54]

### *Limited liability companies*

Like a corporation, investors in a limited liability company (LLC) have limited liability and are not personally liable for losses incurred by management. Unlike corporations, the transfer of shares in an LLC cannot occur without the consent of the other investors.[55]

### *S corporations*

Beginning in 1958, the IRS began allowing shareholders in an S corporation to take income and losses on their personal return. The number of shareholders allowed under an S corporation is limited to 100. Other restrictions include being constrained to one class of stock, while also limiting ownership to non-resident aliens, financial institutions, insurance companies, and domestic international sales corporations.[56]

### *Real estate investment trusts*

In 1960, with the enactment of the Real Estate Investment Trust Act, REITs became a vehicle for the average investor interested in acquiring real estate as in investment. Regulated by the SEC, a REIT allows an individual to acquire a share in the assets of a portfolio of properties. Under a REIT, income and capital gains that are distributed to the investor are not taxed at the corporate level. Other requirements include a minimum number of shareholders (100) and that five or fewer shareholders own no more than 50% of the shares. Also, 75% of the income from a REIT must come from real estate investment activity. Finally, 90% of all net income in a REIT must be paid out to the investors.[57]

## Real estate transactions, agents, and brokers

A real estate agent is often used in the purchase or lease of a property. When another party is authorized to facilitate and act on behalf of the buyer, a relationship of agency is established. Creating an agent can be done by either oral or written agreement. State and provincial statutes govern the appointment of agencies in real estate transactions. A broker is an agent who will receive a commission for their efforts in negotiating and completing a real estate transaction. Individuals, partnerships, and corporations may also act as brokers. All real estate brokers are licensed by the state or province in which they do business. The National Association of Realtors (NAR) in the US and Canadian Real Estate Association (CREA) are trade associations that represent their members on issues of governmental regulations, while promoting professionalism among their membership.[58] Becoming a realtor requires the completion of a training program, the passing of a state or provincial administered exam, and a successful background check.

The signing of a listing agreement establishes the relationship with an agent. The agreement will specify the "listing", or property, the compensation or commission, and the term of the contract. Agency can be an exclusive, right to sell or open.

- With an exclusive listing, the seller will contract with one broker to sell the property, but will reserve the right to sell the property themselves. Under this arrangement, if the owner sells the property, the agent will not receive a commission.
- Under an exclusive–right to sell listing, the broker will receive a commission even if the seller finds a buyer on their own. This relationship ensures that the broker will receive a commission even if another broker should become involved in the sale.[59]
- Under an open listing, the seller is represented by a number of brokers. Only the broker who arranges the sale will receive the commission.
- Under a multiple listing, brokers participate in a pool of exclusive listings. When the sale is completed, the commissions are divided among those negotiating the sale. Multiple listings services are common in the US and Canada.[60]

As the agent, the broker has fiduciary responsibilities to the clients. Built on a relationship of trust, brokers are expected to act in the buyer's or the seller's best interests. As a fiduciary, the broker is expected to work on behalf of their client with professionalism and in strict confidence. The client relies on the broker to disclose all relevant information pertaining to the sale of the property, to obey the law, to follow instructions, to act in a manner that is not to the detriment of the client, and to provide an accurate accounting of all transactions.[61] Agents are expected to provide full disclosure and advise clients in a candid and honest manner. Those who plan to purchase or sell a property will seek the advice of their realtor on a range of issues. As professionals, they are expected to be knowledgeable about the real estate market. They should be able to evaluate and assess the market value of properties. On behalf of the client, they should advise on and disclose any issues or deficiencies that could impact the sale or purchase of a property. Agents should also alert the buyer to issues of compliance with local, state, provincial, and federal regulations. They should always work on the behalf of the client's best interest in negotiating the sale. All offers should be presented to the client in a timely manner. They will also assist the client in drawing up any sales contract and hold any deposits received in trust. In addition, they will advise the client when legal expertise is required.

One of the key responsibilities of the agent is to prepare a "contract of purchase and sale". Every state and province provides standard contract forms that are used in these transactions. In preparing these offers, it is important that all terms and conditions noted in the offer are accurate. Names of all parties, location, terms, dates, price (including deposits), and the date on which the agreement will expire must be included as part of the offer. If there are conditions, including a building inspection or legal review, financing these conditions should also be noted as part of the contract. The realtor should be available to answer any questions during the process of selling or buying a property. After all conditions have been met, the sale is completed. Finally, the agent should help facilitate the closing of the property sale.[62]

***Mortgages***

Though it is possible to purchase a property for cash, financing is generally key in most real estate transactions. In securing financing, a mortgage loan is often used to complete the purchase. Mortgages are loans under which the property is pledged as security. Technically, the mortgage is a lien. In the event that the mortgagor (the party assuming the loan) defaults on the loan repayments, the mortgagee (the lender) has the right to force the sale of the property (foreclose) to repay the debt.[63] Other grounds for foreclosure include non-payment of taxes, failure to maintain the property, and failure to hold insurance. Though breach of any mortgage condition (covenant) could give the mortgagee the right to foreclose, the process of foreclosing is not simple – nor should it be. Missing one payment should not give the mortgagee the right to foreclose on a property. Each state in the US and each province in Canada provides protection to the property. Over half the states in the US provide the right to redeem a property after a foreclosure.

Mortgage documents, like sales agreements, will list the name(s) of the borrower(s), the legal description of the property, the amount borrowed, interest rates, and terms of repayment.[64] Usury laws will restrict the mortgages from charging excessive interest rates. Interest rates may be fixed or variable. Variable rates adjust periodically, generally on a monthly basis. The rate is often fixed to a government Treasury rate plus some predetermined increment. It is common for mortgages to be amortized over their life. For a fully amortized mortgage, each payment represents the interest on the balance plus a portion of the principal. With level payments throughout the mortgage, the borrower will pay proportionally more in interest at the beginning than at the end of the loan. With straight-line amortized loans, the borrower's actual payment varies. For each payment, the amount towards principal is fixed so that, at the beginning of the loan period, interest plus principal will be a larger amount than at the end of mortgage.

With graduate payment mortgages, the borrower's payment actually increases over time. The logic is that, over time, the borrower will earn more as they advance in their career. Partially amortized loans will require an additional principal payment during the course of repayment. This additional "balloon payment" is the balance due before the final payment would have been paid. The borrower who has a 30-year amortization with a balloon payment due in the 15th year is paying off the balance owed in the 15th year. Negative amortization loans apply a negative amortization. Under this case, the borrower does not cover the full interest due with each payment. Over time, the amount owed actually increases, since the interest owed will also accrue.

It is possible to have an interest-only loan. The advantage is a smaller payment than an amortized mortgagee. The disadvantage is that if real estate values do not increase, the owner's equity will not increase in the property. If property values should fall, then the owner may have little incentive to continue making payments, given that the value is less than loan amount.[65]

Mortgages can also be closed or open. When they are open, there is no penalty for early termination. However, for closed mortgages, if the property is sold before the end of term, the mortgagee can charge a penalty – in effect, recovering a portion of the interest the bank would have earned if the mortgage had gone to full term.

The process of underwriting a mortgage by a financial institution will require an appraisal of the property and the submission of financial statements, including income verification for a private home. When the financial institution takes on the full risk of the mortgage, it is considered a conventional loan. With insured or guaranteed loans, a third party will guarantee repayment. In the US, a mortgage can be insured by the Federal Housing Administration (FHA) and the Veterans' Administration (VA). Able to do so under an Act of Congress, the FHA has in the past guaranteed loans up to 97% of the appraised value of the home. A number of options are available for those purchasing their first home, as well as for those who are refinancing. Like all insurance products, there is an additional cost charged for an insured mortgage. It is even possible under the FHA's Energy Efficient Mortgage Program to borrow an amount above what is needed to purchase the home in order to make cost-effective energy improvements that will ultimately lower monthly utility bills. Similarly to the FHA, VA insured mortgages have provided financing to veterans since 1944.[66] In Canada, Canada Mortgage and Housing Corporation (CMHC) provides insured mortgages to home owners. For those looking for assistance financing a small business, the Small Business Administration (SBA) in the US provides guaranteed loans for purchasing land and buildings, making improvements, acquiring machinery, and modernizing facilities.[67]

### *Secondary market*

In underwriting mortgages, the originators are responsible for qualifying the borrower. After the loan has been processed and approved, these loans are often sold off to other investors. When mortgages are pooled together, they can be sold on the secondary market. Creating a market for mortgages is the responsibility of the Federal National Mortgage Association (Fannie Mae), Government National Mortgage Association (Ginnie Mae), and the Federal Home Loan Mortgage Corporation (Freddie Mac) in the US. By buying mortgages, the FHA and VA serve as a source of capital to lenders.[68] In Canada, CMHC will also securitize these mortgages, creating a market for insured loans. Creating a market for these loans relies on the use of mortgage-backed securities (MBSs). By purchasing the mortgages to be sold to investors, the government is expanding credit and providing banks with the incentive to make loans to lower-income households. However, this policy resulted in the financial crisis of 2007–08, when loans were made to individuals who were not creditworthy. Banks and financial institutions knowingly made loans to individuals who would never be in a position to pay back the mortgages. Many properties in cities across Florida, New Mexico, Nevada, and Arizona were purchased as speculative investments. With rapidly escalating home prices, many of these property owners bought with the intention of turning a quick profit. Banks, making most of their profit on origination fees, used mortgages with teaser rates to entice individuals who could barely meet minimum underwriting requirements to purchase properties. After the initial fixed period of a year or two, interest rates on these mortgages became variable, rising significantly and increasing the monthly payments for individuals, many whom were on tight budgets. Some mortgages were interest only, so home owners were not in a position to build up any equity until these loans were converted to a more conventional mortgage. Many of the financial institutions writing these subprime loans had failed to conduct proper due diligence. These MBSs in effect contained varying proportions of subprime mortgages (tranches). In many cases, these tranches could be of very low quality, placed inside an MBS with an investment-grade rating. Unfortunately, these high-risk investments had been purchased by investors seeking high yields and low risk, not understanding that there was a time bomb buried deep within.

In April 2007, the bubble burst, and many of the subprime lenders collapsed, including New Century Financial, leaving Fannie Mae and Freddie Mac, who had issued the debt, on the hook for the all of this bad debt. A bubble in selective real estate markets initiated a price decline, putting many home owners "underwater". With no hope of ever getting out from under a rapidly declining real estate market, many of these borrowers saw their properties end up in foreclosure. For others, the escalating payment schedules under these subprime mortgages became a weight too heavy to bear. With properties worth less than the mortgages, refinancing was not an option, forcing more foreclosures. Lending is based on credit. When markets are responsible, debt instruments are key to an expanding economy. However, in the crisis of 2007–08, the absence of prudent lending practices resulted in a meltdown in the credit markets, forcing the federal government to take drastic action to restore confidence.[69]

With the collapse of the subprime market, under President Obama's Administration, Congress was forced to stem the tide of falling values and a collapsing real estate investment market by lowering short-term rates in 2009 (Figures 4.7 and 5.7). The FHA was also allowed to increase the size of mortgages that could be insured. The US Treasury also began buying many of the MBS investments. Though, eventually, these action would stem the tide, it was not until 2012 that markets began to stabilize. Ultimately, the banks would have to take writedowns in excess of $600 billion.

## Construction loans

When development projects involve new construction, a construction loan will be needed to cover the costs of building. Commercial banks generally make loans to developers for a period of six months to two years to cover the cost of construction. Once the building is completed, construction loans are usually converted to a mortgage or sold to another party. These "construction to permanent" arrangements provide assurances that there will be financing once the project is completed. With stand-alone construction loans, it is possible that, during the period of construction, rates may go up or if the developer, for whatever reason, has difficulties qualifying

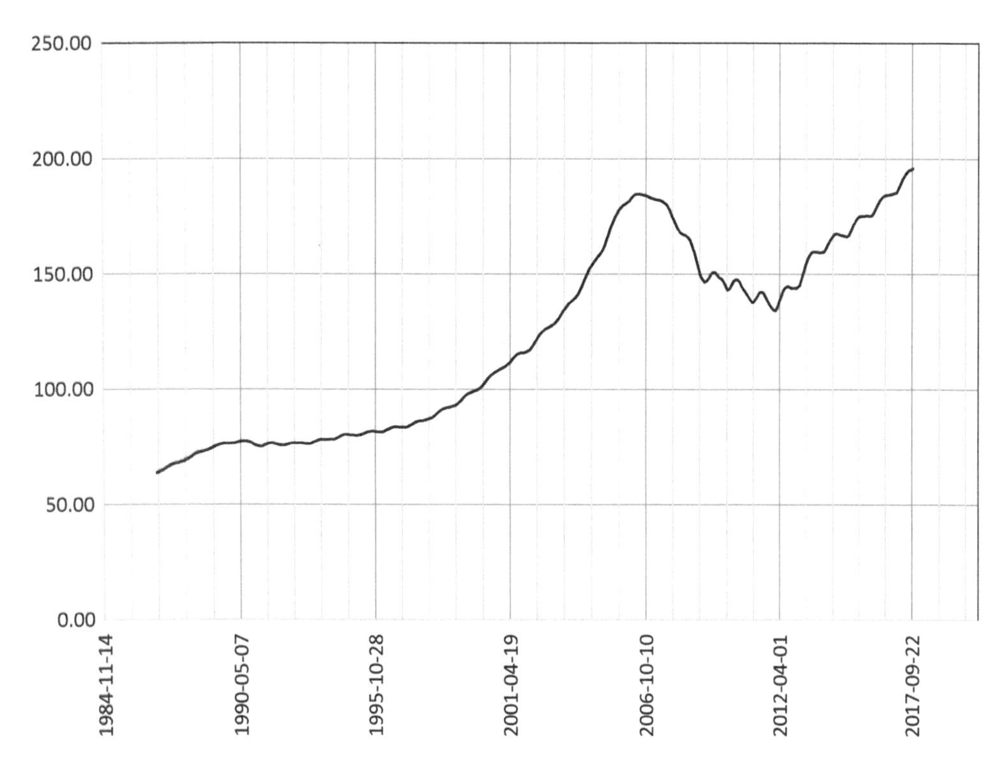


*Figure 4.7* S&P Case-Shiller US national home prices index, 1987–2017 (Jan 2000 = 100) (not seasonally adjusted).

Data Source: Archive, Federal Reserve Economic Data, Federal Reserve Bank of St. Louis.

for a mortgage, then the future of the project could be at risk. To reduce risk, lenders will cover only a certain portion of the costs. The cost of purchasing the land and all associated costs, including legal costs, tile insurance, and commissions, are the developer's equity (borrower's equity) and are not included as part of a construction loan. Hard costs represent those costs associated with construction, and will include equipment, materials, and labor. It may also include a portion of the site preparation costs and the cost of building out the space for the tenants. These costs can include all furniture and fixtures, and communications infrastructure. The soft costs, including fees paid to the planner, architect, and engineering and other consultants, may also be part of the loan amount. It is also possible to include management fees, insurance, and fees paid to governmental authorities as part of the soft loan.[70]

Most construction covers 75% or less of the hard and soft costs. When the builder's equity in the project is insufficient to meet the test, there may also be additional collateral requirements. During the construction loan period, the borrower will be paying monthly interest charges. An interest reserve account, which is funded by the actual construction loan, provides the funds needed to make the interest payment charges. For significant projects, lenders will often be working with a schedule of disbursements. In this way, the lender will know when interest will be paid on the

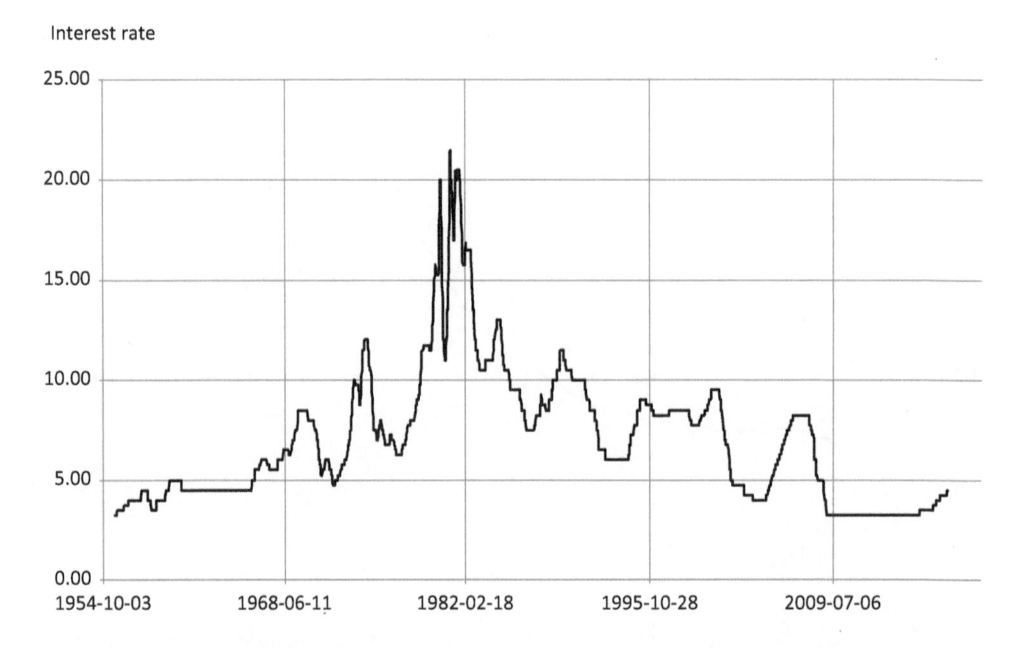


*Figure 5.7* US prime rate, 1955–2018.

Data Source: Federal Reserve, Economic Research.

loan. At each stage, payment is made for work that has been completed. Verification is a requirement for payment at each stage in the project. The lender will also hold back a percentage until the project is completed. This assures the lender that there will be a contingency fund in the event of unforeseen expenses.

In underwriting a construction loan, the lender will need assurances that the project, once completed, will be successful. Extensive review of all financial, legal, and design documents will be part of the underwriting process. The creditworthiness of the project and the track record of both the developer and their general contractor will be reviewed. The financials of the project should also assure the lender that if, for whatever reason, the developer is no longer part of the picture, there will be guarantees that the project will be financially successful. In addition to having a reserve fund, lien wavers, and subordinations, there must be a guarantee that, in the event of a bankruptcy, the lender will be the first to be paid. There may also be assurances in the form of agreements with the architect and engineer that they will complete the project, while surety bonds will provide some financial guarantees. Finally, before a construction loan is finalized, the lender will want to review plans and specifications, building permits, and zoning permits. The developer must also show that there are necessary utilities for the proposed development. When required rights of ways and easements are needed to complete the project, the necessary documentation will have to be submitted to the lender before it will approve the actual loan commitment.[71]

## Real estate closings, land title, and registration

### *Transfer of property*

After the seller approves the purchase agreement and the buyer secures a mortgage, the buyer is in a position to take title of the property. One of the terms noted on the purchase agreement is the date of closing. Though the process of each jurisdiction may vary, at the conclusion of a real estate closing the new owner is in possession of the property. The buyer(s), seller(s), and their attorney(s) will have a number of documents and reports in their possession to review prior to the closing. The new owners will want proof that there are no liens on the property. There should be some assurance that all real estate taxes and fees have been paid and that the property owner is not in arrears. This can include settlement of fire, water, and sewer district taxes. The title abstract also should have been available to all prior to closing.

#### *Property deeds*

One requirement at closing is the transfer of the property deed from the seller to buyer. Signing over the deed does not guarantee that the buyer has title to the property. Having done a title search, a buyer will have information on the transfer of ownership – often back to the original land grant. The goal of this search is to document the chain of title. Title searches will reveal encumbrances recorded against the property. For a fee, title search companies will conduct a search by examining the records kept at the county courthouse or registrar of deeds. Once completed, the actual title abstract or report is then available to the buyer's attorney to review prior to the closing. In many states and provinces, title insurance can be purchased to protect the buyer's interest in a property.

The actual deed is a written contract that is the means of conveyance of ownership. All deeds contain the names of the grantor and grantee, a legal description of the property, and any restrictions placed on the grantor.[72] The deed may take several forms. In warranty deeds, which are commonly used in conveying real estate, the seller provides the following warranties.

- The seller has good title and the right to convey the interest of the property to the buyer.
- The title is not encumbered by any easements, restrictions, unpaid property taxes, mortgages, judgments, and other assessments.
- No other owner has rights that supersede those of the buyer (grantee). In the event that such an owner emerges, the grantor is liable for all expenses associated with defending the title. If another party can prove ownership in the future, the grantor may be sued for an amount equal to the purchase price of the property.[73]

Warranty deeds are also known as "general warranty deeds". Special warranty deeds only provide assurance against title defects during the period for which the seller (grantee) owned the property.[74]

Quitclaim deeds, unlike warranty deeds, make no warranties. Only the interest of the seller (grantor) is conveyed to the buyer, which presents a greater risk to the buyer than a warranty deed. With a warranty deed, the buyer assumes any future liability if another owner should make a claim. That said, quickclaim deeds can be used to remedy both a defect and cloud on the title, which may be the product of an inheritance, dower, or community property right or redemption forced by a foreclosure sale. Quickclaim deeds can also be sued to create or remove an easement.[75]

Bargain and fiduciary deeds convey only title and make no warranty concerning title. This type of deed is used to transfer title in situations in which properties are transferred from one party to another in administrating an estate or as a sale of a foreclosed property.[76]

In transferring title, there may be a requirement for a survey by a registered surveyor. The plat or real property report will show the boundaries of the property and the location of any structures. Sometimes, an updated survey is sufficient and, in some cases, all that is required is a copy of the existing plat (real property report). Any doubts concerning property boundaries should be set aside by a survey. Ultimately, with the transfer of the deed from seller to buyer, the new owner has taken ownership of the property.

The buyer's lender will want to see proof of insurances. The lender will also want to see proof of title insurance. For loans greater than 80% of the assessed value, there may be a general requirement for mortgage insurance. At closing, a "purchaser's statement" and "vendor's statement" will have been prepared showing the financial details of the transaction. These statements show the payments as receipts made by the buyer and seller. Included in these statements are the purchase price, deposits paid, value of any mortgage(s), taxes, legal fees, mortgage insurance, title insurance, warranty fees, transfer taxes, registration fees, prepayment of interest, origination fees, and real estate commissions. For home owners in the US since 2010, the Consumer Financial Protection Bureau has required the completion of a standard settlement form (its Closing Disclosure Form, after October 3, 2015) as part of a "Good Faith Estimate and Truth in Lending" closing disclosure that itemizes all receipts and disbursements.[77] In some jurisdiction, the buyer's funds are held by an escrow agent until the closing is completed, ensuring that all necessary documents are executed prior to the transfer of the property.[78]

At the actual closing, the buyer's agent and attorney will be present. Depending on the jurisdiction, the seller's attorney may be present. Once the transaction is completed, the new owner will receive keys or security codes giving them access to the property. Prior to the actual closing, a walkthrough allows the new owners to inspect the present condition of the property. Registration of the deed may be required before the new owners will be given access to the property.

### *Land registration*

Without the registration of land title, ownership would be difficult to determine. Without this assurance, making improvements to the land is risky. In many parts of the world, land title and registration reform are seriously needed. For example, in many African countries, establishing a land registration system that is fair and affordable among its citizens has been difficult. The conventional approach, first implemented by Europeans in the 19th century, has not been available and thus there has not been the assurance to land holders found in the industrialized world. Without this assurance, it is difficult for landowners to obtain mortgages on properties, greatly limiting capital investment in this part of the world.[79]

Land registration can be traced to ancient times. Land sales and registration appeared in Mesopotamia as early as 2500BC. In ancient Egypt, land records were essential for determining the taxes paid by landholders of agricultural lands.[80] During the Middle Ages, charters showing land ownership were conveyed by a document or deed that would be signed by the grantor. Sealed and witnessed, these documents would be stored in safe place to prove ownership. For the Crown, land ownership was essential to taxation. The Domesday Book of 1086, ordered by William the Conqueror, is a survey of landholdings "by his foresight … . Surveyed so carefully that there was not a hide of land in England of which he did not know who held and how much it is worth".[81] To guard against future false claims, copies of deeds would be made by clerks and stored at separate locations. A technique of tearing the original deed with the word "chirographum" located along the tear line could be used to guard against forgeries. By storing each portion in separate locations, if fraudulent documents were to surface, the two original pieces could be retrieved and matched along the tear line to prove authenticity.[82]

In 1529, Holland created a system of land registration that afforded assurances to those owning property. This new system is credited for increasing property values and making it faster to arrange for mortgages. In England and Wales, it was not until the enactment of the 1677 Statute of Fraud that written deeds were a requirement for land ownership, though it was recognized a century earlier that land registration would have significant benefits. Before 1862, only local registries existed for land registration. With the passage of the Land Registry Act of 1862, registration of land title became statutory. Unfortunately, mismanagement after the enactment of the 1862 Act led to its amendment in 1875 and the Land Transfer Act 1875, which formed the basis of the system now used in England and Wales.[83]

In the US and Canada, registration of time and date of property deeds in the local courthouse completes the transaction. The public record of land sales establishes with a third party the existence of the transaction. The actual process or recording of deeds will differ according to each state's statute. Prior to statutes, common law prevails, in that "first in time, first in right". Recording statutes in each state will determine who has actual right to the property. Under a "notice statute", the owner with the date of a good faith (*bona fide*) transaction determines ownership, while under a "race notice statute", title is determined by date of a good faith purchase. For two states, Louisiana and North Carolina, the only question establishing purchaser is who recorded their deed first.[84]

In the western Canadian provinces of Alberta, Manitoba, and the Northwest Territories, in addition to a few states in the US, the Torrens system is used to register property. Developed by Robert Torrens, a premier of South Australia, it was created as a system that would simplify land registration. The passage of the Australian Real Property Act of 1858 established the Torrens system. Under the Torrens system, the "register of titles" issues "a certificate of title" each time the property is transferred. Under this arrangement, the Crown would cancel the old title and issue a new title. No chain of title is prepared as part of the sale. Under this "curtain principle", no interests from previous owners are of concern. Should there be an error, the government provides compensation from an assurance fund.[85]

## Leases

For investment properties, leases are critical to success of the rental property. They dictate rents, and the rights and responsibilities of the tenant and landlord. Leases are, in legal terms, contracts that convey the right of possession from the property owner to the tenant. The tenant (lessee) has the right to occupy the property during the lease period, after which ownerships reverts back to the landlord (lessor). While the lease is in effect, the landlord has a "reversionary interest" in the property. Binding leases all will include:

- the name of the lessee and lessor;
- the address of the property;
- agreement to grant possession of the property to the lessee;
- the rent and payment period;
- the amount of the security deposit;
- the start and end dates of the lease;
- any cause for termination; and
- the signatures of both parties.

Though leases may be oral, most states and provinces require a written lease for a period of a year or longer. Other tenant responsibilities noted in the lease, including maintenance, will appear as an addendum to the lease. Leases should also note how the tenant will be notified if access to the property is needed. Provisions to sublease all or part of the space should also be clearly outlined in the lease. This provision will be important to those commercial tenants who may want to sublease space if they ever feel the need to downsize their operations.[86] Tenant rights not outlined in the lease may be safeguarded by state and provincial statutes. These rights will include notifications of rent increases, security deposits, landlord's access to the property, requirements that the landlord maintain the property and comply with local zoning and health and safety requirements. Leases should also provide instruction on how communication will be established between tenant and landlord. The lease may also note the penalties for late payment of rent. When tenants are responsible for maintenance, leases will include a list all responsibilities, such as snow removal, repairs, lawn care, and general maintenance of the property. When these responsibilities are not met, the lease should state how these costs will be charged to the tenant. Security deposits, which are often limited to one month's rent, will rarely cover damage and costs accrued by irresponsible tenants.

### *Rents*

Rents on residential contracts are usually fixed for a period of year, with payments generally made monthly. For long-term leases, rents can change. A graduated rent would allow the tenant to start with a lower rent, then see small increases over a period of a few years. When the term of the lease is longer, rent increases can be linked to an index that would be fair to both parties. The consumer price index (CPI) is one index that could be used for this purpose. One issue with the CPI is that it can be biased in favor or one component, such as energy costs, which might favor the landlord where rents have not increased as much as the index.

A participation clause allows the landlord to increase the rent based on an increase in operating costs. To reduce the landlord's risk of having to pay for future increases in taxes, utilities or maintenance, a triple-net lease is a favored arrangement. Under this alternative, the tenant pays for all real estate taxes, insurances, repairs, and utilities. For retail tenants, participation could also include some percentage of gross receipts. Traditionally, agricultural rents were based on a percentage of the output; today, gross sales receipts would be a more likely formula. The logic is that if times are bad, the landlord shares in the downturn, but if there are substantial profits, the landlord shares in the success of the enterprise. In large shopping centers, participating leases also ensures that the landlord has an interest in the success of the tenant, and will more likely uphold their end of bargain by maintaining and making improvements to the site, building, and common areas. Landlords may also make concessions to the tenant as part of signing a lease. Clearly, providing a reduction in the rent for a few months might be an incentive to sign a long-term lease. For commercial and retail tenants, providing a subsidy (a tenant improvement allowance) for building out the space can help a new tenant with the costs of establishing their business.

Establishing the rent for a lease with a duration of 25–99 years is often difficult for both tenant and landlord, given the difficulty of predicting the impact that inflation will have on prices. One solution is to separate the lease for the building from that for the land. Where the tenant plans to construct a building, it is possible to put in place a land lease, with the construction of the building the responsibility of the tenant. Rental increases on the land lease would then be based on an economic index or appraisal.[87]

The rents required by a landlord are not always determined by the marketplace. In some cities, landlords may not have the right to charge the true market rate. In larger metropolitan areas, such as New York, San Francisco, and Washington, DC, rent control laws establish the maximum annual increase in rents. Though establishing maximum increases would at first seem fair, by preventing landlords from practices of rent gauging, ultimately, restricting rents can have a negative impact on the rental market. When rents do not increase enough to cover maintenance and operating expense, properties will deteriorate over time, because landlords will no longer have the cash flow to maintain the property. With low rents and negative profit margins, capital investment in new developments is unlikely, further aggravating housing shortages. Public policy must find a balance between the two extremes if a healthy market is to be sustained.[88]

### *Eviction*

Leases also contain provisions in the event that tenants are late or stop paying rent. Under the terms of the lease, the landlord will have the right to "actual eviction", a legal proceeding that can terminate in a court authorizing a sheriff or marshal to remove the tenant. Landlords also have the right to evict when the tenant has violated a provision of the lease. This could include having individuals live in the property not noted on the lease, or having a dog or cat when none is permitted. When a landlord no longer maintains the premise to meet minimum health and safety requirements, the tenant may use "constructive eviction" to effectively terminate the lease. Under "constructive eviction", the tenant may move out of the property and stop paying rent. The landlord must either fix the problems or terminate the lease.[89]

Eviction, in most jurisdictions, is not a simple matter.[90] In most jurisdictions, getting a date for the hearing can take weeks, if not months. Any delays can result in additional legal fees and expenses. In the interim, tenants will have possession of their apartment while the landlord waits for their day in court. If the tenant has no intention of paying rent or even utilities (if required), the tenant can live rent free for the foreseeable future. During that time period, tenants can become disgruntled by the ongoing legal proceedings and may cause considerable damage to property. Even after an eviction, the process of restitution may be all but impossible. Tenants may move to another state or province. Filing suit in the state or province where the tenant now resides may be difficult if no forwarding address was left. If the landlord proceeds with an action, additional legal fees and expenses will certainly accrue. Even if a judgment is granted by the court, actually receiving restitution may be difficult – especially when tenants are changing jobs, frequently making the process of garnishing wages difficult. The message is clear: tenant selection is critical to the success of a residential real estate. Renting only to tenants with good credit and credible references can help ensure that you have responsible tenants. A property manager with the experience needed to judge prospective tenants can help ensure against renting to "bad tenants".

## Construction contracts

Even for small projects, a contract will be required for building and repairing a new building. Contracts that are enforceable must meet specific basic requirements. Those that sign the contract must be legally competent. Signatures from a minor would violate this requirement. There must be mutual agreement given freely without coercion. To be legal, contracts may not be singed under duress. The contract itself must be lawful. Contracts in violation of the law would not be considered legal in court. There must be consideration, which is usually payment, but could also include land or services. Contracts should also be in writing if they are to be enforceable. Contracts should also address issues of performance and breach.[91]

Construction contracts, like all contracts, establish the work to be done, the price of the work, and when the work will be completed. Because the release of monies during the construction process is fixed to specific milestones, the writing of construction contracts must be done in close coordination with the project management plan. Though it is impossible to foresee events that could jeopardize the success of a project, a risk management plan should be created that directs the risks associated with completing a particular task to those most capable of satisfying this responsibility. Within this framework, the architect and the engineer would be responsible for the design of the building. The general contractor would then be responsible for the overall construction process, with each subcontractor responsible for their particular assigned task. To minimize the risk associated with each task, contractual indemnity provides the assurances needed to assess each phase of the building process. For example, the architects and engineers carry professional liability insurance to guard against specific failures in the design. Contractors and subcontractors will carry general liability insurance and laborer's compensation, while the owner of a building may have owner's protective insurance coverage on large projects.[92] A requirement for a bonding can also lessen the potential of failure and insure project completion. Issued by a surety company, bonds can be required by a private sector client or government. In the event that the general contractor cannot satisfy their contract, moneys will be required to pay judgments resulting from legal action.[93]

The details of any construction contract must address a few key areas. First, there is the question of payment. How will the design professionals, contractors, and subcontractors be paid? For some, payment is a lump sum. On this basis, any miscalculation in costs will be borne by the contractor. Unit price establishes a rate based on a unit cost. For those who are keenly aware of their operating cost, this option minimizes their risk. Cost plus fee, sometimes referred to as "time and materials", will place all of the risk on the client. Under this scenario, the contractor keeps records of all material and labor costs, applies an overhead factor, and sends the bill to the client. In each case, the time frame and verification of completed work is key to managing the construction process.[94]

Contracts should clearly outline the services provided under contract. For architects, the standard American Institute of Architects (AIA) contract and the Royal Architectural Institute of Canada (RAIC) contract each provide an outline of the services that will be provided during each phase of the project. Of course, a developer may want to make additions to this standard lease to meet the specific needs of the project.[95] Traditionally, the architect is responsible for supervising the construction site.[96] However, it is also possible to hire an independent "clerk of the works" as a means of verifying that a work is completed on time according to project deadlines. Alterations in the design will result in both the architect and general contractor requesting changes to the contract and the price to complete the work.

Given that there are always unforeseen events that impact the completion of a project, there may be an advantage to drafting contracts that are considered one-sided, placing all liability on the other party. However, in the event that the project fails, it will be the reputation of the developer that is harmed. Better to have in place a risk management plan that includes insurance, bonding, and money set aside in the budget for contingencies than a court battle where everyone is to blame.[97]

All contracts should be reviewed by the developer's legal team. Attorneys with expertise in construction law will be critical to the success of any development venture, but, as in life, a contract is a just a piece of paper. Successful managers avoid legal battles through skillful negotiations. In the end, getting the job done is what it is all about.

## Notes

1 Hammurabi's Code of Laws (c. 1780BC), www.duhaime.org/LawMuseum/lawarticle-105/1760-bc–hammurabis-code-of-laws.aspx.
2 Karp and Klayman, 2006, 63–64.
3 Ellickson and Tarlock, 1981, 1317.
4 Siedel et al., 2003, 5–6.
5 Ibid.
6 Gans and Kendall, 2010, 643.
7 University of Alberta, A Guide to Property Rights in Alberta, "Are property rights protected in Canadian Law", http://propertyrightsguide.ca/are-property-rights-protected-in-canadian-law/
8 Wex Legal Dictionary, www.law.cornell.edu/wex/tort: "A tort is an act or omission that gives rise to injury or harm to another and amounts to a civil wrong for which courts impose liability. In the context of torts, 'injury' describes the invasion of any legal right, whereas 'harm' describes a loss or detriment in fact that an individual suffers."
9 Siedel et al., op. cit., 2–10.
10 Ibid., 2–19.
11 Ellickson and Tarlock, op. cit., 1315.
12 Karp and Klayman, op. cit., 21–23.
13 Ibid; Hinkel, 2005, 7.
14 *Real Estate Trading Services*, 2011, ch. 1.
15 Siedel et al., op. cit., 101–107.
16 Covenants: Karp and Klayman, op. cit., 359–360; Hinkel, op. cit., 81–82; Benson and Bowden, 1997, 119–121; Ellickson, op. cit.; see Jacobus for Fair Housing Act 1968.
17 Hinkel, op. cit., 5–6.
18 Siedel et al., op. cit., 28–31.
19 Karp and Klayman, op. cit., 38–39.
20 Siedel et al., op. cit., 64; Karp and Klayman, op. cit., 42–45.
21 Benson and Bowden, op. cit., 127; Christensen and Lintner, 2007.
22 Karp and Klayman, op. cit., 40–41.
23 Ibid., 47; Siedel et al., op. cit., 58–59.
24 Benson, op. cit., 125–126.

25 Siedel et al., op. cit., 59–60.
26 FAA News, Summary of small unmanned aircraft rule (Part 107), June 21, 2016.
27 Canada Transports guidelines, www.tc.gc.ca/eng/civilaviation/opssvs/flying-drone-safely-legally.html#regulations
28 Siedel et al., op. cit., 53–58.
29 Benevolo, 112–118.
30 Kontokoska, 2013, 190–198.
31 Siedel et al., op. cit., 56–59.
32 Ibid.; Baker, 1975, 131–158.
33 Karp and Klayman, op. cit., 63.
34 Siedel et al., op. cit., 123.
35 Benson and Bowden, op. cit., 84; Siedel et al., op. cit., 123; Karp and Klayman, op. cit., 66–67.
36 Karp and Klayman, op. cit., 67–73; Seidel et al., op. cit., 126–127.
37 Karp and Klayman, op. cit., 79–87.
38 Ibid., 80–81.
39 Siedel et al., op. cit., 132.
40 Karp and Klayman, op. cit., 81–82; Benson and Bowden, op. cit., 103–105.
41 Siedel et al., op. cit., 134; Benson and Bowden, op. cit., 103.
42 Siedel et al., op. cit., 138–139; Karp and Klayman, op. cit., 85–86.
43 Jacobus, op. cit., 63.
44 Siedel et al., op. cit., 503–505; Karp and Klayman, op. cit., 101–103.
45 Siedel et al., op. cit., 507–508.
46 A History of Housing Cooperatives National Cooperative Law Center, http://nationalcooperativelawcenter.com/national-cooperative-law-center/the-history-of-housing-cooperatives/5/
47 Siedel et al., op. cit., 139–140; Gerstell, "Administrative adaptability: The Dutch East India Company and its rise to power." *Journal of Political Economy* 99, no. 6 (1991): 1307.
48 Karp and Klayman, op. cit., 88–91.
49 https://en.wikipedia.org/wiki/Uniform_Partnership_Act; Siedel et al., op. cit., 140–141.
50 Siedel et al., op. cit.,op.cit., 143.
51 www.inc.com/encyclopedia/joint-ventures.html
52 Karp and Klayman, op. cit., 88–89; Siedel et al., op. cit., 142–143; Jacobus, op. cit., 67.
53 Siedel et al., op. cit., 142–143; Jacobus, op. cit., 66–67; Karp and Klayman, op. cit., 89–90.
54 Samwick, 1996; Samwick tax shelter; IRS, https://apps.irs.gov/app/amt2015/assistant/process?execution=e4s1
55 Siedel et al., op. cit., 143.
56 IRS, www.irs.gov/businesses/small-businesses-self-employed/s-corporations
57 Siedel et al., op. cit., 144–145.
58 Canadian Real Estate Association (CREA), www.crea.ca/about/organization/; National Association of Realtors (NAR), www.nar.realtor/
59 Hinkel, op. cit., 136–137; Karp and Klayman, op. cit., 253–270.
60 Karp and Klayman, op. cit., 257.
61 Ibid., 259–260; Hinkel, op. cit.,127–128.
62 *Real Estate Trading Services*, 2011, §25.4.
63 Siedel et al., op. cit., 283–284.
64 Karp and Klayman, op. cit., 467.
65 Hinkel, op. cit., 256–259.
66 Ibid., 252–253; US Department of Housing and Urban Development (HUD), Energy Efficient Mortgage Program, www.hud.gov/program_offices/housing/sfh/eem/eemhog96, www.hud.gov/program_offices/housing/sfh/eem/energy-r
67 Hinkel, op. cit., 253; US Small Business Administration, www.sba.gov/
68 Siedel et al., op. cit., 295–296.
69 Federal Reserve, History, www.federalreservehistory.org/Events/DetailView/55; Barnes, 2009.
70 Yustein, 2014.
71 Weissman, 2007.
72 Siedel et al., op. cit., 78–79.
73 Jacobus, op. cit., 77; Siedel, op. cit., 332–333.

74 Siedel et al., op. cit., 332.
75 Ibid., 333; Jacobus, op. cit., 83–84.
76 Karp and Klayman, op. cit., 349–350.
77 www.consumerfinance.gov/
78 Siedel et al., op. cit., 343–350.
79 Toulmin, 2006.
80 Ellickson and Tarlock, op. cit., 1377.
81 Mayer and Pemberton, 2010, 3.
82 Hinkel, op. cit., 213–214.
83 Jacobus, op. cit., 94.
84 Siedel et al., op. cit., 246–247, 244–245; US Land Records, www.uslandrecords.com/uslr/UslrApp/index.jsp
85 Benson and Bowden, op. cit., 142–143; Siedel et al., op. cit., 251–252.
86 Siedel et al., op. cit., 375–420; Jacobus, op. cit., 294–312; Karp and Klayman, op. cit., 147–167.
87 Jacobus, op. cit., 299–300.
88 Ibid., 303–304.
89 Ibid., 301–302.
90 Ibid., 304–305.
91 Ibid., 112–124.
92 Bobotek, 2010.
93 Miles, op. cit., 426–427.
94 Epstein, 2004.
95 Miles, op. cit., 417–419.
96 Ibid., 417–419.
97 Epstein, op. cit.

## Bibliography

Baker, Frederick M. "Development Rights Transfer and Landmarks Preservation--Providing a Sense of Orientation." *Urb. L. Ann* 9, (1975): 131–158.

Benevolo, Leonardo. *History of Modern Architecture, the Tradition of Modern Architecture, Volume I*. Cambridge, MA: The MIT Press, 1989.

Benson, Marjorie Lynn and Marie-Ann Bowden. *Understanding Property: A Guide to Canada's Property Law*. Scarborough, ON: Carswell, 1997.

Bobotek, James P. "Construction risk management: Ten issues in construction contracts." *Lexis News Room*, May 13, 2010, www.lexisnexis.com/legalnewsroom/insurance/b/insurance-law-blog/posts/construction-risk-management-ten-issues-in-construction-contracts

Christensen, Randy, and Anastasia M. Lintner. "Trading our common heritage? The debate over water rights transfers in Canada." *Eau Canada: The Future of Canada's Water* (2007): 219–244.

Duca, John V. and Danielle DiMartino. "The rise and fall of subprime mortgages." *Economic Letter, Insights from the Federal Reserve Bank of Dallas, Dallas, Texas* 2, no. 11 (2007): 1–8.

Ellickson, Robert C. "Property in land." *The Yale Law Journal* 102, (1993): 1315–1400.

Ellickson, Robert C. and A. Dan Tarlock. *Land-Use Controls, Cases and Materials*. Boston, MA: Little, Brown and Company, 1981.

Epstein, Robert. "How construction contracts cause litigation." *Alert, Greenberg Traurig* (2004), https://naiopnj.org/resources/Documents/229161085_v%201_RCE%20Article%20–%20How%20Construction%20Contracts%20Cause%20Litigation%20%20%20.pdf

Gans, David H. and Douglas T. Kendall. "A capitalist joker: The strange origins, disturbing past, and uncertain future of corporate personhood in American Law." *The John Marshall Law Review* 44, (2010): 643.

Gerstell, Daniel. "Administrative adaptability: The Dutch East India Company and its rise to power." *Journal of Political Economy* 99, no. 6 (1991): 1307.

Hauben, Jay. "Early history of the Amalgamated housing cooperative, 1926–1932," www.columbia.edu/~hauben/amalgamated/history/Amal-1926–1940.pdf

Hinkel, Daniel F. *Practical Real Estate Law*, 5th ed. Clifton Park, NY: Delmar Cengage Learning, 2005.

Jacobus, Charles J. *Real Estate Principles*, 10th ed. Mason, OH: South-Western Publishing Co., 2006.

Kaiser, Edward J., David R. Godshalk, and F. Stuart Chapin, Jr. *Urban Land Use Planning*, 4th ed. Chicago, IL: University of Illinois Press, 1995.

Karp, James and Elliot Klayman. *Real Estate Law*, 6th ed. Chicago, IL: Dearborn Real Estate Education co, 2006.

Kontokosta, C. E. "Tall buildings and Urban expansion: Tracing the evolution of zoning in the United States." *Leadership and Management in Engineering* 13, (3), (2013): 190–198.

Kratovil, Robert. "Mortgage lender liability-construction loans." *DePaul Law Review* 38, no. 1 (fall, 1988): 43–54.

Levy, John M. *Contemporary Urban Planning*, 5th ed. Upper Saddle River, NJ: Prentice-Hall Inc., 2000.

Masinter, Paul J. and Nicholas J. Wehlen. Arbitration: the good the bad and the Ugly, New Orleans, Louisiana, originally published in the ABA Litigation Section commercial & Business Litigation Committee Newsletter, Vol 5, no. 2 (Summer, 2004).

Mayer, P. and A. Pemberton. *A Short History of Land Registration in England and Wales*. London: HM Land Registry, 2010.

Miles, Mike E., Gayle Berens and Marc A. Weiss. *Real Estate Development, Principles and Process*, 3rd ed., Washington, D.C: Urban Land Institute, 2005.

Real Estate Council of British Columbia. *Real Estate Trading Services, Licensing Course Manual*. Vancouver, BC: Sauder School of Business, University of British Columbia, 2011.

Ryan, Barnes. "The fuel that fed the subprine meltdown," Insights, *Investopedia*, updated February 26, 2009, www.investopedia.com/articles/07/subprime-overview.asp

Samwick, Andrew A. "Tax shelters and passive losses after the tax reform act of 1986." In *Empirical Foundations of Household Taxation*, Martin Feldstein and James Potebaeds. Chicago, IL: University of Chicago Press, 1996, 193–233.

Siedel, George. J., III, Robert J. Aalberts, and Janis K. Cheezem. *Real Estate Law*, 5th ed. Mason, OH: West Legal Studies, 2003.

Toulmin, Camilla. Securing land and property rights in sub-Saharan Africa: The role of local institutions, in *Africa, Improving the Investment Climate Global Competiveness Report, 2005–06*. Switzerland: World Economic Forum, 2006.

Weissmann, David A. "Construction lending from the ground up." *Law Trends & News, Real Estate* 3, no. 1 (May, 2007).

Yustein, Ross L. A primer on construction loan disbursement requests, *Law360, New York* August 12, 2014.

Ziff, Bruce. *Principles of Property Law*, 2nd ed. Scarborough, ON: Carswell, 1996.

## Web resources

American Bar Association
www.americanbar.org/aba.html

Canadian Bar Association
www.cba.org/Home

Canadian Mortgage and Housing Corporation
www.cmhc-schl.gc.ca/en/

Canadian Real Estate Association
www.crea.ca/

King, L. W., *The Code of Hammurabi*, Lillian Goldman Law Library, The Avalon Project
http://avalon.law.yale.edu/ancient/hamframe.asp
National Association of Realtors
www.nar.realtor/
National Cooperative Law Center
http://nationalcooperativelawcenter.com/
US Department of Housing and Urban Development
https://portal.hud.gov/hudportal/HUD
US Department of Housing and Urban Development, Energy Efficiency Programs
www.hud.gov/program_offices/housing/sfh/eem/eemhog96
www.hud.gov/program_offices/housing/sfh/eem/energy-r

# 8 Design and construction

## A team of experts

At some point in the development process, something will be built or renovated – although there are occasions on which little must be done after taking possession of the property. A building that has been recently constructed or is in good repair with good tenants, who are paying reasonable rents, generally demands a premium at the time of sale. But creating additional value is about buying an undervalued property, making some improvements, and ultimately increasing the cash flow out of the property. In these cases, we will need to work with a team of experts if we plan to accomplish our goals. Planners, architects, engineers, contractors, and trades, including skilled artisans, are critical to the success of any project. As the developer, you are responsible for the outcome of the project and your consulting team will look to you for leadership. Though you are not expected to be a registered architect, or a professional engineer, or a certified planner, or a skilled artisan, knowledge and experience of the construction process will help you avoid costly mistakes. Ultimately, this knowledge will reduce the redesign stage and keep the project on time and within budget. Operating on the principle "I'll know it when I see it" is expensive and frustrating for all those involved in the design process.

## Site planning, massing, and space planning

Understanding the process of building must begin by acknowledging the role of urban planners in the design process. In most cities, towns. and rural jurisdictions, a building begins with a concept that conforms to the zoning by-laws, guidelines, master plans. and other constraints established under local, state, and provincial regulation. In Chapter 6, we traced the design approval process from a project's inception. During the concept phase of design, planners and architects must work together to formulate a general plan that conforms to all legal requirements. Although it is sometimes justified to request a variance to make the project work, any request for a relaxation from existing regulations will delay the development process. Besides, there is no guarantee you will receive the variance, which could ultimately add more cost and time to the project. At this early stage, the goal is to develop a massing scheme for the site. Massing studies establish how large a building can be built on the site (site coverage), where it can be located within the setbacks, and its general configuration (building envelope), which must adhere to any limits on height, depth, and width. These factors determine the floor space that can be accommodated in the building. There are times when requirements force a building to have different setbacks in the front, back, and sides. Often, these setbacks are a response to concerns for preserving the character of the neighborhood. To provide a buffer to minimize the impact of shadows of a multifamily apartment building on houses located along the rear property line, an increased setback may be required as the building increases in height. On corner sites, there can also be unique constraints or opportunities. The addition of towers, balconies, and garden terraces can increase the potential value of each unit, while responding to the constraints of the project. In many communities, privacy concerns will dictate the views from windows and balconies, requiring the addition of privacy screens.

Massing studies will also inform many other design considerations, including the location of onsite parking, curb cuts, and door entrances – concerns that can have a direct bearing on interior design issues. For example, if the required onsite parking must be placed in the back of a proposed office building, public entrances will need to be located in the back as well. Fixing the location of the public and service entrances will dictate where the lobby will be located. Consider a building where there is both street and parking access. In this case, the location of the lobby, elevator, and public entrance may have a very different arrangement of space than if public access is limited to the front entrance.

Public buildings in the US and Canada must also respect accessibility regulations providing accommodations for wheelchair access. If the main floor of the building is raised up from the street, long ramps must be incorporated. Without careful planning, these additions will impact the building aesthetics, location of entrances, and cost. Under the Americans with Disabilities Act of 1990 (ADA), interior spaces must include accessible washrooms, hallways, and counters that can accommodate wheelchairs.

Massing studies will also be useful in understanding how orientation will shape interior design, especially in residential buildings. Maximizing the use of natural lighting is always positive. Everyone likes a sunny kitchen and breakfast nook. To the extent that these requirements can be accommodated, additional value can be built into the project, usually with little additional cost.

The requirements of buildings designed to meet Leadership in Energy and Environmental Design (LEED) standards will also need to be acknowledged early in the design process.[1] The LEED scheme offers third-party verification that can apply to all types of building. Silver, Gold, and Platinum certification is awarded to a building if it meets specific criteria in four categories: construction, interior design, building operation, and neighborhood development. In cases in which LEED is a condition of the client, integrating LEED requirements into the design process from the outset will reduce your headaches later. Even for buildings built as speculative projects, LEED certification may be a consideration. LEED buildings have faster lease-up rates and may qualify for tax rebates and zoning bonus. However, capturing all the value associated with energy-efficient buildings may be difficult. Though LEED buildings can exact a higher premium, if a building is built on spec, the developer pays for all the additional costs without any guarantees that a buyer will be willing to pay the associated costs for making the building green.[2] Because the savings on energy accrue to the eventual purchaser, a developer would desire LEED certification only if it helps to market the project and demands a higher sales price for the development.

Massing studies should be done by planners or architects with a good understanding of a community's zoning and by-laws. Once completed, they can form the basis the architect's conceptual plan. The program for the project, based on detailed market studies, surveys, and user studies, will be critical to arriving at the types, sizes, and number of spaces. This data will be useful in arriving at a basic floor plan for the building. In many cases, buildings must satisfy a host of requirements, as is the case for schools, assisted care facilities, nursing homes, and hospitals. At a minimum, individual rooms must be able to accommodate furniture and equipment requirements for each room. Often, market research and health and safety will dictate surface treatments, and the location of heating, ventilation, and air conditioning (HVAC) vents, windows, and doors. Depending on the jurisdiction, requirements for buildings such as hospitals, schools, and government facilities can be accompanied by a very detailed set of specifications. City, state, provincial, and federal requirements should always be reviewed by the architect before the project begins. Because these requirements are not always easily grasped, many architectural firms have specialized expertise in the design of a specific building type. There are firms that are highly focused on nursing homes, hospitals, fire halls, casinos or research labs, for example. An architect that has successfully completed a number of similar projects will bring a wealth of critical experience to the project.

## Mastering the art and science of construction

In the late 19th century, someone who was planning to build a major project such as a university building, railroad station, hospital or government building would have hired an architect.[3] The architect would be responsible for the design of the building, its location on the site, the location of entry and exits, and the placement of windows that can take advantage of views and allow for natural lighting of interior spaces. The contractor, who was responsible for the construction of the building, would have appointed a foreman and hired the laborers needed to build the architect's design. The contractor would then have had to orchestrate a vast army of skilled and unskilled labor. The architect, who was responsible to the client, would make site visits to verify that the building was constructed according to plans and specifications. A "clerk of works", who was the architect's representative, would have had the day-to-day responsibility of verifying that materials were delivered and used properly in the construction of the building according to plans. Inspecting photos from mid-19th-century job sites reveals how little the construction process had changed since ancient time. The art of building in the 19th century would have been intelligible even to an architect of Roman-Greco times. As evident in architectural treatises, very little change in the design and construction of major architectural works has occurred over the course of 2,000 years of architectural history. There has been a continuum in the history of technology that includes the works of Vitruvius, a Roman military architect from 1st century BC and the author of the first-known architectural treatise *Ten Books on Architecture*, Andrea Palladio, an architect of numerous villas in Veneto during the 16th century and author of the *Four Books of Architecture*, and Benjamin Henry Latrobe, the architect of the Baltimore Cathedral in 1804–18.[4] Buildings were largely built by hand, stone by stone, brick by brick, and with wood timbers, using only the simplest of tools. Wood scaffolds, simple hoists, and pulleys were the most complex pieces of equipment found on the site. Any photo or illustration of the construction site from these earlier ages reveals an army of laborers doing tasks that now are done by cranes, excavators, and an array of heavy equipment.

## Technology and the modern building

Changes in technology and the scale of construction since the late 19th century have altered the way we construct buildings today. Before the mid-19th century, structures were largely built from wood, stone or brick. Buildings did not possess sophisticated heating and lighting systems. Indoor plumbing was rare or non-existent. The elevator, first demonstrated at the 1853 New York World's Fair by Elisha Otis, remained a novelty yet to be employed in apartments and office buildings for another 25 years. New iron structural systems adopted from the construction of bridges, mills, and train stations made it possible to build taller buildings and span larger spaces. Wrought iron and cast iron, and later steel and reinforced concrete, enabled engineers and architects to create skyscrapers and spaces of overwhelming proportions. Train stations and exposition halls were great industrial monuments to the Victorian age. With the growing complexity of building systems, the practice of architecture became more specialized and fragmented. Structural engineers now had to be employed to design the structural frame; mechanical engineers, for the heating and ventilation; and electrical engineers, for the electrical, and lighting systems. The architect became largely responsible for the building configuration, floor plans, and building envelope, and then for managing the design process and the communication link between the client, the engineers, and contractors. Not surprisingly, the demands for public safety, including fire safety, in the built environment led to the professionalization of engineering and architecture, as well as specialized trades. Ultimately, the certification and registration of architects and engineers came into existence, specifically to ensure public safety. Without these requirements, it would have been difficult to bar the practices of charlatans and others who assumed titles without the requisite knowledge and experience. Today's highly regulated design profession is partly a response to the demands on practice that began in the 19th century. All designs and renovations, except the smallest buildings, require the professional involvement of licensed architects and engineers.

## Building from the ground up

As an integrated system, the choice of the structural components will play a role in satisfying the space requirements of the project. In the early stages of design, it is important to acknowledge that all structures must have a foundation. Without a firm foundation, even the simplest of structures will be racked by differential settlement over time. Anyone who has visited an old wooden-framed farmhouse on a shallow stone foundation will notice structural problems, such as doors that do not close properly and floors that slope. For example, a marble dropped on a wood floor may start to roll around the room. Of course, this was not the intent of the carpenters who built these homes over a century ago. Soil settlement is the obvious culprit. The geological strata underlying a foundation will have a direct bearing on the stability of the building. Buildings constructed on bedrock or gravel will not settle over time, while those built on clay and other compressive materials can settle at different rates, creating havoc in the superstructure. Clay soils expand and contract with moisture. During the course of a year, water table levels can fluctuate greatly with the seasons, which can affect the stability of a building's foundation. Everyone is familiar with the Leaning Tower of Pisa (*Torre di Pisa*), the foundation stone of which were laid in 1173, and which was built on layers of clay, sand, and gravel. The lean for which the tower is named became a problem shortly after the construction process began. The solution of the architects of the age was to increase the floor heights on the leaning side to compensate for the lean in the tower. Until recently, the whole structure was in jeopardy of self-destructing, but it has now been stabilized through a series of costly foundation improvements. Clearly, tall buildings require stable foundations. Even buildings of modest proportions can self-destruct over time when placed on poor foundations.

Based on geological surveys, the foundation engineer, working in conjunction with a structural engineer, should be able to develop a foundation design that ensures stability. Many projects may require additional field studies to investigate how soil conditions will respond to the weight of the imposing structure. Pits and borings are two common approaches to learning more about geological formations under building sites. Core samples taken from the site can be analyzed in the lab, providing the engineer with critical data to determine the type of foundation needed to support the proposed building. Ultimately, what is required is knowledge of the allowable stress levels. Soils that are overstressed can see substantial settlements and even failure. It is also important to know how the geological conditions vary over the site. In these cases, the foundation must adapt to the variation in bearing capacity. On special occasions, where projects are significant in scale and size, actual field testing of loads can be used to verify values arrived at in the lab.

The most common foundation is constructed out of reinforced concrete. Though the Romans created structures that have lasted 2,000 years, concrete can deteriorate over time. Sulfates in the soil in certain regions can attack concrete, causing it to crack and lose its strength. However, taking special precautions by adding calcium hydroxide and calcium silicate hydrate to the cement can counter the effect of sulfates in the soil. If a foundation is not built below the frost line, the freezing and thawing of soils can be a destructive force, lifting the foundation in winter only for it to settle in spring. If the foundation rests on a layer of gravel, this makes it immune to the effects of temperature and water, and it can be built at a much lower cost. Bedrock will also provide a stable foundation. However, where the bedrock is close to the surface, the cost of excavation may bar the construction of a basement. A building built without a basement may be the solution to the problems posed by high water tables and even potential flooding. Where there is a requirement to build on unstable materials that are below the water line, costs can be considerable. For example, waterfront development in Boston begins only after extensive foundations are completed. However, given the high cost of land in Boston, the additional expense in development costs was justified.[5] Like the foundation stones of Gothic cathedrals, a firm foundation is never seen or appreciated. Estimating the projected cost for foundations early in the development process may help you determine whether to abandon the project. Given that the added expense cannot be retrieved in terms of higher rents, it may be impossible to make the project profitable if the foundation costs escalate beyond a reasonable value.

### *Foundation design*

The simplest of foundations are spread footings. Used for centuries, spread footings take the loading from the columns and distribute it over a larger surface area (Figure 1.8). In this way, the foundation does not overburden the substrate beyond its bearing capacity or safe limit. For example, a building with column loads at the foundation of 400 kips (equivalent to 400,000 lbs) may be easily handled by a concrete column measuring 10 × 10 inches (a cross-sectional area of 100 sq. inches). With a working stress level of approximately 4,000 psi, there is a significant safety factor designed into these columns. However, if the soil substructure is able to support a maximum of only 100 psi (including an appropriate safety factor), a column placed directly on the soil will exceed its safe limit and we can expect the building to settle or possibly collapse over time. To solve this problem, we will need to spread this load over a greater area. In this case, we need at least a ratio of 40:1 in

*Figure 1.8* Column footing calculations.

the size of the foundation footing to the cross-sectional area of the column. In engineering design, an appropriate safety factor, typically 300%, is included in our calculation to insure against the unforeseen. Soils, unlike manufactured and engineered materials, are hardly homogenous in nature. In foundation design, safety factors can be several multiples. In our example, we will require a cross-sectional area of 4,000 sq. inches of surface to distribute the column loading of 400,000 lbs. In this example, a column footing of 63.25 × 63.25 inches should be adequate. The engineering solution is to provide a mechanism for distributing our load over a greater area. In the past, a pyramid form was used to distribute this load (Figure 2.8). During the 19th century, rafts made from used railroad ties buried in concrete were used to distribute the column loading; today, we can use reinforced concrete to solve this problem. Formwork, with carefully placed steel reinforcing bars, creates an homogeneous and continuous structure that transmits the column loads to the bearing material just below the foundation (Figure 2.8).

Continuous footings can also be used to support an entire bearing wall. Usually found at the perimeter of the basement wall, a continuous footing has a cross-section similar to the reinforced concrete footing (Figure 2.8). Spread footings are useful when the bearing wall supports a significant portion of the weight from the floors and roof structure. When

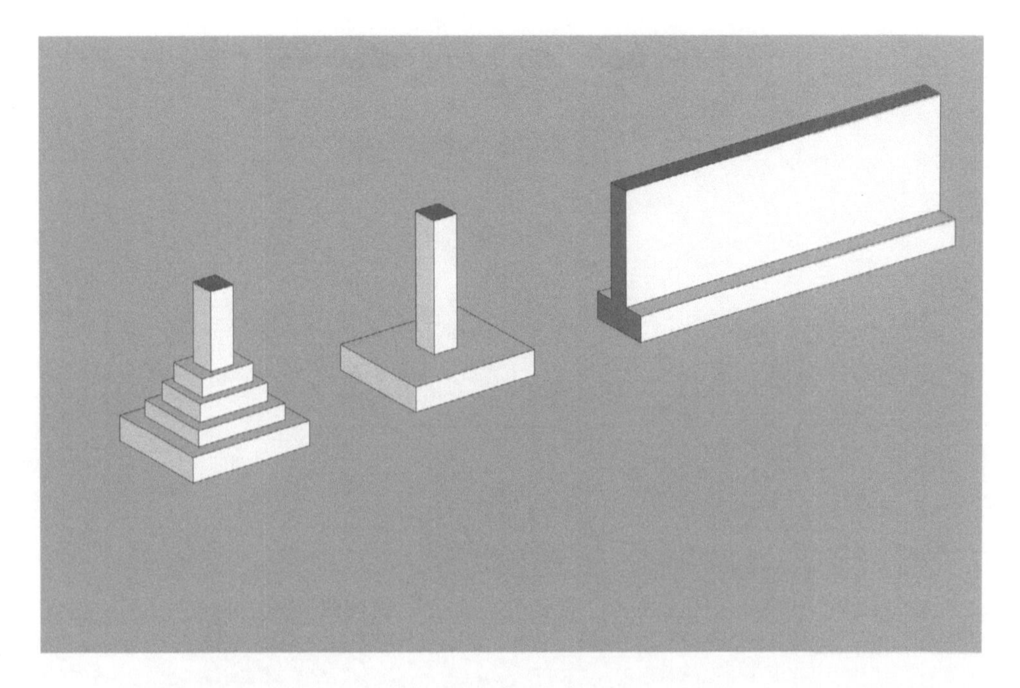

*Figure 2.8* Typical spread column footing, reinforced concrete column footing, and continuous footing for supporting a wall.

there are interior columns in the building, a matt or raft foundation can be used to support the structural frame. With a matt foundation, the columns bear directly on a heavily reinforced basement floor (Figure 3.8). Matt foundations can be an effective strategy in areas in which settlement is expected during the first few years of the project. If the building settles as a single entity, then it is possible to design the building to accommodate its ultimate elevation. If, for example, it is known that the building will settle several inches during the first few decades of its life, then underground parkades will need to incorporate architectural transitions between the building and the adjacent ramps to accommodate the slow, but inevitable, settlement in the foundations. Similarly, the steps, ramps, and entrances will ultimately need to align with sidewalks and steps that are not attached to the building.

There are occasions on which the material directly below the footings is unable to support the loads from the building. When bedrock is located below the surface, piles can be an effective means of supporting a building (Figure 4.8). In Roman times, a wood trunk stripped of its bark and branches was driven into the ground by human power. Dropping a heavy weight suspended from a pulley and wood frame could be used to hammer piles into the ground.

In his treatise *Ten Books on Architecture*, Vitruvius specifies that if piles from alder, olive or oak are charred before being driven into the ground, they will last for many years.[6] The charring process makes the wood pile toxic to both insects and mold spores, prolonging its life. Later, human strength was replaced by horsepower. In the 19th century, steam power became the dominant force of mechanical pile drivers. Today, pile drivers will use either a falling weight or an exploding piston, like in a gas or diesel

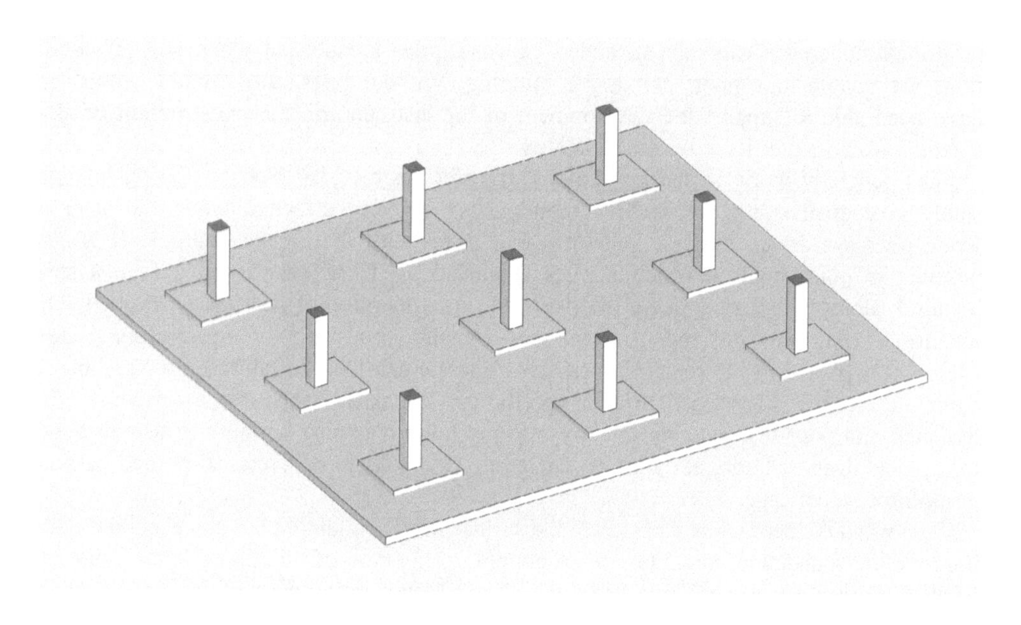

*Figure 3.8* Matt foundations.

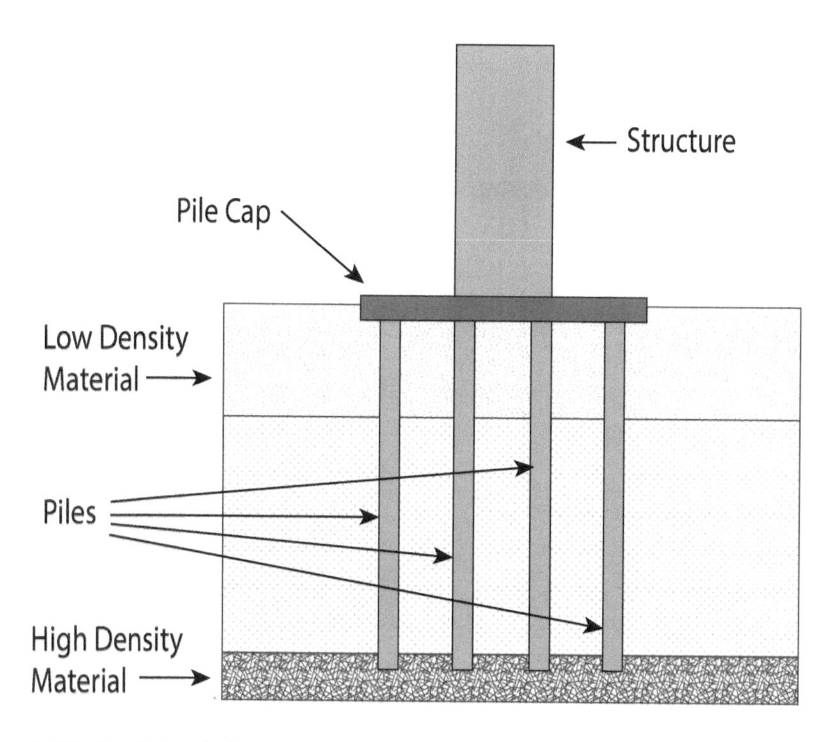


*Figure 4.8* Pile foundation design.

engine, to drive piles into the ground. The actual pile can be made from wood, steel, and reinforced concrete. Although an added expense, piles are needed when subsoil conditions are unable to support a massive building. Without piles, many cities would not have been able to support the development of the last century: the construction of skyscrapers would never have become a reality.[7]

Piles generally work on the principle of transferring the load from a footing to more stable geological strata below. This strategy can be very effective when the steel or wood piles are driven through loose material down until they reach bedrock. It is also possible to pour in place concrete piles (Figure 5.8). To accomplish this feat, a steel mandrel, along with a steel shell, is driven into the ground until bedrock is reached. The mandrel is then removed and the steel shell remains as a form for the reinforced steel cage that is placed in the remaining cavity. Concrete can then be poured into the hole to complete the pile (Figure 5.8). Under specific conditions, it is possible to dispense with the shell and pour the concrete directly into the hole. It is also possible to use an auger to core a deep cylindrical hole in the earth into which concrete is poured around a reinforced steel cage.

Even when the bedrock is far beneath the surface of the building, piles can also be used to improve the foundation conditions for a building. This type of "friction pile" depends on frictional forces between the surface of the pile and the surrounding soil. Like a nail being

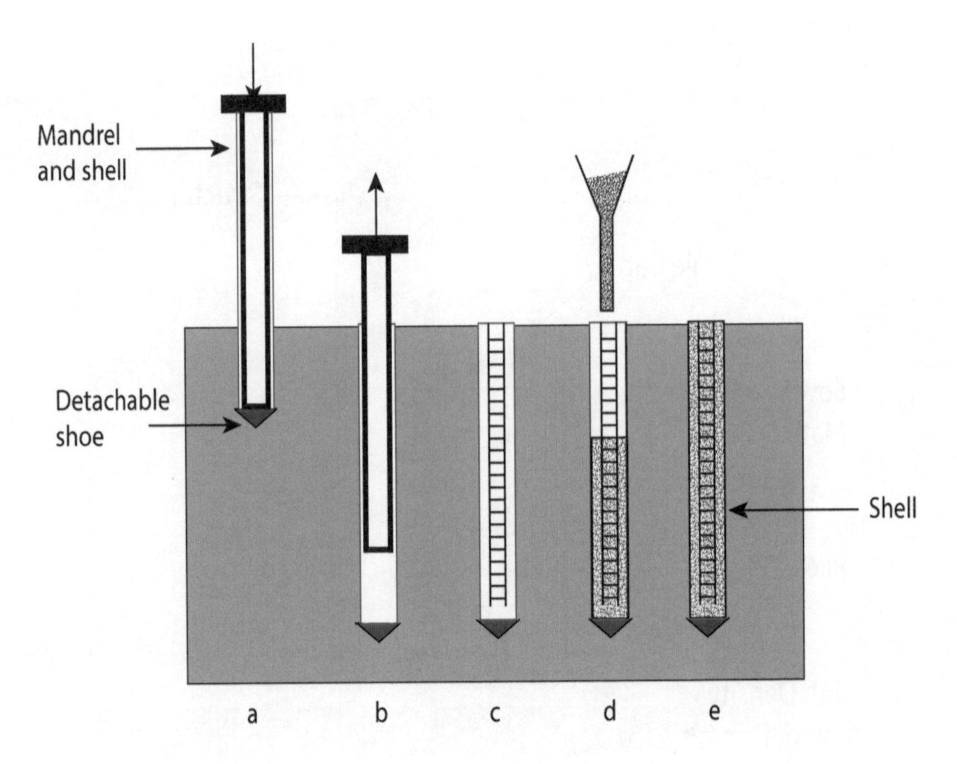


*Figure 5.8* (a) Cast in place pile; (b) casing is driven into the soil; (c) final position reached and the mandrel is removed; (d) a reinforced cage is placed in the form; (e) concrete is poured into the completed pile.

driven into a block of wood, it is the frictional forces that provide the support for the column footings above. Caution must be exercised when using frictional piles in earthquake-prone areas, where soil subject to liquefaction can result in building settlement. To offset this potential loss in bearing capacity, piles need to be designed with greater bearing capacity at their heads to insure against settlement.

In marine environments, caissons can be found for bridge piers and other deep-water structures. For buildings, they offer designers a strategy for supporting great loads on bearing material sometimes over 100 ft. below the surface of the water. John and Washington Roebling used such a strategy to support the massive piers for the Brooklyn Bridge, completed in 1883 (Figure 6.8). In this case, steel caissons were kept from flooding by air pressure. At several atmospheres of pressures, workers who were removing the material below the caisson were subjected to Caisson disease (also known as the bends). John Roebling, inspecting the foundation in 1872, succumbed to the condition, which left him in a weakened condition for the rest of his life. For much of the time the bridge was under construction, Roebling would be imprisoned in an apartment with a view of the site. Too ill to visit the site in person, a pair of binoculars and visits to the job site by his wife Emily provided the critical link in overseeing the completion of the project.[8]

### *Structural systems*

During the conceptual design phase of the project, the engineer may consider several potential design concepts for the structural system of a building. At this early stage, the advantages and disadvantages of each system should be studied and compared. For example, it is not uncommon, during the conceptual design phase, to consider the

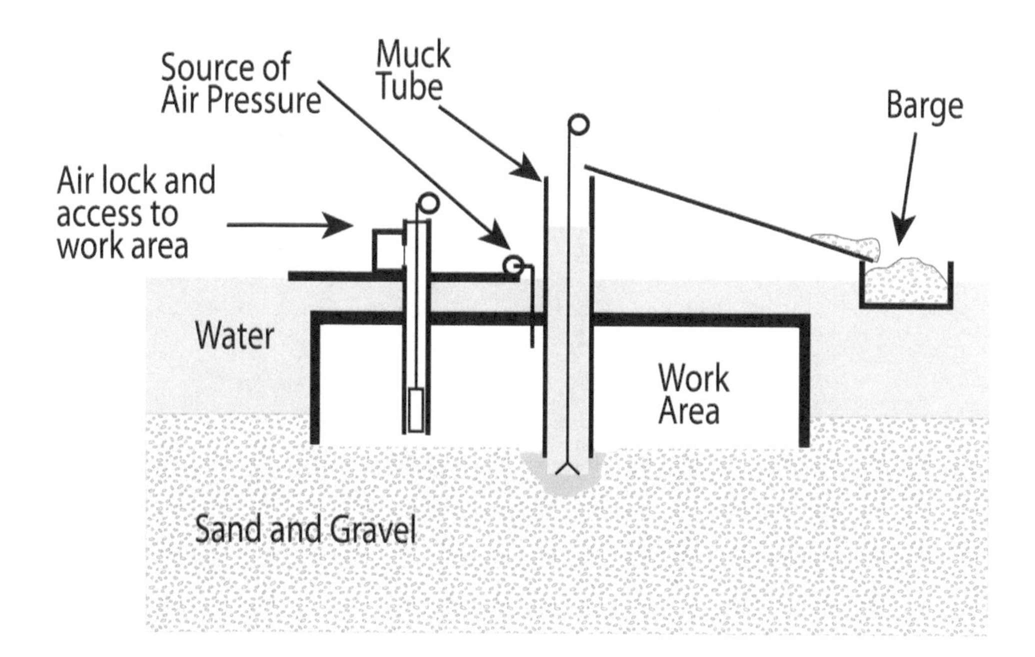


*Figure 6.8* Pressurized caisson, cross-section.

relative merits of reinforced concrete vs steel framing. This choice should weigh a number of factors, including cost of labor, the availability of specialized skills, the relative cost of materials, and the architectural requirements of the design. In addition, the contractor's experience may impact the choice of the building system.

To better understand how a structural grid works, consider the simple case of beam and columns, illustrated in Figure 7.8. In this example, the weight of the beam and loads placed above them is transmitted to the columns below, which are supported by the foundation. There are two types of loading: dead and live loads. The dead loads, or constant loads, include the weight of the structure, floors, ceiling, curtain wall, mechanical equipment, and anything permanently attached to the structure. Live loads, or dynamic loads, will change over time. They can include wind loads, snow, pedestrians, cars, and other vehicles. Live loads can also include cranes and forklifts.

In our example, the cross-sectional area of each column and beam is sized to withstand live and dead loads, while factoring in a margin of safety. To calculate the size of the columns, it is important to consider the strength of the material. Materials that can resist compression, such as stone, have been traditionally used for columns (Figure 1.8). Concrete is also good under compression. The working stress or the stress that a material can be subjected to when in use is often expressed in lbs/sq. inch, tons/sq. ft. or kgs/sq. m. A measure of safety provides for a margin of error. This margin can vary from 1.5 to 2.5 for concrete buildings. The allowable working stress will then be a proportionally lower number. In Figure 1.8, if we assume a working stress of 4,000 psi, it should be possible to support a dead load of 200 tons (400,000 lbs) at each column's base. In this simple case, a column of 10 × 10 inches should easily suffice. However, when applying an additional safety factor of 2, the cross-sectional area will double.

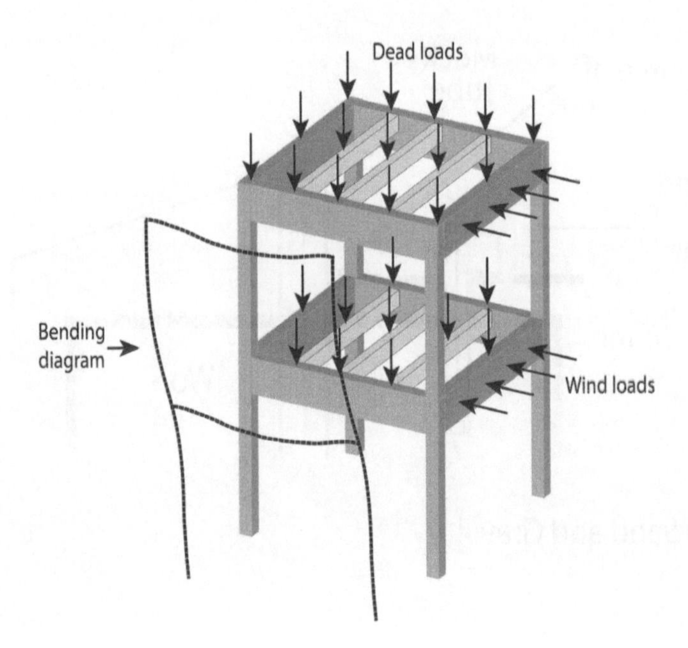


*Figure 7.8* Structural frame with dead and live loadings.

**Example: calculation of column dimensions**
Allowable stress = 4,000 psi
Cross-sectional area = Load/Allowable stress
= 400,000 lbs/4,000 psi = 100 sq. in.
Safety factor = 2
Required cross-sectional area = Cross-sectional area × Safety factor = 100 sq. in. × 2
= 200 sq. in.

If the column cross-section is square, then the dimensions will increase to 14.142 × 14.142 inches.

With the compressive force being applied to the column, we would expect the column to foreshorten slightly. The relationship between stress and foreshortening is referred to as strain. Under tension, materials will elongate, but concrete is not an elastic material; we can depend only on its compressive strength in building design, rather than its tension strength.

To compensate for this lack of strength in tension, steel reinforcing rods are placed inside the formwork prior to pouring concrete. The additional strength provided by the steel reinforcing bars makes it possible to use reinforced concrete for the entire structural grid of a building, including floors and beams. In sizing the beam, it is important to consider its bending action (Figure 8.8). In this case, the concrete above the neutral line is being placed in compression, while the material below the neutral axis is actually in tension. Tension is when the forces are pulling from both ends, like stretching a rubber band. From practical experiences, we know that materials such as concrete and stone are good in compression, but not in tension. This is why the distances between columns for Greek and Roman temples facades were not very expansive. To compensate for the low allowable compressive stress in concrete, we can use steel rebars placed in the concrete (Figure 8.8). Placed at the bottom of the form before pouring, these steel bars will act to compensate for the low allowable tensile stress levels for concrete. Next time you walk

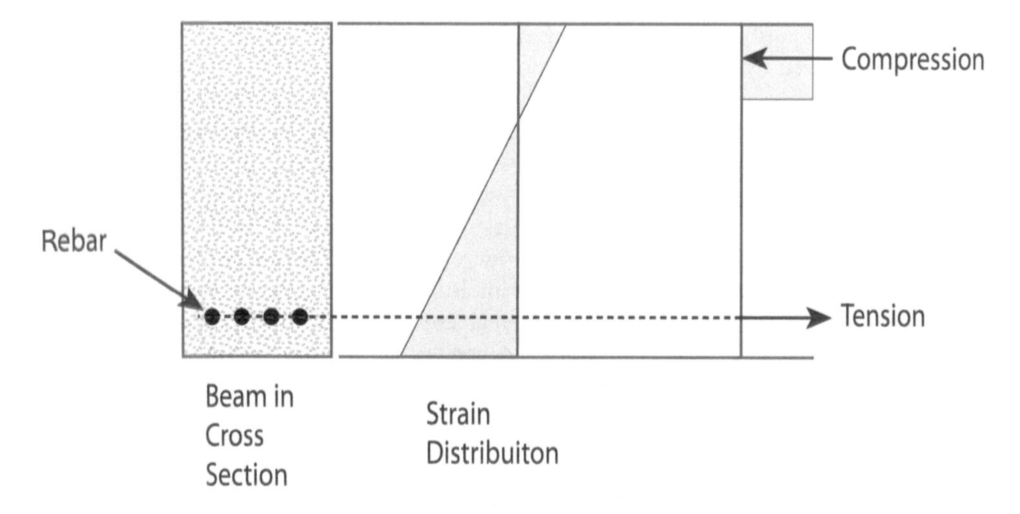


*Figure 8.8* Typical design for a reinforced concrete beam.

under a neglected concrete bridge in disrepair, you may see cracks and even some exposed rebars. The rebars are taking on the tensile loads as a result of the bending action of the beam.

Most buildings are not simple beams resting on columns; instead, each beam is firmly attached to the column, forming a structural grid, making it possible to withstand the dead and live loads (Figure 7.8). The use of structural grids has freed architects from the constraints of an exterior bearing wall. Until the creation of the modern structural grids used in skyscrapers and industrial buildings, the exterior walls, along with the interior columns, supported the weight of upper floors and roofs. With the invention of the metal structural frame, the exterior bearing wall was no longer essential to the stability of the building. Though the inner columns of 19th-century mill buildings were required to support the floor loadings, much of the dead load was still assumed by the heavy stone masonry walls. Numerous examples of this type of industrial building exist throughout the northeast (Figure 9.8). With the invention of the structural grid, the curtain wall became an architectural design element, with large expansive glass windows. Some of the best-known examples were designed by the Chicago architects Burnham & Root, Sullivan & Adler, and William Le Baron Jenney. Built between 1880 and 1910, these skyscrapers and department stores, characterized by their structural grid, strip windows, and terracotta detailing, came to be known as the "Chicago style". Examples of this style can still be found in many major American cities, including Chicago itself, New York, Buffalo, St. Louis, and San Francisco.[9]

In New York City, where the skyscraper assumed towering proportions in the 20th century, buildings including the Chrysler Building and the Empire State Building are a reflection of new zoning regulations and an evolving construction technology. With the introduction of the 1921 zoning regulations, buildings were required to be set back from the street at the higher elevations. Both the Chrysler Building and the Empire State Building are examples of this new stepped architectural form. The Empire State Building, clad in Indiana limestone and completed in 1931 in the record time of three years,[10] would be the tallest building in the world for the next 55 years, until the construction of the Sears Tower in 1973. Upon close inspection, we see that the steel frames of these towering giants incorporated cross-bracing and knee joints to withstand the wind loads of gale forces. A curious fact about the Empire State Building is that the structural stability was improved by the addition of a stone facade, which provided a damping action against wind.

In 1921, Ludwig Mies van der Rohe asserted a new aesthetic in steel and glass in his revolutionary skyscraper design. Though it would be decades before Mies would have the opportunity to design a glass-and-steel skyscraper, his drawing of 1921 would inspire generations of architects. In 1961, for the design of the Seagram Building, Mies was able to give his ideas physical expression (Figure 10.8). With its curtain walls of glass and black mullions, the design of the Seagram Building made use of non-load-bearing glass curtain walls supported by a steel frame, creating expansive interior views that permit natural light into the interior spaces. A century earlier, the extensive use of glass first appeared in the construction of conservatories and expositions. In the late 19th century, skylights and clearstory windows were a prominent feature in railroad stations, and industrial and commercial buildings. One disadvantage in the use of glass was the interior heat gains from occupants, electronic equipment, and lighting. Anyone who has been in a greenhouse on a summer day will understand the problem of using glass in hot, sunny climates. Reflective films, double walls, computer-controlled vents, and louvers

*Figure 9.8* Wilkinson Mill, Pawtucket, RI, built between 1810 and 1811.

Source: Author's collection.

*Figure 10.8* Toronto-Dominion Centre, Ludwig Mies van der Rohe with John B. Parkin Associates and German & Hamann, 1964–74, 1985–71. NB The Seagram Building in New York City, completed in 1958, served as a model for this complex of buildings built in Toronto.

Sources: Patricia McHugh, Toronto, Architecture, Toronto, Ontario: McClelland & Stewart Inc., Author's Collection.

are some of the strategies architects can use to compensate for the heat gains of glass walls and skylights. However, it is air conditioning that made these iconic building a reality.

In making curtain walls work for taller buildings, architects and engineers must give special consideration to wind loads, which have, on occasion, resulted in glass panes bursting out under unusual circumstances. In 1972, a design flaw in the John Hancock Building resulted in the redesign and replacement of the glass curtain wall at a cost of over $7.7 million (about 10% of the original building cost of $75 million or about 4.5% of the final total costs of $175 million) (Figure 11.8). John Hancock is a slender building that actually bends in the wind, placing stress on the frame. In addition, the thermal loads created from the expansion and contraction of the metal mullion placed excessive strain on the glass. Thermal stresses have been known to break the seals holding the glass windows in place. Once the seals are broken, water and air penetration can pose a serious problem to the integrity to the building.[11] The result was the loss of many windows, especially on the windward side of the building, where the stresses were the greatest.

Though tall glass buildings are not without their associated environmental and structural risks, the view from upper floors has so intoxicated tenants, critics, and owners that they look beyond many of the design issues that surround these modern symbols of commercial power. However, with such height requirements, architects and engineers are confronted with the realities of making these buildings safe. Wind velocity increases with elevation, creating greater stresses on the structural frame of tall buildings. On a windy day, upper floors can be subjected to considerable sway, making occupants uneasy and sometimes even nauseous. The addition of cross-bracing and dampeners can help in resisting these lateral loads by countering the sway in these cantilevers (Figure 12.8). In areas that experience earthquakes, achieving structural stability can be a challenging additional exercise. Specially designed structural frames and foundations are needed to allow the structure to react to an earthquake without destroying the integrity of the building. Structural engineers have been able to solve these problems because of the computer revolution that began in the 1960s. In particular, with the advances in computing, the analysis of the stresses and strains in complex steel frames has become a reality.

Prior to the development of advanced structural design software, it would have been impossible to accurately determine how a 3D grid would behave under different loading conditions. Skyscrapers are highly redundant structures, which, in simple terms, means that the failure of one beam in the structure will not result in a catastrophic failure. Analyzing the stresses and strains in these structural grids requires the solution of equations with hundreds, if not thousands, of unknowns. In the early history of skyscraper design, engineers had to rely on mechanical calculators and slide rules, which limited the techniques that could be used to analyze a structural frame. Only the simplest of analysis could be done with these hand calculators. Accounting accurately for the live action of wind loads would have been all but impossible given the mechanical calculators of the time. Because these solutions were crude by today's standards, engineers increased the margin of safety to be assured of the structural integrity of their designs. A comparison of skyscrapers of today with structures like the Empire State Building reveals a lightness of form and elegance that has been possible only with advances in computing technology and materials science. It is now possible to design tall skyscrapers that can withstand high gale-force winds and earthquakes. In 1956, Frank Lloyd Wright's proposal for a "mile-high skyscraper design" was viewed as

*Figure 11.8* John Hancock Building, 2018.

Source: Author's collection.

*Figure 12.8* X-Bracing, steel frame construction, Auckland, NZ, 2018.

Source: Author's collection.

a fantasy.[12] With the completion of the Burj Khalifa in Dubai, at a height of 2,710 ft., slender skyscrapers of towering heights have become possible. With the planned completion of the Jeddah Tower in 2020, at a height of 3,280 ft. (1 km), an altitude will be reached that could only be imagined in 1956.[13]

Architects of tall buildings must give special consideration to escape during fire and other disasters. In the Triangle Shirtwaist Factory fire of 1911, 146 employees died when exits were locked, making egress impossible.[14] This eight-story building, built in a time of horse-drawn fire equipment, experienced one of the most horrific fires in New York City's history. Planning for disaster must be a concern in today's political environment. The attack on the World Trade Center in 2001 will always speak to the need for continued improvements in fire protection and evacuation technology. Most buildings are rated between one and two hours before their structural integrity will be jeopardized. Smoke and heat also present serious impediments to the successful exit of inhabitants in the event of a fire or explosion. Sprinkling and fire suppression systems, as well as backup electric power systems, are just a few of the design considerations needed in all tall buildings. With elevators out of commission during a major disaster, planning the evacuation of thousands down staircases in buildings that can exceed heights of over 1,000 ft. will always present a challenge.

### *Optimal solutions*

Most major commercial, industrial, retail, and residential projects will be built of concrete or steel. Ultimately, an experienced architect and engineer will be able to guide the developer on the most appropriate structural material. Material costs will be of prime concern in making this decision. For most projects, over 75% of the materials used to construct a building are sourced within a radius of less than 150 miles. Transportation costs figure heavily in the choice of building systems. Where sand, gravel, and cement are readily available, concrete construction may be a cost-effective solution for major projects. The associated labor costs for formwork will also figure into the final cost calculation. One advantage of concrete is for sites where there is little room for storing materials: poured concrete construction may offer a space-saving solution.

In developing the structural solution for a building, consideration should always be given to the cost of construction in relation to the cost of land and the rents that are to be paid by the tenants. Clearly, in cities with high rents, the cost of land, site development, and construction can justify constructing a tall building. The history of the skyscraper shows that higher rents and land costs support higher densities. Taller buildings distribute the high cost of land over a larger area of rentable space. Architect Cass Gilbert, designer of the Woolworth Building, the tallest skyscraper in the world when it was completed in 1913, is credited with saying in 1900: "A skyscraper is a machine that makes the land pay."[15] However, taller buildings demand more expensive structural and mechanical solutions. One of those expenses is the need for vertical transportation, the elevator, not needed in a single-story building. As a building grows in height, an increasing number of elevators are needed to accommodate the growing number of employees and visitors. A building of moderate height, six stories, might have two elevators; however, as the building advances in floors, banks of elevators will be required to service the occupants. With the addition of each elevator, the amount of rentable space is reduced on each floor. It is not unusual for a tall office building to have more than 40 elevators, which will occupy 30% of the floor space. So as not to consume a larger and larger percentage of the space on each floor with elevators, engineers have resorted to the use of high-speed elevators, banks of express elevators, and sky lobbies. Like taking an express train and making a connection to a locale, the use of sky lobbies allows visitors to travel quickly to an upper portion of the building, where they take an elevator that services only part of the building.

Arriving at the optimal solution that balances costs of land, structure, elevators, and rent will vary by city. In cities such as Hong Kong, Shanghai, and New York, it is clear that high rents and land prices have made tall buildings, even with small floor plates, economical.[16] In these cities, the high cost of land would make building anything lower than 40 stories unprofitable. Few cities in the world have this set of unique economic constraints. In rural areas, it may be difficult to support more than one story. Within suburban communities, it is not unusual to find office buildings with between five and eight stories, and apartment buildings and condos in the three- or four-story range. Only in the downtown core of major cities in North America do you find anything that would be considered a skyscraper. New York, Toronto, Chicago, San Francisco, Los Angeles, Vancouver, Seattle, Los Angeles, Houston, Calgary, and Denver are the cities in North America where economics support new skyscraper construction in the 21st century. Next time you see a budget hotel of four stories being constructed, note the use of wood or metal studs, and count the number of floors and elevators. Count the number of rooms, and subtract the amount of space used for halls and public areas. Going through this calculation will reveal how developers economize on materials to build profitable design solutions. There is a reason why budget hotels all look alike.

## Structural systems: concrete, steel, or wood

### *Wood construction*

Wood framing is a general all-purpose solution for most residential building in North America. It is relatively inexpensive, commonly available, and easy to transport and work with on the job site. With only modest change since it was invented in Chicago in the 1850s, "balloon framing" offers a versatile form for constructing buildings of one, two, and three stories. Using standard milled forms of nominal size, 2 × 4 inches, 2 × 6 inches, 2 × 8 inches or 2 × 10 inches, these building elements can be assembled into exterior and interior walls floors, floors, and roof trusses to create houses of varying dimensions and size.[17] If a framer were transported in time from the late 19th century to a job site today, they would easily adapt to the construction site of a residential development project (Figure 13.8).

Residential house construction begins with the foundation. Today, the stone foundation found in many older homes of the 19th century has been replaced by a poured concrete foundation, using wood formwork or concrete poured in foam forms. Once the concrete cures, the framing begins. Walls today are still constructed with studs connected to top and bottom sill plates to form wall panels. With the additional of the roof and floor joists or trusses, the house takes shape (Figure 13.8). Windows and doors that have been premanufactured for over a century are ordered from a catalog, delivered to the site, and placed within the framed openings. Today, even the doors and jambs come as a unit already mounted on hinges ready to be installed by the framer. One development that has increased the productivity of wood framing construction has been the introduction of

*Figure 13.8* Modern housing framing.

lightweight power tools. The hammer has been replaced by the nail gun and the hand saw has been supplanted by reciprocating radial and miter saws that make cutting and assembling dimensioned lumber much quicker to fashion a house or small building. Preassembled engineered floor and roof joists have replaced the built-up wood beams used just a few decades ago. They are ordered from the manufacturer, delivered to the job site on a flatbed truck, lifted into place by small cranes, and nailed, stapled or bolted into place. When residential construction is engaged on a mass scale, even wall sections can be framed offsite. Plywood and composite 4 × 8 ft. sheets attached to the floor joist with high-strength glue and screws results in a strong, but light, floor system. Other components, including staircases, banisters, and balustrades, can come preassembled and ready for installation. Once the framing is completed, glass fiber insulation placed between the wood studs, a vapor barrier, and a siding complete the exterior of the house.

With the house framing completed, plumbers and electricians can install the heating, cooling, plumbing, and electrical systems. Unlike houses of the past, in which cast iron and copper pipes were assembled using solder and flux, plumbing today is dependent on plastic pipe, which is easily cut and glued together. These plastic pipes placed in wall and floor cavities can withstand hot water, and can easily be designed to bend around obstacles. Houses using high-efficiency furnaces reduce energy costs in comparison with those from a few decades ago. By eliminating the chimneys found with conventional furnaces, considerable interior space can be recovered on each floor. The addition of solar energy systems for space heating, hot water, and electricity will only further reduce the operating cost for the home owner. Interior wallboard, molding, and trim completes the interior. Only paint, carpet, engineered floorings, and tile are needed to finish the interior spaces. With the installation of kitchen and bathroom cabinets, sinks, tubs, and toilets, the house is ready for flooring, including carpet, vinyl tiles or engineered hardwood, and paint. For more expensive residential properties, the use of costly finishes and surface treatments, custom cabinetry, saunas, wine cellars, and entertainment and security systems can add significant costs to the final product.

Though wood construction is largely associated with single-family residential construction, it can also be found used in multifamily designs. In many jurisdictions, the building code may restrict heights to four stories, though the construction of tall wood-framed buildings of 8–12 stories is now becoming more common in North America and Europe. With the introduction of composite beams (joists made of wood glued together), an engineered product has been created that is both strong and lightweight. When these components are assembled into prefabricated panels and floors, construction is greatly simplified. Unlike the construction of a single-family home where framers work with dimensional lumber nailed together, the construction of tall wood-framed buildings can resemble a kit of parts, which components can be bolted together on the job site. With fire ratings that are equal to or better than steel, engineered wood systems can be used to create aesthetically pleasing architectural spaces. Growing interest in the use of renewable materials in the construction process in Europe and North America will only promote greater reliance on wood construction techniques in the future.

### *Concrete*

Reinforced concrete buildings are integrated systems composed of columns, beams, and floors. The floor system can be designed as a flat plate with reinforcing mesh and bars. In some cases, to make the floor structure lighter, large tubes called hollow core concrete slabs can be placed in the floor form. Waffle slabs can be made by placing metal tubs in the forms prior to pouring the concrete. They can also be designed with spandrel beams running the length of the slab. Reinforced concrete can be used to create a diversity of architectural forms. Using forms made of plywood, metal, and foam, and even packed earth, concrete can be shaped into beams, columns, domes, and shells, limited only by the designer's imagination.

The history of reinforced concrete must reference the industrial buildings of the early 20th century with their mushroom capital columns and open floor plans. They inspired a generation of architects, including Le Corbusier, the architect of the iconic Villa Savoye (1923). A house for modern living, stripped of ornament, featuring an open plan and roof garden, it is a work still studied by students for its innovative approach to design.[18] In the early 20th century, the bridge designs of Robert Maillart established that great elegant spans could be created using reinforced concrete. Later in the century, Salvador Levi created air hangars, sports coliseums, and industrial buildings that would inspire a generation of architects and engineers, including Santiago Calatrava, the Spanish architect best known for his design of spacious lightweight structures in steel and concrete. His design for the World Trade Center Transportation Hub in New York City is one of many designs that won awards for excellence in concrete construction.

Concrete elements can also be precast offsite and assembled like giant toy blocks. Brought to the job site on flatbed trucks and lifted by cranes, these structural elements can be assembled into tall buildings, highway overpasses, bridges, and large-span enclosed spaces, including sports arenas and rail stations. To counteract large dead and live loads, steel cables are placed within the formwork to offset the tensile forces in long-span beams. These cables can be pretensioned while the concrete is curing or post-tensioned after the concrete cures. Though more commonly used in bridge construction, this approach can be used in the construction of architectural forms requiring large spans. In the 20th century, this strategy was used to create iconic rail stations, air hangars, stadiums, and bridges. As with any steel used in reinforcing concrete, water will cause rusting. Due to the high stresses in pretensioned or post-tensioned beams, the cables can snap, creating a hazardous situation.

Concrete can also be poured onsite into forms and tilted up to form walls. The tilt-up system can result in a finished product that contains holes for windows, doors, and connectors. In the lift-slab system, concrete is being poured at ground level and elevated using hydraulic jacks. Using this technique, the building literally grows out of the ground.

Though steel is generally associated with skyscraper design, many of the tallest buildings in the world actually have been built from reinforced concrete, including the Trump International Hotel & Tower and the China International Trust and Investment (CITIC) Plaza in China. In Canada, two of the tallest structures, the CP and CN Towers in Calgary and Toronto, constructed in the 1970s, demonstrate how concrete construction can be used in building tall structures (Figure 14.8). The completion of the Jeddah Tower (3,280 ft.) in 2020 will demonstrate the versatility of reinforced concrete as a design solution (Figure 14.8).

In many cities in the US, the cost advantage of concrete is more evident because of the price of materials and the cost of labor. One factor that will influence the design of a building is the floor-to-floor heights. Having a waffle floor or using a series of T-beams may be a more efficient use of material, but can also increase the distance between floors. This may not always be an issue, but in the case of a multistory building designed to fit within maximum

| | Building | Height | Country | Overall World Rank |
|---|---|---|---|---|
| 1 | Trump International Hotel & Tower | 423 | USA | 9 |
| 2 | CITIC Plaza | 390 | China | 12 |
| 3 | Central Plaza | 374 | China | 15 |
| 4 | Almas Tower | 360 | UAE | 18 |
| 5 | Shimao International Plaza | 333 | China | 24 |
| 6 | Q1 | 323 | Australia | 28 |
| 7 | Wenzhou Trade Centre | 322 | China | 29 |
| 8 | Nina Tower | 319 | China | 32 |
| 9 | HHHR Tower | 318 | UAE | 34 |
| 10 | Sky Tower | 312 | UAE | 336 |

*Figure 14.8* The world's ten tallest concrete structures.

height constraints, a slab design can make a project more profitable. For example, using a solid slab over a waffle or T-beam it is possible to reduce the height between floors by a few inches per floor. When this difference is multiplied by the number of floors, the saving may allow for the addition of one or more floors. As long as the foundation is capable of supporting the additional loads, the additional space may be the deciding factor in making a project profitable. This is especially true in cities with high land costs. Again, by dividing the land costs over a larger area of rentable space, we have improved the margins of the project.[19]

### *Steel*

Steel, with its high strength-to-weight ratio, offers architects and engineers a material well suited to the design of tall buildings. William Le Barron Jenney is credited with the first use of steel in a tall building. In his innovative design of the Home Insurance Building in Chicago, completed in 1885, he used steel and wrought iron plates and angles in the assembly of structural beams, with wrought iron and cast iron in the columns of this ten-story skyscraper. Less expensive than steel, these more brittle cast iron columns would be unacceptable in modern buildings today. In the 19th century, steel was reserved for the beams that would experience higher tensile stress levels.[20] Masonry was used to protect the metal frame from the calamities of fire. The Home Insurance Building marked the early beginning of modern fireproof construction in tall buildings. Like buildings today, this ten-story building with elevators provided office space for a growing class of office workers.

Today's architects and engineers rely on standard rolled sections to create commercial buildings that would have towered over Jenney's original design. Many innovations in material science and structural analysis are responsible for the increased height of steel-framed towers. Using standard mill sections, a 3D array of columns and beams are assembled with welded or bolted connections. In a modern steel frame structure, these joints are designed as an integrated structure to resist both dead loads and wind loads. Often, to increase the resistance to lateral loads, cross-bracing, or "X" bracing, is added to the structure. In buildings like the John Hancock in Chicago, the "X" bracing is adopted as an architectural element, though in most building designs the bracing is hidden from view (Figure 12.8). In cities where earthquakes are known to be severe, structural frames are designed not only to resist lateral loads, but also to act as dampers that absorb earthquake loads. Using tuned mass dampers, the energy from an earthquake is absorbed by the action of a massive pendulum.

Multistory steel structures built today are structural systems in which the weight of the concrete floors is transmitted to the supporting steel frame. Floor slab systems can use a metal deck welded to a steel floor beam. Concrete is then poured directly onto the metal deck, creating a strong and versatile floor system. For long spans where the floor heights are not critical, steel joists manufactured to specification and delivered to the site ready for installation offer an economical solution. With steel joist, spans of 60 ft. can be achieved for both single-story and multistory structures. In special cases in which greater spans are needed, trusses that can be a story in height can be used to create large open spaces for both commercial and industrial buildings. Attached to steel columns or masonry walls, steel joists are commonly used as part of the roof structure for large box retail outlets and warehouses. In addition to the floors and the roof of the building, the frame of a multistory building must also support the exterior curtain walls, made from an assembly of glass, metal, masonry, and composite materials. These curtain wall systems are attached to the frame with vertical mullions. In special cases, mullions can be replaced by truss-like structures that act as an armature for a wall of glass.

Finally, the structural grid must be designed to support elevators, stair towers, and the HVAC system necessary for the operation of the building. The inner core of a skyscraper that houses the elevators and mechanical systems can add significant structural stability to the frame. In many designs, the structural core acts like an internal tower that supports the floor joists that span the space from the curtain wall to the elevators. Once the frame is completed and sprayed with insulation of vermiculite or perlite, water sprinkling systems can be added to protect the structure against fire.

### *Exotic structures*

Though it is customary to feature in architectural periodicals the work of architects who create imaginative structures that can cantilever great spans or resemble a piece of sculpture, such as the Guggenheim Museum Bilbao, architectural masterpieces are not the subject for a book on real estate development. Though noteworthy for their daring approach to the use of form to create imaginative spaces, they do not serve as a precedent for projects that must turn a profit. Structures, including stadiums, built to massive proportions require a commitment of vast sums of money to complete and are generally not pursued without public support. Likewise, designs for concert halls, museums, libraries, and other public structures, created as icons and serving as major tourist attractions, are costly beyond any rational rate of return. The high cost often results from the use of exotic materials and complex construction techniques generally not applicable to projects that must promise a reasonable return to its investors. Similarly, new and largely untested technology, including 3D printing and robotic masons, has yet to offer any commercial advantages over more traditional approaches. On occasion, architectural elements inspired by contemporary architecture are introduced into more commercial buildings. Though adding to the cost of the project, they may be justified for creating a unique signature location.

## Historic restoration and preservation projects

Traditionally, masonry composed of cut stone and brick was used for most of the public buildings in the 19th century. In the US, a major incentive in the preservation and reuse of these older buildings has been a tax credit program. With the passage of the 1986 Tax Reform Act, developers received a 20% credit for renovating certified historic structures and 10% credit for buildings built before 1936.[21] With the passage of tax reform legislation in 2017, lower tax rates and a requirement that the credit is to be distributed over a five-year period have made this provision of the tax code less attractive to developers.[22] As important as the credit has been, having a specialized knowledge of historic architecture is critical to preserving the integrity of these buildings. Today, finding the skilled artisans needed in the renovation of historic architecture can be a challenge. Badly done restoration attempts can actually damage the buildings, making a bad situation worst. For example, brick buildings of the 19th century were constructed using a soft lime mortar. Over time, the mortar that binds the bricks will need to be renewed. Referred to as “repointing”, the process relies on the use of a special chisel to remove the old mortar from the masonry joints. After the removal of the old mortar joints, a mason must repoint the masonry with fresh mortar. If modern cement mix is used to complete this task, the mortar will, over time, actually place enough stress on the softer brickwork to crack and destroy it. Experts in the field of historic preservation will be able to specify the proper mix that should be used to renew these older masonry structures. Historic preservation experts can also advise developers on which contractors and tradespeople in a community are capable of completing each part of a restoration project. Carpenters, plasters, and masons who work in building preservation are undoubtedly artisans. Without their skill, a project is in jeopardy. Learning their trade is not easy, often requiring years of apprenticeship and training before they are qualified to restore a significant work of historic architecture. For these reasons, it is important to consult with established experts in the field of historic preservation during the early stages of a project.

Buildings that are not restored according to the guidelines and rules established for historic preservation may not receive the grants and tax advantages available under various historic tax preservation programs. These rules and guidelines will specify how wood, brick, plaster, and stone must be treated. Other rules will address questions on the percentage of the original facade that can be modified, what types of window are allowed, and what types of paint can be used on wood surfaces. In cases in which balustrades and other architectural detailing are lost, procedures for reproducing these lost features are recommended. For noteworthy historic buildings, renovation and preservation of the interior spaces can be both costly and time-consuming. In addition, if interior walls cannot be moved or changed, it may be all but impossible to accommodate new uses. Even under the best of circumstances, the cost of renovating these older structures may make the project unprofitable. In these cases, additional funding from grants may be critical to underpinning the cost of the project. Much planning and negotiation always precedes the renovation and reuse of historic buildings. Generally, developers who take on these projects have the experience to know how best to make these projects a success.

## Heating, cooling, and other building systems

Ultimately, a building is more than a structural system composed of columns, beams, floor slabs, and curtain walls. As an integrated system, building design must consider a number of other systems, including heating and lighting, water and sewer, security, and communication systems. Getting this right means that the structural engineers and architect must work in concert with the mechanical engineer and other consultants brought into the project. For example, if a requirement is for a cooling system to be located on the roof, this additional load must be a consideration when sizing the columns and beams that comprise the structural system. Noise and vibration, which can travel along the structure, must be isolated. Column and grid size must also take into account the use of spaces within the structure. If an auditorium or theatre is planned for the first floor of a skyscraper, the column grid will need to support massive trusses that can withstand the loads from the floors above while creating a column-free space for the auditorium. The Chicago Opera House (1885–95), designed Cobb & Frost, and the Auditorium Building (1889) by Adler & Sullivan are just two early examples of innovative design incorporating the new technologies of steel-frame construction, electric light, and mechanical air-handling systems.[23] With its massive trusses supporting the upper floors above an auditorium lit only by incandescent lights, the Chicago Opera House was a marvel of engineering and architectural design.[24]

In today's environment, energy costs associated with the operation of residential and commercial constitute approximate 39% of all energy consumption in the US (Figure 15.8).[25] New technology provides opportunities to create integrated environmental systems for reusing waste heat, grey water, and other waste products. To the extent that a building can be designed to reduce operating costs, these savings will be reflected in its bottom line. In many parts of the world, a growing interest in energy efficiency and sustainability has the attention of the public and business community. In the future, it may be common for buildings to be able to generate their electric power and recycle their water. Buildings that are less expensive to operate will have a higher resale value. Though economic return will always be the prime driver of investment in green technology, governments will continue to take positive action to provide incentives for reducing the ecological footprint of our cities.[26]

Finally, in many urban environments, security has also become a concern of both tenants and owners. To the extent that these concerns can be addressed during the initial conceptual design for a building, there will be less dependency on the use of surveillance technology to provide security for its occupants. An area of study, crime prevention through environmental design (CPTED), has emerged in recent years to address the concerns of public and private security. Many of the principles governing CPTED are focused on physical design, increasing visibility, and removing obstructions from sight lines, while limiting access to pedestrian spaces through the use of bollards, walls, and other features. Architectural and planning firms with CPTED expertise can provide critical input into the design of buildings that require sensitivity to security.

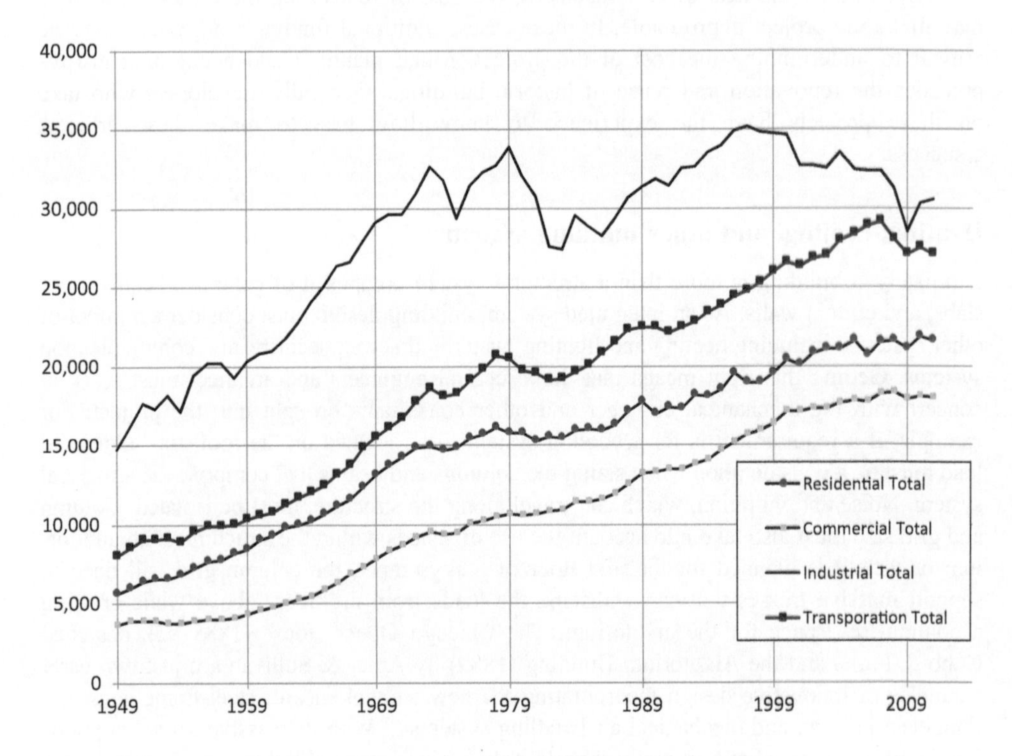


*Figure 15.8* Energy consumption estimates by sector, 1949–2011 (Trillion Btu).

Data Source: US Energy Information Administration.

## Managing the design process

Design software is an important tool in design. For simple buildings, the use of simple graphic techniques, including bubble diagrams, can help during the conceptual phase of design. In today's design environment, computer-aided design (CAD) software has become an essential tool in the design of architectural space. Unlike the CAD tools of the early 1980s, which could support only 2D design, CAD applications have evolved into sophisticated expert design tools that enable the architect to track space requirements from concept to final working drawings. Building information modeling (BIM), as a design tool, offers many advantages over traditional paper and pen.[27] Built on top of extensive databases, these applications greatly facilitate the planning, design, construction, and operation of buildings. Within this environment, while an architect is drawing a line that represents a wall, the application is creating the actual wall section for the building. As the designer develops the floor plans with exterior and interior walls, the program links this information to a database that tracks the materials and specifications on how the wall should be assembled on the construction site. At any point in the design process, it is possible to query the database and to calculate the quantity and type of materials needed to build every wall in the building.

For example, a wall represented by a double line on a computer display represents an actual wall system. Even a simple exterior wall may include wallboard, wood studs, insulation, particle board sheathing, vapor barrier, and vinyl siding. Having an application that not only tracks the required materials, but also has the ability to generate detailed drawings for the general contractor can help eliminate confusion and simplify the construction process. A library of standard wall sections built into all BIM applications can be modified at any point to meet specific code requirements. By merely clicking on a wall section, changes can be made easily and quickly to the design. If there is a need to substitute a different wall section, this can be done by a replacement command rather than by redrawing the entire building. Once this substitution is made, it is possible to obtain updated itemized costs for the project. With the walls in place, windows and doors are added to the building, and the application adjusts the quantity of materials needed to construct the wall.

In the process of designing, building elements are modified and updated. Facilitating the design process, many of these elements are available from the manufacturer as parametric blocks. By merely clicking on a window or door, another choice in the database may be substituted. If there is a need to detail an office space, workstations, chairs, bookcases or conference tables can be added and tracked by the interior designer. Using this same approach, the structure, HVAC and electrical systems are integrated into the design. The structural engineer can size the columns and beams, while the mechanical engineer can design the HVAC for heating and cooling the building. At each step in the process, the design can be evaluated for both cost and performance.

Building information modeling offers many time-saving advantages over design approaches used a decade ago. Design changes can be easily updated without the need to manually redraw with pen and ink. Finally, alternative design solutions can be compared and evaluated for their efficiency in their use of space and cost. Finally, even as the designer works on the design in plan view, the application can represent a plan view as a 3D model. The benefits of having a 3D model over a 2D plan can be essential in visually inspecting spaces as you would in a walkthrough of a finished building. In these virtual spaces, it is possible to ask questions such as: "Does the furniture fit in the spaces? Are site lines appropriate? Are the dimensions and scale appropriate for the type of use? Does the proposed structural grid of columns and beams work well for the room sizes?" Having columns placed in the center of classroom or seminar space is a signal that the proposed structural grid is not compatible with the division of the architectural space.

It is even possible at this early stage to test the proposed layout with future users of the space, thus providing critical feedback early in the design.

- Can pedestrians find where they are going?
- Are office workers spending too much time walking back and forth from meeting spaces to workspaces?
- Have they properly considered accessibility and accommodation for those with physical limitations?

These are all questions that could be tested in a virtual environment of the building prior to construction. Though some architects may begin the discussion with their clients over a cup of coffee, sketching in a notebook, ultimately, getting more accurate numbers will require the use of more precise tools.

Using BIM will also facilitate the design process by coordinating the input of engineers, contractors, and suppliers. At critical stages in the design cycle, the architect will need input from structural mechanical, electrical, and acoustic engineers. By using a BIM approach, it is possible for the architect, engineers, and general contractor to engage in a collaborative design process. By having a single BIM database for a project, the software can help eliminate costly mistakes arising from miscommunication. Engineers can test the compatibility of complementary systems. For example, heating and lighting are linked to daylight and the design of the curtain wall. With BIM software, it is possible to create solutions that are pretested in a virtual environment before they are built.

Building information modeling also offers an opportunity to link the design directly to project management software. Many software tools are now available to monitor time and expenses at each stage in the development process. Project management tools can offer an integrated view of the development process, including tools for cost estimation, budgeting, scheduling, resource management, and evaluation. The decision of whether to use project management software is a function of complexity and scale of the project. Clearly, a small renovation project may not require a computer application to track the progress of a few trades on the site at any one time, in contrast to major commercial development project, which will require the monitoring of numerous contractors and subcontractors on the site at any one time. Having accurate as-built drawings is invaluable when the inevitable modifications are made in the future.

Finally, a word must be said about the use of consultants. Nobody likes to be managed by their consultants. Though most developers are not experts in architecture, planning, and engineering, to be successful, knowledge of the design and construction process is essential. Ultimately, it is the registered architect and engineer that will assume responsibility for the design and construction before a building permit is issued. Though all developers need advisors with specialized expertise, it is the developer who must ultimately make critical design choices. Your consulting engineers and architects are not in the position to make these critical financial decisions. Aesthetics can be an important issue in both the sale and rental of real estate. However, aspiring young architects with a desire for recognition among their peer group may not always be concerned with the profit and loss of the project. Getting a photo of the finished building into a leading architectural publication is not bad for business, but it should not be the prime concern. In some cases, the additional costs of achieving the architect's goals may do little to help the bottom line. Real estate development is not about ego; it is about building value.

## Notes

1 Information on Leadership in Energy and Environmental Design (LEED) can be found at https://new.usgbc.org/leed
2 Youngheng, 2014; Eichholtz et al., 2012; Eichholtz et al., 2010.
3 Levy, 1980.
4 Vitruvius, 1960; Palladio, 1965; Pierson, 1976, 360–368.
5 Houghton and Dietz, 1990.
6 Vitruvius, op. cit., 88.
7 Aldrich and Lambrechts, 1986; Lambrechts, 2008a, 2008b.
8 McCullough, 1972.
9 Condit, 1973.
10 Condit, 1961.
11 *New York Times*, 1988; Goldberger, 1988; Durgin, 2004.
12 Mostoller, 1985.
13 Post, 2012; Stephens, 2012.
14 Berger, 2011.
15 Irish, 1989.
16 Tse et al., 2009; Wong et al., 2011.
17 Note that the notation used for milled lumber varies from actual size. A 2 × 4 actually measures 1½ by 3½ in. and a 2 × 6 measures 1½ by 5½ in.
18 von Moos, 1983.
19 Construction Week Online Construction, "Week lists the world's tallest concrete buildings," September 27, 2019. www.constructionweekonline.com/article-9184-top-10-worlds-tallest-concrete-buildings/1
20 Condit, op. cit., 1973, 88; Larson, 1987.
21 See Tax Reform Act of 1986 (PL 99–514; Internal Revenue Code Section 47 [formerly Section 48(g)]) for references to the 20% tax credit for the certified rehabilitation of certified historic structures and a 10% tax credit for the rehabilitation of non-historic, non-residential buildings built before 1936.
22 Kamin, 2017.
23 Condit, op. cit., 1964, 59.
24 Ibid.
25 www.eia.gov/Energyexplained/index.cfm?page=us_energy_use
26 Kramers et al., 2014; US Energy Information Administration, "Use of Energy in the United States Explained, Energy Use in Commercial Buildings," last updated September 28, 2018, www.eia.gov/Energyexplained/index.cfm?page=us_energy_commercial; US Energy Information Administration, "Energy Consumption by sector", 2012, www.eia.gov/totalenergy/data/annual/showtext.php?t=ptb0201a
27 Azhar, 2011. For information on BIM and CAD software, see: www.autodesk.com/; www.bentley.com/en/products/brands/microstation

## Bibliography

Aldrich, Harl P. and James R. Lambrechts. "Back Bay Boston, part II, groundwater levels, man-made structures that permanently lower groundwater levels can have adverse effects on buildings with water table sensitive foundations." *Civil Engineering Practice* 1, no. 2 (Fall 1986): 31–64.

Azhar, Salman. "Building information modeling (BIM): Trends, benefits, risks and challenges for the AEC industry." *Leadership and Management in Engineering* 11, no. 3 (July 2011): 241–252.

Banham, Reyner. *The Architecture of the Well-tempered Environment*. London: The Architectural Press, 1969.

Banham, Reyner. *Theory and Design in the First Machine Age*. New York: Prager Publishers, Inc, 1970.
Berger, Joseph. "Triangle fire: A half-hour of horror." *The New York Times* (March 21 2011).
Kamin, Blair. "Historic preservation tax credit is saved, but weakened," *Chicago Tribune*, December 20 2017.
Burg, David F. *Chicago's White City of 1893*. Lexington, KY: The University Press of Kentucky, 1976.
Condit, Carl W. *American Building Art, The Twentieth Century*. New York: Oxford University Press, 1961.
Condit, Carl W. *The Chicago School of Architecture, A History of Commercial and Public Buildings in the Chicago Area, 1987–1925*. Chicago: The University of Chicago Press, 1973.
Cowan, Henry J. *Architectural Structures, an Introduction to Structural Mechanics*. New York: American Elsevier Publishing Company, Inc., 1971.
Deng, Youngheng and J. Wu. "Economic returns to residential green building investments: The developers' perspective." *Regional Science and Urban Economics* 47 (2014): 35–44.
Durgin, Frank H. "Lessons from studies of cladding pressures at the John Hancock Tower in Boston structures." *Congress 2000: Advanced Technology in Structural Engineering* 103 (2004).
Eichholtz, Piet, Nils Kok, and John M. Quigley. "Doing well by doing good? Green office buildings." *American Economic Review* 100, no. 5 (2010): 2492–2509.
Eichholtz, Piet, Nils Kok, and Erkan Yonder. "Portfolio greenness and financial performance of REITs." *Journal of International Money and Finance* 31, no. 7 (2012): 1911–1929.
Fitch, James Marston. *American Building and the Environmental Forces that Shape It*. New York: Schocken Books, 1972.
Frampton, Kenneth. *Modern Architecture, A Critical History*. New York: Thames and Hudson Inc, 1992.
Goldberger, Paul. "The limits of urban growth." In *Architecture and Design in the Postmodern Age, on the Rise*. New York: Penguin Books, 1985.
Houghton, Robert C. and Deborah L. Dietz. "Design and performance of a deep excavation support system in Boston, Massachusetts." *Geotechnical Special Publication* 25 (1990): 795–816.
Irish, Sharon. "A 'machine that makes the land pay': The West Street building in New York." *Technology and Culture* 30 (April 1989): 376–397.
Jordy, Willliam. H. *American Buildings and Their Architects, the Impact of European Modernism in the Mid-Twentieth Century*. Garden City, NY: Anchor Books: Doubleday, 1972.
Kostof, Spiro, ed. *The Architect, Chapters in the History of the Profession*. New York: Oxford University Press, 1977.
Kramers, Anna, Mattias Höjer, Nina Lövehagen, and Josefin Wangel. "Smart sustainable cities: Exploring ICT solutions for reduced energy use in cities." *Environmental Modelling & Software* 56 (2014): 52–62.
Lambrechts, James. "The problem of groundwater and wood piles in Boston, an unending need for vigilant surveillance." Master's thesis, Wentworth Institute (2008a).
Lambrechts, James. "The problem of groundwater and wood piles in Boston, an unending need for vigilant surveillance." *ASEE Annual Conference and Exposition, Conference Proceedings* (June, 2008b).
Larson, Gerald and Roula Mouroudellis Geraniotis. "Toward a better understanding of the evolution of the iron skeleton frame in Chicago." *Journal of the Society of Architectural Historians* 46, no. 1 (March 1987): 29–48.
Levy, Richard M. *The Professionalization of American Architects and Civil Engineers, 1865–1917, Ph.D. thesis*. Berkeley: University of California, 1980.
Levy, Richard M. "Data or image: The influence of professional culture on computing in design." *Acadia* (Fall 1997): 8–11, 22–23.
McCullough, David. *The Great Bridge*. New York: Avon Books, 1972.
Mostoller, Michael. "The towers of Frank Lloyd Wright." *Journal of Architectural Education* 38, no. 2 (Winter 1985): 13–17.
Palladio, Andrea, Isaac Ware edition, intro by A. Placzek. *Four Books of Architecture*. New York: Dover Publications, 1965.

Parker, Harry and James Ambrose. *Simplified Engineering for Architects and Builders*. New York: John Wiley and Sons, Inc., 1993.

Paul, Goldberger. "Architecture view: A novel design and its rescue from near disaster." *New York Times* (April 24 1988).

Pierson Jr., William H. *American Buildings and Their Architects, the Colonial and Neo-Classical Styles*. New York: Anchor Press, 1976.

Post, Nadine M. "Designers apply lessons from world's tallest tower to improve future 'megatallest'." *Engineering News-Record* (March 5, 2012).

Salmon, Charles G. and John E. Johnson. *Steel Structures: Design and Behavior*, 4th ed. Upper Saddle River, NJ: Prentice Hall, 1997.

Salvadori, Mario. *Why Buildings Stand Up*. New York: W. W. Norton & Company, Inc. 1990.

Sauder School of Business. *Commercial Property Analysis*. Vancouver, BC: University of British Columbia, Real Estate Division, 2009.

Stephens, Suzanne. "Kingdom come." *Architectural Record* 200, no. 5 (May, 2012): 160.

Stringer, M.E. and S. P. G. Madabhushi. "Effect of liquefaction on pile shaft friction capacity." *Fifth International Conference on Recent Advances on Geotechnial Earthquate Engineering and Soil Dynamics and Symposium*, Paper No. 515a (May 2010).

Timoshenko, S. and D.H. Young. *Elements of the Strength of Materials*. Princeton, NJ: D. Van Nostrand Company, Inc., 1968.

Tse, K.T., P.A. Hitchcock, K.C.S. Kwok, S. Thepmongkorn, and C.M. Chan. "Economic perspectives of area dynamic treatments of square tall buildings." *Journal of Wind Engineering and Industrial Aerodynamics* 97 (2009): 455–467.

Vitruvius, Pollio. *The Ten Books on Architecture, Trans. Morris Morgan*. New York: Avon Books: Dover Publications, 1960.

von Moos, Stanislaus. *Le Corbusier, Elements of a Synthesis*, 2nd ed. Cambridge, MA: The MIT Press, 1983.

Wisely, William H. *The American Civil Engineer 1852–1974: The History, Traditions and Development of the American Society of Civil Engineers*. New York: ASCE, 1974.

Wong, S.K., K.W. Chau, Y. Yau, and A.K.C. Cheung. "Property price gradients the vertical dimension." *Journal of Housing and Built Environment* 26 (2011): 33–45.

## Web resources

Autodesk, Inc.
www.autodesk.com/

Bentley's Microstation
www.bentley.com/en/products/brands/microstation

Leadership in Energy and Environmental Design (LEED) programme
https://new.usgbc.org/leed

# 9 Project management

## Introduction: why do I need project management skills?

Project management is a critical skill in real estate development. As in many endeavors, showing up a "day late and a dollar short" never works out very well. Project management is taught as part of many professional programs. In engineering, architecture, and real estate development, a significant component of success is good planning and its execution. In real estate development, working with planners, architects, engineers, and contractors requires close attention to scheduling. Everyone has a schedule that must be accommodated if the work is to get done on time. Even for those who will never hire a contractor, or engage an engineer or an architect, project management skills can be useful in everyday life. Many of life's challenges require good project management skills. Whether it is planning a wedding, applying to college, securing your first job, buying a house, starting a business, or preparing for retirement, you will need a plan. Just as importantly, you will need to execute your plan to achieve success.

Project management techniques are essential to the successful outcome of projects, even those of modest size. Without these skills, success may depend on just plain dumb luck. Even with some knowledge of project management techniques, success in life is never assured. However, if things turn in the wrong direction, you will have some strategies that may help you avoid the risk of total disaster. If nothing else, you will have a rational framework to help keep your emotions in check when a disaster occurs. That said, project management is not a panacea. It cannot replace good judgment or creativity. Also, project management does require time. Most importantly, to be effective, the tools must actually be used and incorporated into business operations. However, devotion to project management activities will not guarantee success in business. Spending critical time tracking business activity should never be an excuse for not paying attention to tasks that need attention. In real estate development, like any other business endeavor, it is the relationship you have with your partners, staff, agents, consultants, suppliers, and bankers that will keep you in business. Project management should be viewed as just another tool to keep a business on track.

### *Avoiding failure*

Perhaps the most important reason for taking the time to learn about project management is that it may help you avoid failure. The reasons why you may have not achieved success in business or life should be considered. Perhaps you had no plan or, if you had a plan, you were unable to actually execute this plan. Not having a plan may be the fundamental reason for not achieving success. Many of us believe we have a plan when there is not even one document that provides proof of a plan. It can be argued, in this case, that you really do not have a plan. A written plan can be evaluated, shared, and revised. Without a plan, you are never quite sure if you are meeting your goals and objectives. Without a plan, it is likely that decisions made one day may be changed the next, without rhyme or reason. Resources may not be available when they are needed and the schedule may be a bit unrealistic. Without a schedule, consultants and subcontractors may lack deadlines; their deliverables could be hit or miss at best. Most importantly, without a clear and well-defined schedule, there may not be sufficient cash flow to cover expenses when they come due. Clearly, a plan is a critical part of managing any project.

## Time, budget, and performance

Achieving success requires attention to three important elements: time, budget, and performance. Clearly, there is a relationship that exists between these three constraints. First, time is always limited. Not being able to complete a project on time will have dire consequences. For a development project, penalties may be incurred because a tenant cannot move in on the start date of the lease or financing will not be available as planned because a certificate of occupancy has been held up by a building inspection. There are major milestones that, if not met, can ruin the financial stability of a project. Often, failing a major objective was not the consequence of missing a single deadline, but of failing to make progress on a series of related tasks. With each missed deadline, more time is needed to complete the project. With a plan and proper project management techniques, the art of managing time can keep a project on schedule.

The budget is always a key constraint. Work is done either for a fixed fee or on an hourly basis plus material cost. Anyone who has hired a plumber on an hourly basis plus parts knows that, without a fixed price, costs can quickly escalate. Unfortunately, it is often difficult to get an accurate estimate of cost. In renovating an older building, often new problems are discovered only after the work begins, even though an inspection was completed by a registered architect or engineer. Now, the only choices are to complete the required work, which will increase the cost of the project, or to sell outright to another investor, possibly at a loss. Even when work is done on a fixed-fee basis, changes are inevitable. A turn in the market may require the addition of new features to the project to make the product salable. Negotiating these additional costs after the project has started can create delays and increase costs, while reducing the profit margins.

Performance is key to success. Merely completing a job on time and within budget is meaningless if basic requirements of the project have not been satisfied. If a foundation is poured one day only for you to discover major cracks and settlements several months later, the work has not been completed satisfactorily and now requires costly repair. If the framing has already been completed, the ramifications are wide-reaching, placing the entire project at risk. Clearly written specifications outlining how the work is to be executed must be part of any contract. However, ensuring that the specifications have been met is also critical. Having time-tested procedures to ensure that work has been completed according to specifications is key.

## Project management defined

In simple terms, project management involves the application of knowledge, processes, methods, and techniques to achieving expressed objectives. Within a business framework, reaching your project goals will require many different skill sets, including strategic planning, team building, appropriating resources, assigning tasks, budgeting, monitoring, and tracking progress. Most courses and textbooks on the subject begin with a discussion of the difference between projects and processes generally found in a manufacturing environment in which the same item is produced over and over again. Baking cookies or crackers would be a clear example of such a process. It is unlikely that the process changes very much from one batch to the next. In fact, changing the recipe or process might result in some very disappointed customers. Differences between processes and projects can be summarized as in Figure 1.9. In our example of the bakery producing crackers or loaves of white bread, the same jobs must be performed every day by trained staff using tried-and-true practices. Their goal is bread that is ready for delivery to the supermarket every morning. Projects are less predictable. In the world of baking, creating a wedding cake would be a project. Each cake will be made to suit a different customer, each with unique demands and preferences. Depending on the size of the wedding party, what is currently in fashion, the budget, and the date of the wedding will determine how the baker will plan and execute this work of art in cake and icing. To arrive at the final design of this masterpiece, the baker will hold meetings with the client, to show them pictures and examples of cakes. Sample cakes will be tasted, delivery dates will be discussed, and contracts will be reviewed and signed. The seriousness of this agreement, of course, is understandable: a wedding without the requisite cake would be a major disappointment.

In real estate development, each new endeavor is unique. No two projects are ever alike. Sites will differ in location, size, and acquisition cost. Even for a developer that

| Process | Project |
|---|---|
| Process or product is a constant | Process or product is unique |
| Ongoing | One time, limited life |
| People have dedicated responsibilities | Responsibilities can change with product or process |
| Well-established systems | System must adapt to the product or process |
| Based on established practices | Responsive to changing practice |
| Supports the status quo | Upsets the status quo |

*Figure 1.9* Differences between process and project management: basic components.

Source: Pinto, 2007:7.

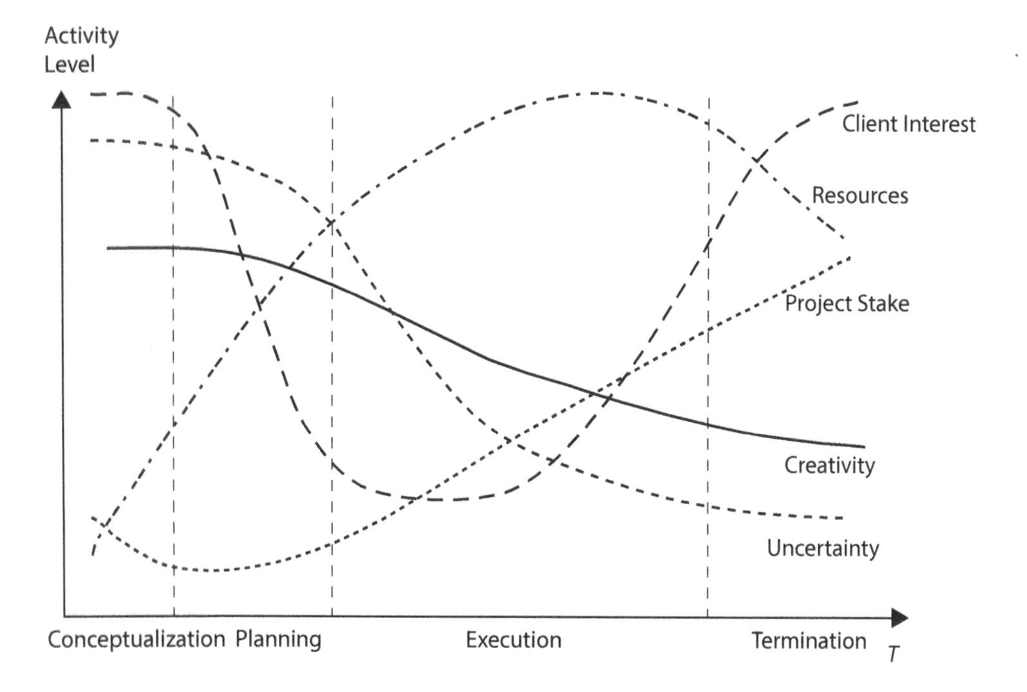


*Figure 2.9* Project management process: life cycles of project management.

Source: Sohmen, 2002.

specializes in one type of development, for example single-family homes for the first-time home buyer, changes in the market or the location of the site will have a bearing on the design of the product. With varying lot size and proportions, adjustments to even a standard plan are needed to make the basic design fit the lot. Zoning requirements may differ from one locale to the next, forcing developers to consider house designs that accommodate a narrow lot, or conform to more restrictive height requirements or larger setbacks. Changes in consumer preferences will also compel the developer to make changes to floor plans and other features in the design, including the kitchen and bathroom layouts, choice of appliances, paint colors, and flooring materials. Changes in an economy in which incomes are falling may force the developer to find ways of lowering their costs so that prospective buyers can afford their product. Not being flexible and adaptive could result in building a house nobody wants to buy.

Every project has a beginning and an end (Figure 2.9). Four stages or phases characterize every project: conceptualization, planning, execution, and termination.

- During the *conceptualization* stage, the goals and objectives of the project are determined. Also, an initial budget for capital and expenses is estimated, along with the needs for professional expertise. Stakeholders are identified that may have a direct bearing on the outcome of the project.
- During the *planning* stage, a set of specifications and a preliminary schedule are established, with the goal of assigning tasks to individual team members and consultants.

- During the *execution* stage, the actual work is completed and the actual project objectives are achieved.
- Finally, in the *termination* stage, the project comes to a close, with any outstanding responsibilities assigned to the client or third party.

Clearly, in the first stage of the project, resources are much fewer than in either the planning or execution stage. Depending on the project, planning and execution will consume the lion's share of the resources assigned to the project. As the project progresses, the level of commitment increases. With more and more resources assigned to the project, any risk of failure will generate potentially larger and larger losses. Early on in the project cycle, if the project should suddenly come to a close, the expenditure of resources may be very modest. In real estate development, the conceptualization stage will require site surveys and some preliminary calculations. At this stage, the costs can be quite small relative to the development continuing through to construction. Real estate development projects may involve different levels of financial commitment depending on the project objectives. If the project involves only the assemblage of properties and rezoning, the level of financial resources will be much less than a project involving new construction.

## Application of project management to real estate development

Real estate development encompasses projects of different scales and complexity. Obviously, a shopping center, office tower, and subdivision will require years of planning before construction can even begin. In comparison, acquiring a small apartment building and making some repairs and modest improvements can be accomplished with much less time and effort. Project types may be usefully divided into those that are already built and those that will need to be designed and constructed. For projects involving the purchase of a property such as an apartment building, you will need to conduct your due diligence and secure financing. Leases will need to be reviewed. In this case, project management involves scheduling meetings with all those involved in the project, including the real estate agent, banker, attorney, architect, engineer, general contractor, and urban planner. Each will need sufficient time to review any pertinent documents, including leases, permits, and city by-laws. If properties need to be inspected, arrangements will need to be made with the existing owner for onsite access. During this period of investigation, it is important to discover any outstanding or potential code violations. The goal is to uncover potential failures in the physical integrity of the building. Telltale cracks and leaks should alert the buyer to a need for more investigatory work by the architect and engineer. It is likely that some improvements and repairs will be needed. As with many older buildings, there will be a list of deferred maintenance items that will need to be addressed. As part of any sales negotiation, there may be items that the seller will need to rectify before the buyer takes ownership.

Time will be needed to secure financing for the project. In acquiring an investment property, financing will be the most important consideration. Potential financing options, including interest rates and terms, should be reviewed. As part of this review, some consideration should be given to the potential risks; the future is always uncertain. A property may look like a real winner on paper, generating positive cash flows when the economy is doing well, only to need a cash infusion to break even when there is a slight downturn in the economy. Rents may not go up as projected. During bad economic times, tenants may not renew their leases or bankruptcy may force them to vacate even before the end of their terms, leaving your building partly empty. To keep existing tenants, you may be forced to lower rents, so as not to lose them to another landlord offering a sweeter deal. If interest rates should rise during this period, the consequences can be financially devastating. A variable mortgage even with caps could push up expenses into a range where the project is no longer profitable. In examining most likely, worst, and best cases, reasonable estimates on rents and expenses need to be established. As in any financial decision, having reliable data is essential, which all takes time to collect and review.

Where there are significant changes planned for the building, new construction or renovations are always major undertakings. For modest-sized projects, your architect will be responsible for the planning, design, and construction supervision. Clearly, scale matters. A renovation or construction of a single-family home will require the expertise of an architect or house designer. Architectural drawings, including plan views and elevations, are part of the initial concept phase. In the past, ink and pencil drawings, sketches, and desktop models were indispensable to the design process. Since the 1980s, computer-aided design (CAD) software has transformed all aspects of practice. Now, all drawings from concept to final working documents are computer outputs from a 3D virtual model.

Having the ability within a CAD program to generate views on the fly allows the architect or designer to more easily convey their ideas. In the past, when changes to the building design required architects and drafters to produce a new set of documents; today, once the model is updated, producing any view of the model is incidental. In the past, if the client requested a change, a new meeting would need to be scheduled to review design changes; today, much of the communication can occur electronically, without the need to physically meet. Updates to the design, whether they are different window types or perhaps skinning the building in stucco rather than wood siding, can be done as part of an interactive design session. Also, unlike drawings, when there are changes, it is easy to get a quick list of materials. In the past, the list of the number and type of doors, windows, and fixtures, along with an estimation of materials, including brick concrete and lumber, would require the attention of an estimator. The estimator's job was to read architectural drawings and specifications, compile lists, and tabulate costs. In using the table output from the CAD application, it is possible to select the manufacturer for major elements specified in the plan, including windows, doors, and cabinets and fixtures. In addition, specifications of surface treatments, including carpet, tile, and paint, may cover much of the details of the project. If an interior designer is working on a project, actual samples are submitted to the client for approval.

For projects of modest size, a contractor will review the document and provide an estimate of costs. Contracts should also specify that costs are not to exceed a specific amount and dates for the completion of major phases of the project. If a new heating, ventilation or air conditioning (HVAC) system is required, you may get several quotes from competing contractors or you may have the contractor take care of the installation. Anyone who has done a kitchen renovation knows that even simple projects can quickly get out of hand when changes are made during the renovation process. If you had specified linoleum and now decide to substitute tile, the cost is now not only for the tile, but for the additional expense for the new subfloor. It is important to remember that any changes made after the contract has been awarded can result in additional charges and delays. A contractor may bid low on designs that are poorly documented, knowing that their profit margin will increase with each new requested change and addition.

On major projects, the architect will be responsible for site survey, the design of the building, and the supervision of the trades. On large projects, a hierarchy composed of junior architects, clerks, inspectors, and supervisors would report to the architect on the status and progress of the building process. What is noteworthy about projects up till the early 20th century is the lack of of documents by today's standards. Plans, elevations, and perhaps a section were all the drawings needed, suggesting that much of an architect's time was taken up by onsite supervision in order to ensure that designs were properly interpreted. Specifications that outline all aspects of the construction process are the creation of the modern age. Today, a large-scale project requires careful planning and the management of a complex multistage process.

Building information management (BIM) software is a significant aid in overseeing the construction process. Even the early stages of estimating a project's cost and completion date can take time. Though it may be easy to update a computer database, negotiating with contractors and subcontractors on price and schedule may require time. Estimates during the concept phase are often based on average costs per square foot. Using numbers that are based on historical data for the type and scale of construction should be considered as a rough first approximation. More detailed estimates will accompany each stage in the planning process. It is important to work with your planner, market analysts, architect, and engineers during the conceptualization stage. Drawings and estimates may appear crude, but should provide sufficient detail to decide if the project is worthy of proceeding to the planning phase.

In the planning stage, the design will need to be finalized. Based on final drawings and specifications, more accurate costing will establish the required financial commitments for the project. It is not uncommon that your initial offer can change as the lender learns more about your project. As they acquire more knowledge about your project, banks and lending institutions may re-evaluate the risks that have emerged since your initial proposal. With greater risk, the terms of the loan may change. A larger down-payment with higher points may be required, forcing you to look for additional investors or perhaps even another source of financing. This can create delays or additional costs. If an additional lender is needed to plug the gap, that subordinate lender may demand higher interest rates. If financing looks uncertain, it may be the time to bail on the project before going any further.

A goal of the planning phase of the work will be the completion of those detailed drawings and specifications critically needed for the bidding process. Again, BIM software and CAD software have been an important development for managing the construction process. Without the need of estimators, who traditionally would be required to read the architect's drawings to determine the quantities of materials needed for the project, estimates can be managed by merely updating the database. One important feature of BIM software is the ability to compare the costs associated with different building strategies. A change to the wall section of a building or the substitution of a window from a different manufacturer can be compared as part of the total strategy for cost reduction. One other feature of BIM is the ability to examine the construction process. Estimating the sequence of tasks and time during the scheduling process is essential if milestones are to be met. Building information modeling software can also be used to visualize the entire construction process. For example, when employing tower cranes for multistory buildings, it may be necessary to test for interference with other cranes and surrounding buildings.

Depending on the scope of the project, at every stage experts may need to assist the efforts of architects, planners, engineers, and others, such as geologists or soil scientists. Managing this process can be an extremely demanding exercise. For example, the architect must be aware of any changes made by the city. The architect must keep the engineer abreast of any changes that will impact the structural, mechanical or electrical design. For example, if the number of floors for a commercial office building changes, the HVAC system will need to be redesigned. In some cases, if the number of floors should change dramatically, it could impact the design of the building itself, requiring more elevators. Once the drawings and specifications are finalized, they will go out to bid. General contracting firms capable of managing the construction process are key to reaching the goal of making the architectural design a reality. General contractors (GCs) have the professional expertise and experience essential in sourcing the needed materials and coordinating the activities of the numerous trades that will be needed to construct the type of building you are planning. Clearly, if you are under contract to build a hospital, you will want to hire a GC with experience on similar projects to complete your design. Professional publications including *Engineering News Record*[1] can provide useful data on the size and expertise of general contracting firms in your area.

### *Bidding process*

Once architectural drawings and specifications (the bid documents) are prepared, they are available to go out to bid. In the example of a house renovation, you may work with an architect or designer to secure estimates on the proposed work. In many cases, based on the recommendations from your design professional, you will review the estimate, make some decisions on various options such as the type of fixtures, appliances, and cabinets, and then sign a contract with a price not to be exceeded by more than a certain percentage. There will also be a date on which the project is to be completed, often with late penalties. On small projects, the work may proceed on good faith, given your personal relationship with both the designer and contractor. However, if you talk to anyone who has done a renovation to their home, you will hear countless horror stories. One common complaint is the length of time the project took to complete. When contractors are juggling several jobs, it is not uncommon for a project without any late penalties to be delayed. Not all of the delays are the fault of the contractor; clients who are unclear or change their mind from day to day can easily create headaches for all concerned. Also, undiscovered issues such as mold behind wallboards and rotting floorboards cannot always be anticipated. However, once they are discovered, solutions must be found, adding time and expense to the project.

In new construction, good management will be essential to the success of the project. Though the tendency is to award the contract to the lowest bidder, performance must also enter into this decision. To guard against giving the contract to a GC who will try to subvert best practices, it is important to examine other of their completed projects. Site visits and interviews with former clients will provide some reference points for determining if your low bidder is a good best choice. In reviewing the bids if all the bids have come in higher than expected, it may be necessary to revisit the original design. Costly detailing may be simplified or eliminated. Finishes that require higher-than-average expenditure in materials and labor may be substituted with ones that are less costly. Decisions at this stage may place project objectives in jeopardy. For example, if a commercial office building is intended to be "Class A" space, using lower-quality materials for the walls and floor surfaces in the lobby may elicit a negative reaction from prospective tenants. The result may be a cheaper-looking building that will not demand top rents. However, if the preference was purely to satisfy an architectural impulse that only members of the design profession will appreciate, finding a less costly solution is demanded. Working with an architect who can create value will be essential in keeping costs from escalating out of control.

In reviewing each bid submission, it is important to consider who will actually do the work. Each phase of the construction will be dominated by a few subcontractors: foundation, framing, windows, HVAC, and roofing. Work that is done late or substandard will be more difficult to rectify and can easily delay important milestones. During the course of reviewing contracts, specific warranties and penalties for work not completed on time and according to specifications will need to be included. If the choice of a particular supplier or subcontractor is known to be problematic, the discussion should focus on finding another subcontractor. Good communication between all those responsible for the success of the project may help avoid future problems. Having one member of the team make a change without the approval of the architect, engineer or GC can result in misunderstandings and more costly additions later.

#### *Design–build: a different approach*

Over the last few decades, design–build firms have become an alternative to a GC who completes the work designed by the architect and engineer. Under this arrangement, a single firm will design and build your project. Clearly, this arrangement can offer some advantages. Experience with both design and construction should reduce the risk of having a building that is difficult to construct. However, with design–build, there will be less control over the details of the design. Once the basic concept has been approved, contracts may specify that control over final details is in the hands of the design–build firm. If this design comes early in the process, it is possible that the final product may not be to the liking of the client.[2] Also, meeting performance standards defined by the client will require carefully drawn contracts that can ensure that each phase of the project satisfies all conditions. However, for generic building types such as warehouses, a design–build firm may be a very cost-effective option.

### ***Construction phase***

Once the contract has been awarded, the implementation stage of the project commences. Now that the contractor has taken over as the project leader, your role and responsibilities do not fade away. Daily progress reports must be reviewed. Weekly meetings with the contractor and design team may help avoid the major crises that often feel inevitable during the construction phase. Hindsight is always 20/20, but it is not always possible to see into the future. Material costs can change with global demand and temporary shortages of specific components may require adjustments to the plan. Innovative designs can result in assembly issues that will require modification and may result in delays. Also, no one can predict a labor union strike or boycott. Perhaps the message is "acknowledge that there will be crises that will emerge at some time during the construction process and you won't be surprised when they do". That is why all projects have a line item for contingencies that can range from 10% to 20%, depending on the type of building. Also, having a bit of slack in a schedule can provide some insurance that the project can make its important milestones.

### *Scheduling and resource management*

Scheduling is at the core of managing any project. Without a deadline, it would be impossible to coordinate the multitude of subcontractors needed to complete any construction project. Consider our simple kitchen renovation. A specific order to the process must be followed if good progress is to be made towards completing the project. In renovating our kitchen, the tasks include the removal of existing cabinets, putting down a new floor, the placement of the cabinets, new counter tops, new tile backsplashes, painting, plumbing, the installation of appliances, and the installation of new lighting. Clearly, it would be impossible to have everyone complete their task at the same time. It is also imperative that certain tasks be completed before others begin. The old cabinets and floor need to come out before the new flooring and cabinets can be installed. But if you plan to paint the walls a new color, it would be easier to complete this task before installing the new cabinets. Each task may be done by a different trade, requiring careful scheduling of their time. Coordination is required if the entire project is to run smoothly. It is the responsibility of the GC to make sure that the entire job gets done by the deadline. If there are delays in any of the required tasks, there should a strategy for recovering some of the lost time, otherwise the deadline will not be met. It is generally unwise to sign a contract without a date for completing the project, even though the contractor may be reluctant to accept a date. Nobody wants to live without their kitchen, though I suspect many home owners will find that they are going out to dinner and ordering takeout for many more nights than they had planned before their dream kitchens are ready to use.

In putting together a schedule, the first step is determining the scope of the project, which includes a list of all the activities, products, and resources needed for the job. To begin the process, it is useful to make a list of all the tasks. In project management jargon, this is often referred to as the work breakdown structure (WBS). The WBS defines the task or activities as a hierarchical list. At the lowest level in the WBS are the work packages. Each work package has an assigned time required for completion and an associated cost. It is also important to know the order needed to complete these tasks. In our kitchen renovation example, it was important to know which tasks must be completed before others can begin. In project management terminology, a task that must begin before another is referred to as the predecessor, while those that come after a task are the successors. To visualize the relationships among all of the work packages, it is customary to use graphical diagrams. Though these network diagrams can be constructed with the help of software, for simple projects they could also be done on a piece of graph paper.

#### *Project management terminology*

Being acquainted with the terminology used by project managers can be helpful in understanding the output of project management programs. It will also be important in communications with all parties involved in project management.

**Scope** A description of the products, results, resources, and time associated with a project or a phase of the project.

**Work breakdown structure (WBS)** A set of tasks organized in a hierarchical structure.

**Work project** The deliverable associated with a specific task; can include specifications, time, and cost.

**Activity descriptor** Task description.
**Alphanumeric code** Code assigned by the software application to track each task.
**Path** An ordered set of activities describing a set of tasks completed over time.
**Node** Junction point along a path.
**Predecessor** A task that must be completed prior to the completion of the current task.
**Successor** A task that will be completed after the current task is completed.
**Activity duration** Time between the start and finish activity.
**Early start (ES) date** The earliest possible date a task can begin without creating overall delays in the project.
**Late start (LS) date** The latest possible date a task can begin without creating overall delays in the project.
**Early finish (EF)** Earliest possible date by which the task or activity can be completed.
**Late finish (LF)** Latest possible date by which a task or activity can be completed without delaying the project.
**Forward pass** Calculation of early start and early finish (ES–EF) dates for a network.
**Backward pass** Calculation of late finish and late start (LS–LF) dates for a network.
**Merge activity** An activity for which two or more successor activities must be completed.
**Burst activity** An activity for which two or more activities must be completed.
**Serial path** Path on which one task is completed before another begins.
**Parallel path** Path on which two or more tasks take place at the same time.
**Lag** Logical relationship between the start and finish of two tasks.
**Float** The amount of extra time that is available before there will be an overall delay in the project; normally calculated as the difference between the LS and ES or LF and EF.
**Critical path** In network analysis, the path that must be followed if the project is to be completed on schedule; any extension required for any of the tasks on this path will result in a delay in the overall project.
**Critical path method (CPM)** A series of analytical techniques that results in the minimum time needed for a project.

*Network analysis*

To visualize our home renovation project as a network diagram, we could start with the list of activities and draw a box on our graph paper showing each task (Figure 3.9). Going from left to right, we can establish a timeline showing how each task is related to the next.

In Figure 3.9, we have established the relationship between the first four tasks. The project begins with the removal of cabinets and countertops. Once the old cabinets and countertops are gone, we can remove the old floor. Then, we will call in the plumber and electrician to begin their work. It would be useful to add to our simple diagram the start date and how much time is estimated for each task. If we add this information to our network diagram, we would now have a more complete view of our construction process (Figure 4.9). If I have the start date for the project and the time required for each task, we can now look at our calendar and schedule when each of the trades should start their work.

A simple calculation is required to determine the start and end dates for each node. In our example, if the start date of the project is November 21, 2016, and the first task (removal of cabinets and kitchen countertops) takes one day, then the earliest this part of the project can finish is November 22, 2016 (Figure 5.9):

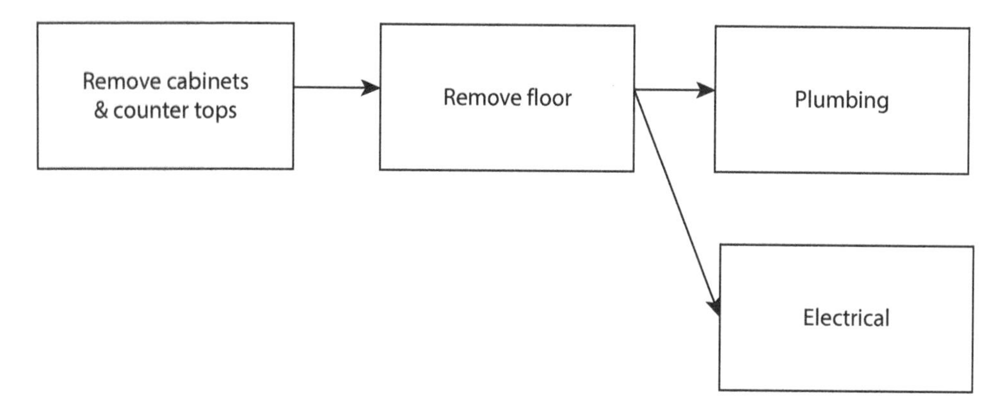


*Figure 3.9* Network diagram.

| Early Start | ID Number | Late Finish |
|---|---|---|
| | Activity Descriptor | |
| Late Start | Duration | Late Finish |

*Figure 4.9* Network node defined.

| Nov 21 2016 | ID 1 | Nov 22 2016 |
|---|---|---|
| | Remove cabinets and counter tops | |
| Late Start | 1 | Late Finish |

*Figure 5.9* Network node, early start, early finish.

$$\text{Early Start} + \text{Duration} = \text{Early Finish}$$
$$\text{ES} + \text{Duration} = \text{EF}$$
$$\text{Nov 21} + 1\text{ Day} = \text{Nov 22}$$

Similarly, if the second task of removing the floor takes one day, the earliest this task can be completed is one day later, given that this task cannot start until the cabinets and countertops are completed on November 22, 2016 (Figure 6.9):

$$\text{ES} + \text{Duration} = \text{EF}$$
$$\text{Nov 22} + 1\text{ Day} = \text{Nov 23}$$

Applying network analysis to a critical path analysis is referred to as a forward pass. When we use popular project management software Microsoft Project, our network analysis gives the start and ES and EF dates for our entire project (Figure 7.9).

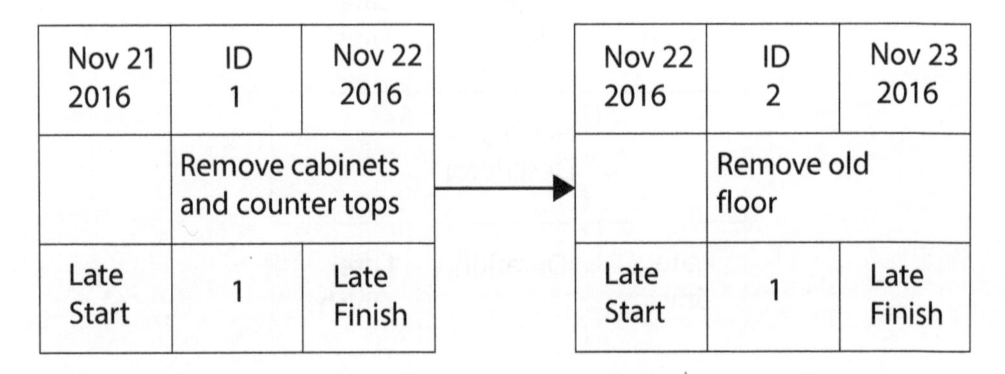


*Figure 6.9* Network node, early start, early finish; removal of cabinets, countertops, and old floor.

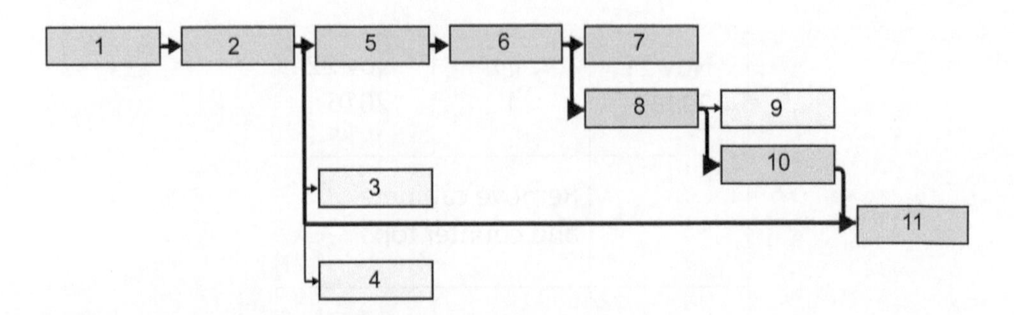


*Figure 7.9* Network analysis using Microsoft Project showing all the tasks for a kitchen renovation: 1) remove cabinets and countertops; 2) remove flooring; 3) plumbing; 4) electrical; 5) install new subfloor; 6) tile floor; 7) paint wall; 8) install new cabinets; 9) tile backsplash; 10) install counters; 11) install lighting. NB The gray boxes and darker lines show the critical path through the network diagram.

Once we have completed the forward pass analysis on our network diagram, it is also possible to conduct what is referred to as a backward pass. In this case, we begin with the last task and the date by which we must complete the job and work backward. Using the following relationship, we can now determine the latest date on which each task can begin and still meet our final deadline:

$$\text{Late finish} - \text{Duration} = \text{Late Start}$$

or

$$\text{Lf} - \text{Dur} = \text{LS}$$

If for example, we know that the contract calls for completion of all work on December 8, 2016, then we know that the final task of wiring the new lighting must end on that date. If this job takes two days, then in order to complete the final contract on time, this last task must begin on December 7, 2017, as long as the contract gives you the entire day to complete all work. If the contract specifies a time of 8am or 12 noon, then we will need to start the final task of wiring on December 6. Of course, having a bit of slack in your project is always a good idea in all projects working to a tight deadline.

In analyzing any network, it will become clear that there are many tasks that can begin at the same time. These parallel paths are illustrated in our home renovation project. In this example, we have the installation of the new subfloor, plumbing, and electrical beginning at the same time just after the old floor is being removed (Figure 8.9). If you do not want your plumber and electrician tripping over the laborers who are putting down the new subfloor, then you will have to wait until they complete their work (Figure 9.9).

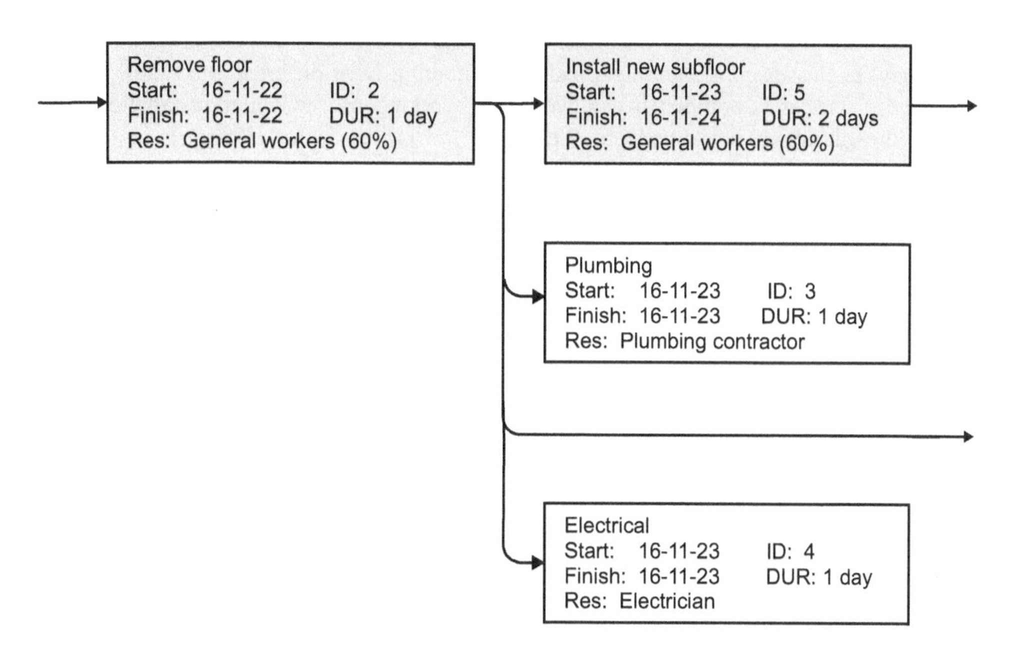


*Figure 8.9* Network diagram: installing new subfloor, plumbing, and electrical all begin on the same date.

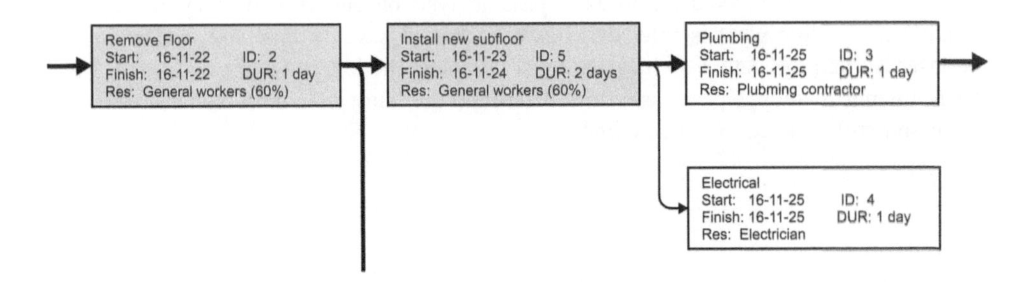


*Figure 9.9* Network diagram: plumbing and electrical begin after the new subfloor is completed.

*Lags and network analysis*

"Finish to start" is a common logical relationship between a set of tasks. In this case, we need to complete our task before moving on to the next. In our example, it might be possible that, before we have the counters and cabinets out of the way, we can remove the old floors. In the example, the end date of ID1 is the same as the start date of ID2. If we are not sure how long the removal of the old cabinets and floors will take, we can always add some lag or extra time – perhaps another day between these two tasks (Figure 10.9). This delay between when one task ends and another begins is referred to as "lag". In that case, the finish date of ID1 would be November 22 and the end date of ID2 would be November 24.

"Finish to finish" is the case in which two or more tasks must be completed before we can move on to the next (Figure 11.9). We need the plumbing, electrical, and a new boiler for the heating system to be installed before moving on to the installation of radiators. Not all tasks will take the same amount of time nor will they all begin on the same day. In this example, the electrical and plumbing have different end dates, so only after the electrical is completed can we begin working on the boiler installation. This makes the plumbing work node to the boiler the critical path. Having a lag for the

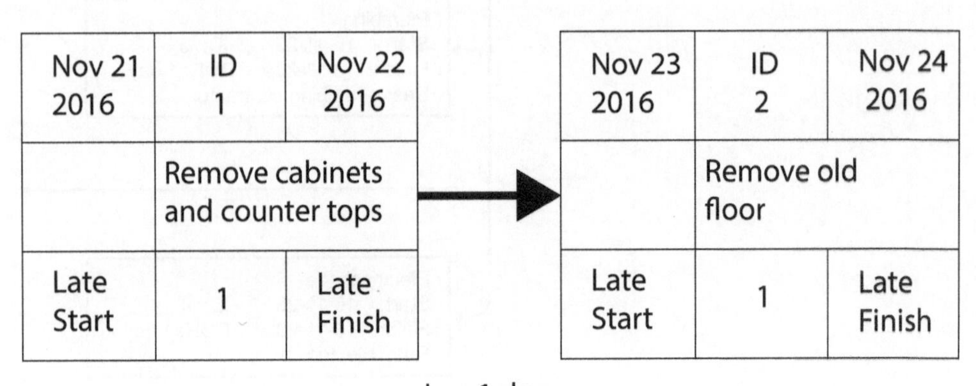


*Figure 10.9* Kitchen renovation: start/end with a lag of one day.

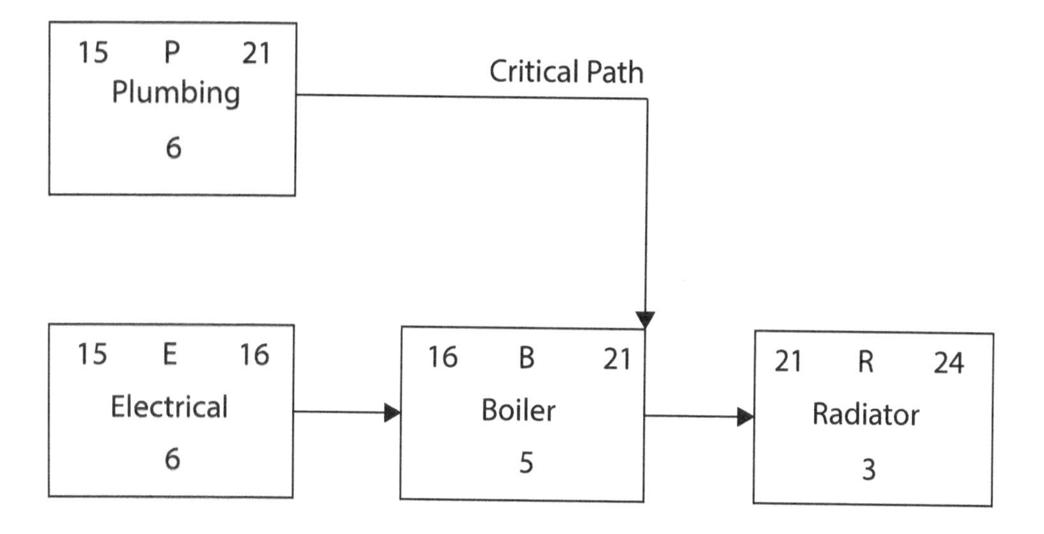


*Figure 11.9* (a) Finish to Finish;

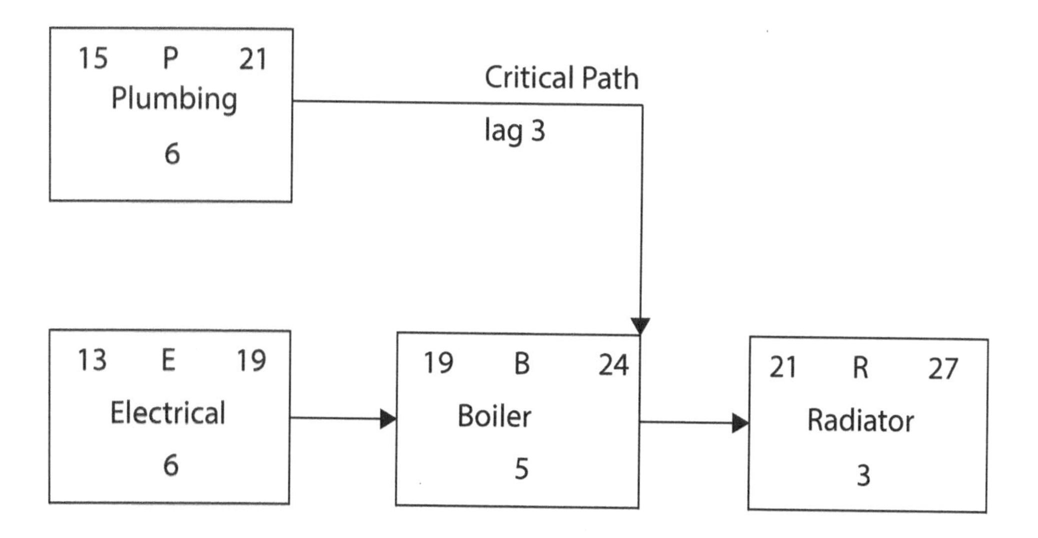


*Figure 11.9* (b) Finish to finish with a lag period of three days.

plumbing may not be a bad idea, given that we may have unforeseen issues with our old house. However, if the plumber finishes their work a day earlier, the work on the radiators could begin a day sooner.

There are also tasks that can start on the same day. This is referred to as a "start to start" relationship (Figure 12.9). In this case, both plumbing and the boiler installation will have some overlap in their scheduling. Start to start must maintain their logic in

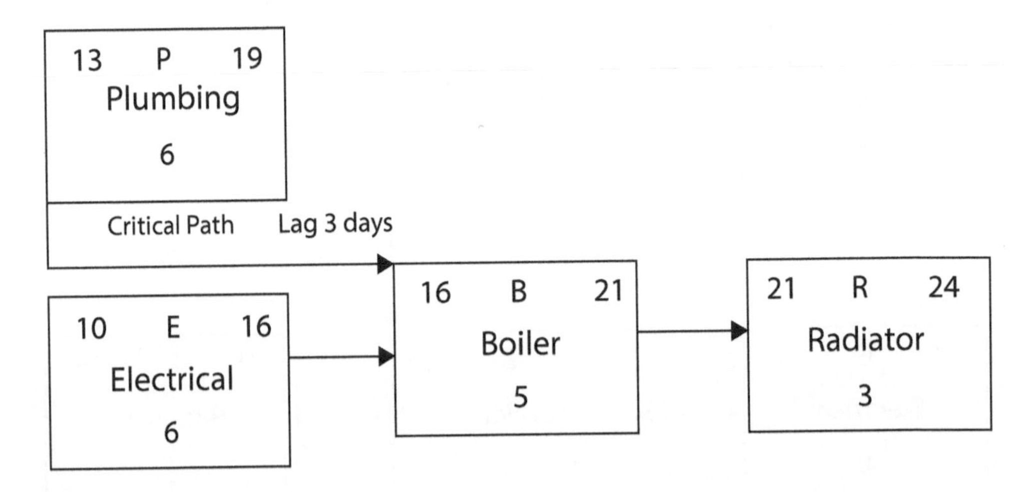


*Figure 12.9* Start to start.

a forward and backward pass through the network. Good examples of this relationship can be found when fast-tracking the design and construction of a building. You may not have all the details of the interior design nailed down when the foundations are being poured or even when the structural steel is being erected. Given the experience of the architect, with knowledge of the constraints and specifications, this may not be a problem. In this case, past experience indicates that there is little connection between interior design issues and the foundation, so it is possible to start pouring the foundation long before interior design decisions are made.

### *Gantt charts*

Anyone who has worked with a project manager has seen a Gantt chart. Often, they appear on a wall, so that anyone who visits the office can see the progress being made on the project. These timelines show each phase of the project and the dates on which each task was completed. These charts, devised by Henry Gantt, were created during the early years of the scientific management movement. Gantt was a graduate of Johns Hopkins University and Stevens Institute, where he earned a Master's degree in mechanical engineering, which brought him in contact with Frederick Taylor at Midvale Steel in 1887. New management practices of Taylor and Gantt would revolutionize the factory floor by replacing the foreman's control over production with that of the industrial engineer. In 1903, Gantt published an article entitled "A graphical daily balance in manufacture", which outlined this new approach to tracking work.[3]

Using the scientific management approach, every operation would be dissected and reduced to tasks that could be analyzed and improved with an eye towards efficiency. As part of this approach, time and motion studies would eliminate any wasted effort by workers. Levels of production would be based on these experiments and bonuses would be paid when these minimums were surpassed. Gantt's charts were first used in military production by the US Ordnance Department during World War I. In the days before computers, these charts were created with paper and pen as a means of establishing the process of production on an assembly line floor. Also shown in these charts was the level of resources committed to each task, along with the responsible individuals.[4] Today, with personal computing power at everyone's fingertips, it is possible to produce Gantt charts using software that can analyze and display the status of a project. When data comes from mobile devices in the field, these charts can be a powerful tool in analyzing a project in real time. However, when there is a lack of attention given to updating the progress made in the field, inappropriate or bad decisions can be made on the basis of old data. Management procedures should be in place to guarantee that field data is always entered into the database in a timely fashion.

Gantt charts are simply timelines showing the order of execution of all the tasks that are scheduled for a project. In the output shown in Figure 13.9, all of the tasks for the kitchen renovation appear, along with their duration, start date, finish date, predecessor, and who is responsible for each the task (resource). On the right-hand side along the top of the chart is a scale, with the days of the month. A bar representing each task shows the start and end dates. As is customary, the bars in gray show the critical path through the project.

### *Project management software*

Creating a simple Gantt chart could be done in a spreadsheet program or on a piece of graph paper. But unless the project is more than only a few tasks that could be managed with a simple calendar program, project management software can greatly simplify the process. The selected software should work seamlessly with other core applications. Standalone systems that require your staff to keep separate databases in sync will always be subject to error. Besides introducing error into your accounting systems, having independent systems will waste valuable staff time that could have been spent on other tasks. Depending on the focus and range of your operations, it may be desirable to use BIM systems, which track the design and construction management process. An application developed by Autodesk, a CAD software developer, allows managers to monitor all aspects of the design process from concept design to final drawings. Programs such as Rivet, a parametric design application developed by Autodesk, is one of several applications that have transformed architectural practice. Unlike early CAD programs, which

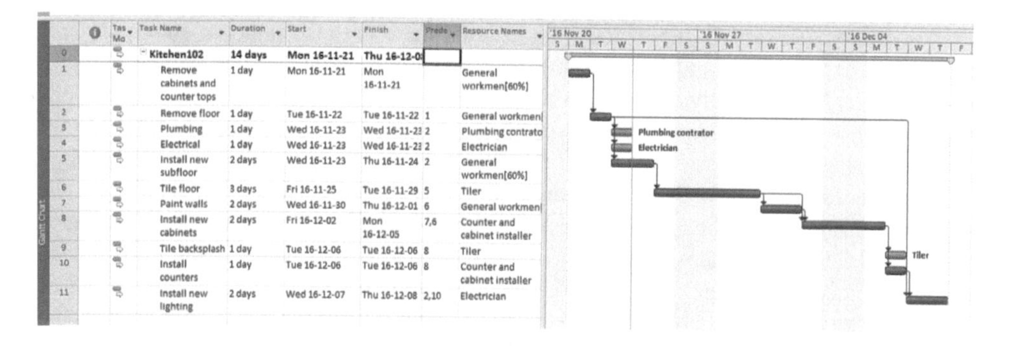

| | Task Name | Duration | Start | Finish | Prede | Resource Names |
|---|---|---|---|---|---|---|
| 0 | Kitchen102 | 14 days | Mon 16-11-21 | Thu 16-12-0 | | |
| 1 | Remove cabinets and counter tops | 1 day | Mon 16-11-21 | Mon 16-11-21 | | General workmen[60%] |
| 2 | Remove floor | 1 day | Tue 16-11-22 | Tue 16-11-22 | 1 | General workmen |
| 3 | Plumbing | 1 day | Wed 16-11-23 | Wed 16-11-2 | 2 | Plumbing contrato |
| 4 | Electrical | 1 day | Wed 16-11-23 | Wed 16-11-2 | 2 | Electrician |
| 5 | Install new subfloor | 2 days | Wed 16-11-23 | Thu 16-11-24 | 2 | General workmen[60%] |
| 6 | Tile floor | 3 days | Fri 16-11-25 | Tue 16-11-29 | 5 | Tiler |
| 7 | Paint walls | 2 days | Wed 16-11-30 | Thu 16-12-01 | 6 | General workmen |
| 8 | Install new cabinets | 2 days | Fri 16-12-02 | Mon 16-12-05 | 7,6 | Counter and cabinet installer |
| 9 | Tile backsplash | 1 day | Tue 16-12-06 | Tue 16-12-06 | 8 | Tiler |
| 10 | Install counters | 1 day | Tue 16-12-06 | Tue 16-12-06 | 8 | Counter and cabinet installer |
| 11 | Install new lighting | 2 days | Wed 16-12-07 | Thu 16-12-08 | 2,10 | Electrician |



*Figure 13.9* Gantt chart: kitchen renovation.

Source: Microsoft Project

replaced hand drawing with computer graphics, today's CAD applications focus on creating a 3D database for a building. By using a cloud-based application, all contractors and subcontractors can monitor any design or schedule change. By relying on tablets in the field, any issue that can arise on site can be documented, with updates sent to the architect and project manager.

For those who are not involved directly in the design and construction, Microsoft Project offers an easy-to-use software solution for most undertakings. A tool for conducting network analysis, creating Gantt charts, managing resources, and monitoring the budget, its power lies in its ability to provide an integrated approach to project management. When these applications are linked into existing accounting software, they can also serve as the front end for invoicing, bill paying, and general accounting functions. For those who are newcomers to project management software, Microsoft Project provides sample templates that can be used as a basis for creating and managing many aspects of the development process, including concept development, design, and construction.

### *Project management with Microsoft Project: renovation of an office building*

Even for modest projects, using project management software such as Microsoft Project can greatly simplify the scheduling process. On complex projects, project management software will be indispensable. As an example of how Microsoft Project might be used to assist in managing the renovation of a commercial building, our case study involves the acquisition of an older building office built during the 1920s. Like many buildings from this period, the fireproof structural frame is steel, with a masonry facade. Some repairs have been made over the years to the roof and the brick facade. Though the building is in generally good shape, we will need to make some repairs and improvements. The inspection reveals that the old tar-and-gravel roof will need to be replaced and small repairs to parapet wall are likely. Apparently, some of the flashings is less than perfect and a few water stains have revealed themselves on the ceiling of rooms located on the upper floor. Also, there are plans to improve the efficiency of the building by replacing all of the single-pane windows with modern double-glazed windows. The masonry will need to be reappointed in several places, and some of the brickwork and terracotta detail may need to be replaced or repaired. The electrical, plumbing, and heating systems are in working order. However, in order to suit the needs of the new tenants, there will be some upgrades to the bathrooms. An additional bathroom on the ground floor is needed. To satisfy one of the new tenants, there are also plans to subdivide several of the larger spaces and install some new lighting throughout. The carpet will also need to be replaced. Finally, there are a few cracks in the sidewalk along the front of the building that will need attention. However, this last item is not a high priority.

Given that we took possession in the fall, the greatest concern is making the building watertight before the snow flies. Once the building is sealed up, we can worry about the interior renovation. In scheduling the tasks, estimating the hours and wages paid to each worker will be the first steps. We begin by entering the list of tasks and the time needed to complete each. If we are also acting as the GC, we will need reliable quotes for all of the planned work. If we are using subcontractors with whom we have worked in the past, their experience should help us assess the estimated costs for all the work that needs to be done. For example, in this case, a trusted subcontractor has provided a reliable estimate of the time needed to replace all the windows. However, when it comes to the roof, we might have less confidence in the estimates that were submitted. Though an inspection revealed that only the old tar paper and gravel need to be removed before installing the new roof, once into the project we may find other issues that will need to be addressed. It is always useful to consider adding some slack to tasks that are more likely to require more time. The critical path shown in red will help us plan the work that needs to happen before the onslaught of winter (Figure 14.9).

When we review our first Gantt chart for this project, the roof is not scheduled to be completed till December 6, which may be a concern, especially if this is the year in which we have an early snowstorm (Figure 15.9). In completing our scheduling for the work, we have to remember that we have only entered the number of hours in total for each task. Since each task may require small crews, we need to enter this pertinent information into the application. For this project, the roofing and painting will be done by crews of three. To assign the actual number of workers to each task in Microsoft Project, we can enter the data about each worker into the resource sheet, including their hourly and overtime rate. We can also enter the percentage of their work week that each will devote to this project. It is not uncommon for someone to be splitting their time between more than one job (Figure 16.9).

When this data is added to our table (under the column "Resource name"), we can now see that the names of all those engaged in the project (Figure 15.9). With three roofers working full-time on this project from day 1, the repairs on the roof will be completed by December 6, with the entire project to be completed by January 19 (Figure 17.9). However if we are still concerned, we could add another crew to the job. In adding additional resources to the project, we need to remember that tasks are completed by small teams. If we want to speed up the process of installing the windows, we may need to add another team of three workers. Similarly, we cannot add only one more roofer to the job if we are concerned about an early winter; instead, we will need to add another team of three.

There may be occasions on which we cannot find additional trades to work on our project. When the construction industry is doing well, finding skilled trades can be very difficult. In Microsoft Project, like many project management software, the default is for your workers not to work overtime during the week or on weekends. In the resource table, however, we have the ability to enter overtime rates. Paying overtime may be required when there are significant penalties involved for completing the contract after a specific date (Figure 16.9).

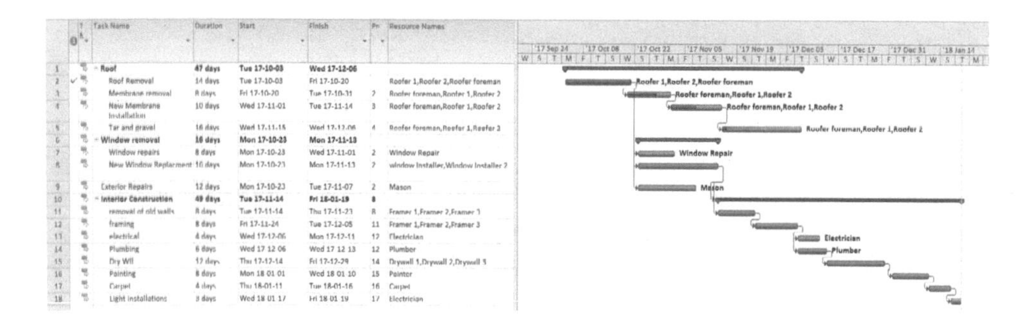

| | Task Name | Duration | Start | Finish | Pr | Resource Names |
|---|---|---|---|---|---|---|
| 1 | Roof | 47 days | Tue 17-10-03 | Wed 17-12-06 | | |
| 2 | Roof Removal | 14 days | Tue 17-10-03 | Fri 17-10-20 | | Roofer 1,Roofer 2,Roofer foreman |
| 3 | Membrane removal | 8 days | Fri 17-10-20 | Tue 17-10-31 | 2 | Roofer foreman,Roofer 1,Roofer 2 |
| 4 | New Membrane Installation | 10 days | Wed 17-11-01 | Tue 17-11-14 | 3 | Roofer foreman,Roofer 1,Roofer 2 |
| 5 | Tar and gravel | 16 days | Wed 17-11-15 | Wed 17-12-06 | 4 | Roofer foreman,Roofer 1,Roofer 2 |
| 6 | Window removal | 16 days | Mon 17-10-23 | Mon 17-11-13 | | |
| 7 | Window repairs | 8 days | Mon 17-10-23 | Wed 17-11-01 | 2 | Window Repair |
| 8 | New Window Replacment | 16 days | Mon 17-10-23 | Mon 17-11-13 | 2 | window Installer,Window Installer 2 |
| 9 | Exterior Repairs | 12 days | Mon 17-10-23 | Tue 17-11-07 | 2 | Mason |
| 10 | Interior Construction | 49 days | Tue 17-11-14 | Fri 18-01-19 | 8 | |
| 11 | removal of old walls | 8 days | Tue 17-11-14 | Thu 17-11-23 | 8 | Framer 1,Framer 2,Framer 3 |
| 12 | framing | 8 days | Fri 17-11-24 | Tue 17-12-05 | 11 | Framer 1,Framer 2,Framer 3 |
| 13 | electrical | 4 days | Wed 17-12-06 | Mon 17-12-11 | 12 | Electrician |
| 14 | Plumbing | 6 days | Wed 17 12 06 | Wed 17 12 13 | 12 | Plumber |
| 15 | Dry Wll | 12 days | Thu 17-12-14 | Fri 17-12-29 | 14 | Drywall 1,Drywall 2,Drywall 3 |
| 16 | Painting | 8 days | Mon 18 01 01 | Wed 18 01 10 | 15 | Painter |
| 17 | Carpet | 4 days | Thu 18-01-11 | Tue 18-01-16 | 16 | Carpet |
| 18 | Light installations | 3 days | Wed 18 01 17 | Fri 18 01 19 | 17 | Electrician |

*Figure 14.9* Gantt chart: renovation of a historic commercial building.

| | | Task Mode | Task Name | Duration | Start | Finish | Pre | Resource Names |
|---|---|---|---|---|---|---|---|---|
| 1 | | | **Roof** | **47 days** | **Tue 17-10-03** | **Wed 17-12-06** | | |
| 2 | ✓ | | Roof Removal | 14 days | Tue 17-10-03 | Fri 17-10-20 | | Roofer 1,Roofer 2,Roofer foreman |
| 3 | | | Membrane removal | 8 days | Fri 17-10-20 | Tue 17-10-31 | 2 | Roofer foreman,Roofer 1,Roofer 2 |
| 4 | | | New Membrane Installation | 10 days | Wed 17-11-01 | Tue 17-11-14 | 3 | Roofer foreman,Roofer 1,Roofer 2 |
| 5 | | | Tar and gravel | 16 days | Wed 17-11-15 | Wed 17-12-06 | 4 | Roofer foreman,Roofer 1,Roofer 2 |
| 6 | | | **Window removal** | **16 days** | **Mon 17-10-23** | **Mon 17-11-13** | | |
| 7 | | | Window repairs | 8 days | Mon 17-10-23 | Wed 17-11-01 | 2 | Window Repair |
| 8 | | | New Window Replacement | 16 days | Mon 17-10-23 | Mon 17-11-13 | 2 | window Installer,Window Installer 2 |
| 9 | | | Exterior Repairs | 12 days | Mon 17-10-23 | Tue 17-11-07 | 2 | Mason |
| 10 | | | **Interior Construction** | **49 days** | **Tue 17-11-14** | **Fri 18-01-19** | **8** | |
| 11 | | | removal of old walls | 8 days | Tue 17-11-14 | Thu 17-11-23 | 8 | Framer 1,Framer 2,Framer 3 |
| 12 | | | framing | 8 days | Fri 17-11-24 | Tue 17-12-05 | 11 | Framer 1,Framer 2,Framer 3 |
| 13 | | | electrical | 4 days | Wed 17-12-06 | Mon 17-12-11 | 12 | Electrician |
| 14 | | | Plumbing | 6 days | Wed 17-12-06 | Wed 17-12-13 | 12 | Plumber |
| 15 | | | Dry Wall | 12 days | Thu 17-12-14 | Fri 17-12-29 | 14 | Drywall 1,Drywall 2,Drywall 3 |
| 16 | | | Painting | 8 days | Mon 18-01-01 | Wed 18-01-10 | 15 | Painter |
| 17 | | | Carpet | 4 days | Thu 18-01-11 | Tue 18-01-16 | 16 | Carpet |
| 18 | | | Light installations | 3 days | Wed 18-01-17 | Fri 18-01-19 | 17 | Electrician |

*Figure 15.9* Resource allocation in Microsoft Project.

| | | Resource Name | Type | Material Label | Initials | Group | Max. Units | Std. Rate | Ovt. Rate |
|---|---|---|---|---|---|---|---|---|---|
| 1 | | Roofer foreman | Work | | R | | 100% | $65.00/hr | $95.00/hr |
| 2 | | Roofer 1 | Work | | R | | 100% | $65.00/hr | $95.00/hr |
| 3 | | Roofer 2 | Work | | R | | 100% | $65.00/hr | $95.00/hr |
| 4 | | Plumber | Work | | P | | 100% | $100.00/hr | $150.00/hr |
| 5 | | Electrician | Work | | E | | 100% | $100.00/hr | $150.00/hr |
| 6 | | Painter | Work | | P | | 100% | $60.00/hr | $75.00/hr |
| 7 | | Carpet | Work | | C | | 100% | $55.00/hr | $75.00/hr |
| 8 | | General Contractor | Work | | G | | 100% | $200.00/hr | $0.00/hr |
| 9 | | Project Manager | Work | | P | | 100% | $150.00/hr | $0.00/hr |
| 10 | | Window Repair | Work | | W | | 100% | $85.00/hr | $120.00/hr |

*Figure 16.9* Renovation of a historic commercial building, resource allocation.

Source: Microsoft Project

## *Crashing a project*

Accelerating, or crashing, the project can involve a combination of several strategies. In examining our schedule, if we can shorten the time along the critical path, we can reduce the time to completion. Adding another crew to help remove the old gravel-and-tar roof might be a good first step in shortening our time. If we assign additional resources to the project, we should be certain that these new workers will not be in each other's way. Space limitations on the job site may restrict how many workers can be added to do a specific set of tasks. Having people trip over each other is never a good

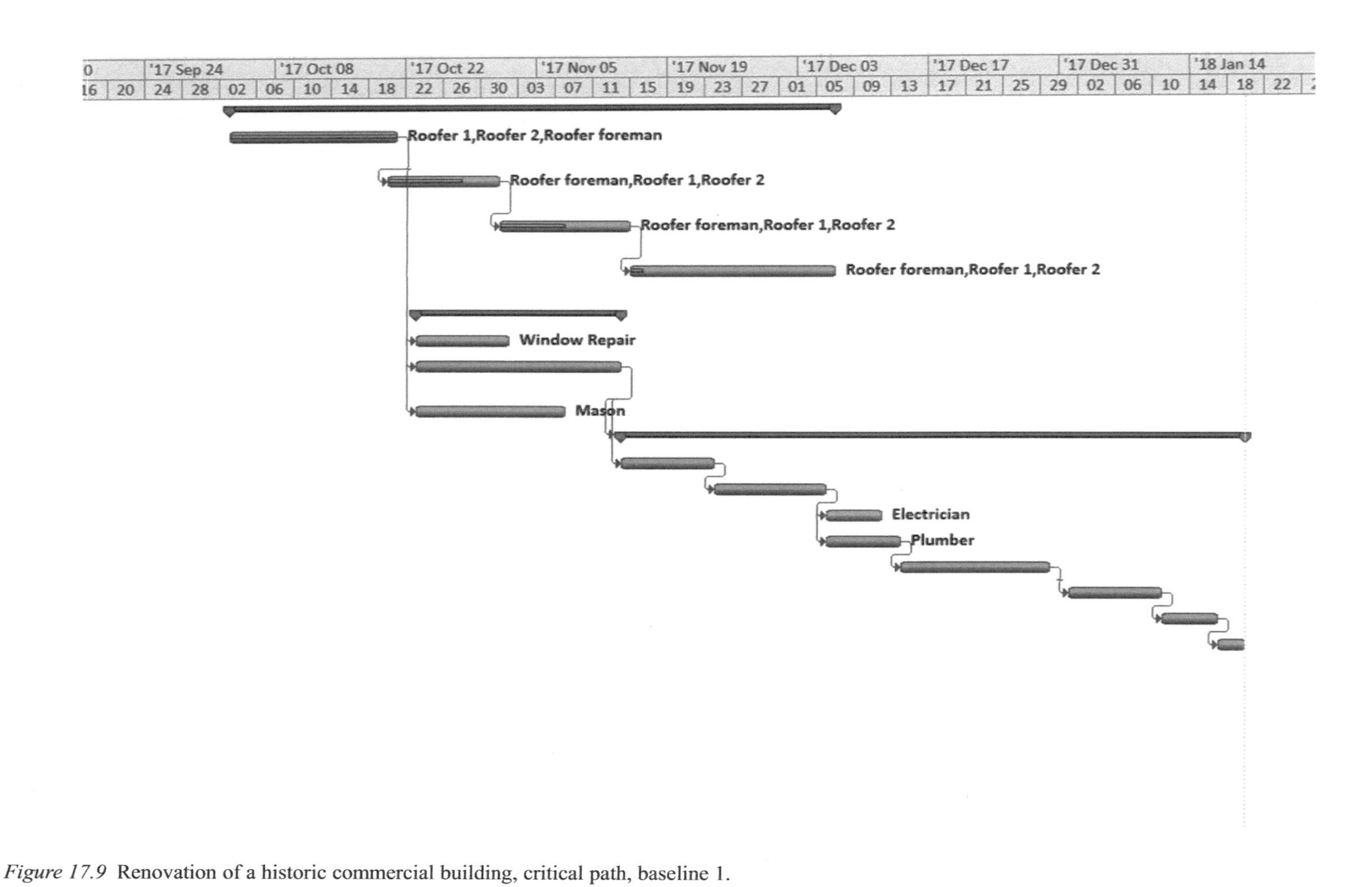


*Figure 17.9* Renovation of a historic commercial building, critical path, baseline 1.

Source: Microsoft Project

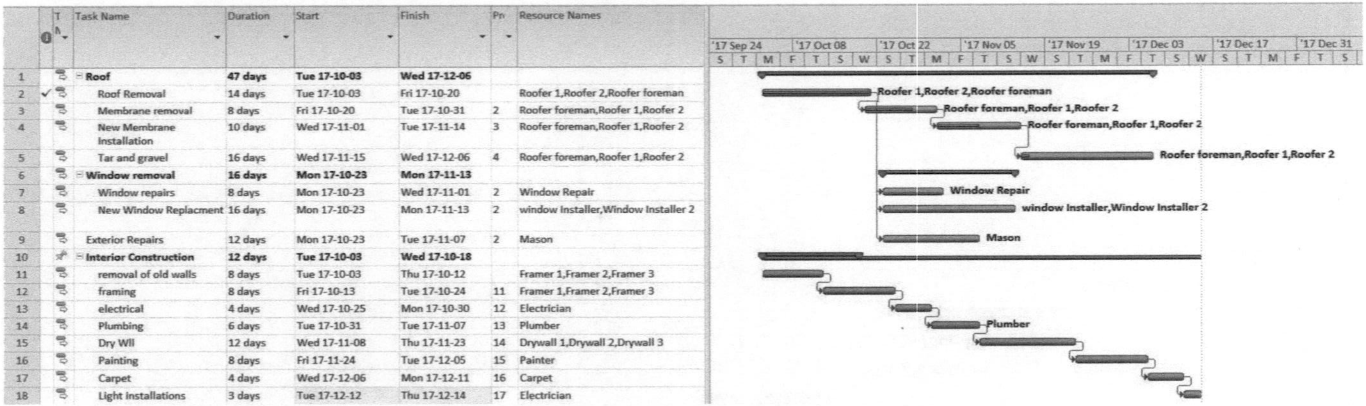

| | Task Name | Duration | Start | Finish | Pr | Resource Names |
|---|---|---|---|---|---|---|
| 1 | **Roof** | **47 days** | **Tue 17-10-03** | **Wed 17-12-06** | | |
| 2 ✓ | Roof Removal | 14 days | Tue 17-10-03 | Fri 17-10-20 | | Roofer 1,Roofer 2,Roofer foreman |
| 3 | Membrane removal | 8 days | Fri 17-10-20 | Tue 17-10-31 | 2 | Roofer foreman,Roofer 1,Roofer 2 |
| 4 | New Membrane Installation | 10 days | Wed 17-11-01 | Tue 17-11-14 | 3 | Roofer foreman,Roofer 1,Roofer 2 |
| 5 | Tar and gravel | 16 days | Wed 17-11-15 | Wed 17-12-06 | 4 | Roofer foreman,Roofer 1,Roofer 2 |
| 6 | **Window removal** | **16 days** | **Mon 17-10-23** | **Mon 17-11-13** | | |
| 7 | Window repairs | 8 days | Mon 17-10-23 | Wed 17-11-01 | 2 | Window Repair |
| 8 | New Window Replacment | 16 days | Mon 17-10-23 | Mon 17-11-13 | 2 | window Installer,Window Installer 2 |
| 9 | Exterior Repairs | 12 days | Mon 17-10-23 | Tue 17-11-07 | 2 | Mason |
| 10 | **Interior Construction** | **12 days** | **Tue 17-10-03** | **Wed 17-10-18** | | |
| 11 | removal of old walls | 8 days | Tue 17-10-03 | Thu 17-10-12 | | Framer 1,Framer 2,Framer 3 |
| 12 | framing | 8 days | Fri 17-10-13 | Tue 17-10-24 | 11 | Framer 1,Framer 2,Framer 3 |
| 13 | electrical | 4 days | Wed 17-10-25 | Mon 17-10-30 | 12 | Electrician |
| 14 | Plumbing | 6 days | Tue 17-10-31 | Tue 17-11-07 | 13 | Plumber |
| 15 | Dry Wll | 12 days | Wed 17-11-08 | Thu 17-11-23 | 14 | Drywall 1,Drywall 2,Drywall 3 |
| 16 | Painting | 8 days | Fri 17-11-24 | Tue 17-12-05 | 15 | Painter |
| 17 | Carpet | 4 days | Wed 17-12-06 | Mon 17-12-11 | 16 | Carpet |
| 18 | Light Installations | 3 days | Tue 17-12-12 | Thu 17-12-14 | 17 | Electrician |

'17 Sep 24 | '17 Oct 08 | '17 Oct 22 | '17 Nov 05 | '17 Nov 19 | '17 Dec 03 | '17 Dec 17 | '17 Dec 31

S T M F T S W S T M F T S W S T M F T S W S T M F T S

Roofer 1,Roofer 2,Roofer foreman
Roofer foreman,Roofer 1,Roofer 2
Roofer foreman,Roofer 1,Roofer 2
Roofer foreman,Roofer 1,Roofer 2
Window Repair
window Installer,Window Installer 2
Mason
Plumber

*Figure 18.9* Renovation of a historic commercial building, critical path, baseline 2.

Source Microsoft Project

| | Task Name | Baseline Dur. | Baseline Start | Baseline Finish | Baseline Work | Baseline Cost | Baseline2 Cost |
|---|---|---|---|---|---|---|---|
| 1 | **Roof** | **14 days** | **Sun 17-10-01** | **Wed 17-10-18** | **896 hrs** | **$58,240.00** | **$58,240.00** |
| 2 | Roof Removal | 14 days | Mon 17-10-02 | Thu 17-10-19 | 336 hrs | $21,840.00 | $21,840.00 |
| 3 | Membrane removal | 8 days | Fri 17-10-20 | Tue 17-10-31 | 192 hrs | $12,480.00 | $12,480.00 |
| 4 | New Membrane Installation | 10 days | Wed 17-11-01 | Tue 17-11-14 | 240 hrs | $15,600.00 | $15,600.00 |
| 5 | Tar and gravel | 5.33 days | Wed 17-11-15 | Wed 17-11-22 | 128 hrs | $8,320.00 | $8,320.00 |
| 6 | **Window removal** | **16 days** | **Fri 17-10-20** | **Fri 17-11-10** | **320 hrs** | **$24,640.00** | **$24,640.00** |
| 7 | Window repairs | 8 days | Fri 17-10-20 | Tue 17-10-31 | 64 hrs | $5,440.00 | $5,440.00 |
| 8 | New Window Replacement | 16 days | Fri 17-10-20 | Fri 17-11-10 | 256 hrs | $19,200.00 | $19,200.00 |
| 9 | Exterior Repairs | 12 days | Fri 17-10-20 | Mon 17-11-06 | 96 hrs | $8,640.00 | $8,640.00 |
| 10 | **Interior Construction** | **55 days** | **Mon 17-11-13** | **Fri 18-01-26** | **920 hrs** | **$55,760.00** | **$55,760.00** |
| 11 | removal of old walls | 8 days | Mon 17-11-13 | Wed 17-11-22 | 192 hrs | $10,560.00 | $10,560.00 |
| 12 | framing | 8 days | Thu 17-11-23 | Mon 17-12-04 | 192 hrs | $10,560.00 | $10,560.00 |
| 13 | electrical | 4 days | Tue 17-12-05 | Fri 17-12-08 | 32 hrs | $3,200.00 | $3,200.00 |
| 14 | Plumbing | 6 days | Tue 17-12-05 | Tue 17-12-12 | 48 hrs | $4,800.00 | $4,800.00 |
| 15 | Dry wall | 12 days | Wed 17-12-13 | Thu 17-12-28 | 288 hrs | $15,840.00 | $15,840.00 |
| 16 | Painting | 12 days | Fri 17-12-29 | Mon 18-01-15 | 96 hrs | $5,760.00 | $5,760.00 |
| 17 | Carpet | 6 days | Tue 18-01-16 | Tue 18-01-23 | 48 hrs | $2,640.00 | $2,640.00 |
| 18 | Light installations | 3 days | Wed 18-01-24 | Fri 18-01-26 | 24 hrs | $2,400.00 | $2,400.00 |

*Figure 19.9* Renovation of a historic commercial building, projected costs for baseline 1 and 2.

Source: Microsoft Project

strategy when it comes to increasing productivity. Another possibility would be to spend additional resources on overtime. If the overtime rates are not excessive, we could have our crews work over the weekend. However, having people work without a break may reduce their overall productivity. In any case, we may not want to increase the budget beyond our standard rates even if there are penalties involved for late completion. Calculating the cost of penalties vs overtime expenditures has to be considered part of the equation when considering crashing a project.

Looking at the Gantt chart, it may be possible to begin some of the tasks earlier than first considered. When we began the project, we had planned to begin the interior work only after all the windows were in place (Figure 14.9). If we are under pressure to complete this project before January 1, we could fast track the project and begin the interior work sooner after the project's start date (Figure 18.9). Again, it is important that we do not jeopardize the progress made on earlier tasks. In our example, we may want to inspect the Gantt chart to make sure that the drywall, painting, and carpet do not happen before the roof and windows are finished. In this way, we can sleep at night knowing that bad weather will not destroy any of the interior renovations. Having it rain in on the new carpet would be a disaster. Besides, having the window installers track dirt and mud on the new carpets would be ill-advised (Figure 18.9). As part of our efforts to balance time against cost, we can use our project management software to calculate the projected costs of our accelerated plan against our original baseline. In this case, there were no increased costs, so starting the interior work earlier might be a strategy worth considering. (For Baseline 1 and Baseline 2, the costs are the same: $58,240 – see Figure 19.9.)

In looking for opportunities to accelerate our project, we may want to revisit the construction process with our GC. Are there new techniques or technology we could employ to speed up the project? Perhaps we could rent a piece of specialized equipment that might make our crews more productive? Anyone who has worked on a home improvement project knows having the right tools for the job is essential. Of course, if there are

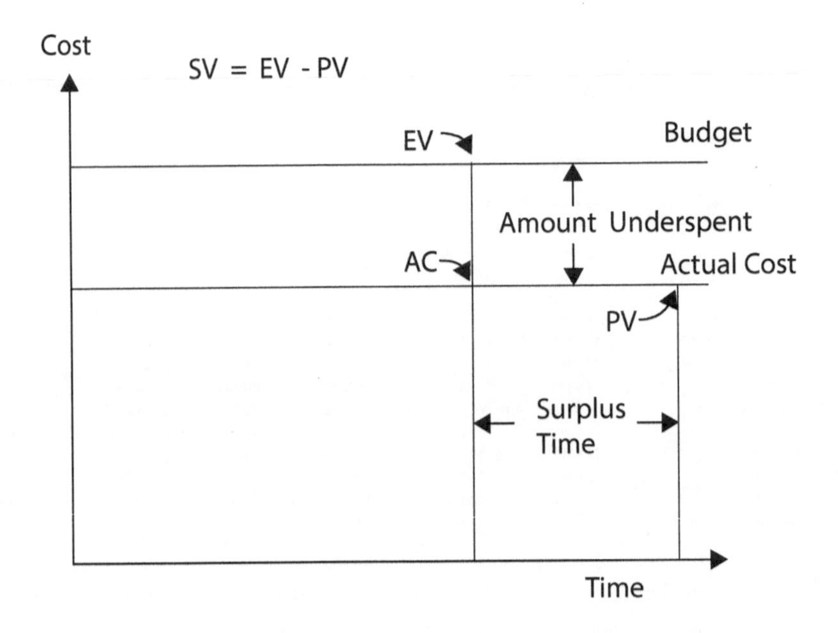


*Figure 20.9* Earned value.

plans to employ a more productive piece of technology, our crews must be trained to use it effectively. Training costs may also have to be added to our proposed plan. In making each decision, the cost of accelerating a project can be examined within the project management software by considering one alternative against another (Figure 19.9).

*Project evaluation and control*

One central responsibility of every project manager is tracking progress. As Henry Gantt stated in his 1903 article on scientific management, it is important to know "at a glance each day just what has been done and what remains to be done, in order to enable us to lay out the work for the next day in the most economical manner".[5] Essential to this management practice is to understand what are the deliverables associated with each task. How and what is being measured becomes critical to this exercise. For example, if we need to tile a concrete floor and we know it is going to take four days in total, then it might be easy enough to count the number of tiles laid down each day. If we complete 25% of the tiles each day, then we know we are on target to get the job done on time. But if the project requires removal of the old floor before we begin the tiling, then, in order to measure progress, it might be important to make floor removal a separate task for tracking purposes. How much detail we put into our plan may in part be a function of our comfort level and experience. In either case, having a clear set of milestones for each task is essential if we are going to track our progress.

There are a few terms that may be useful to know in relation to project, evaluation, and control. First, we must be clear about the work packages or significant deliverables that we need to track. We will also need some parameters for tracking our progress. Major milestones need to be defined for our project that align with important financial and construction deadlines. Often, the release of funds is linked to the completion of

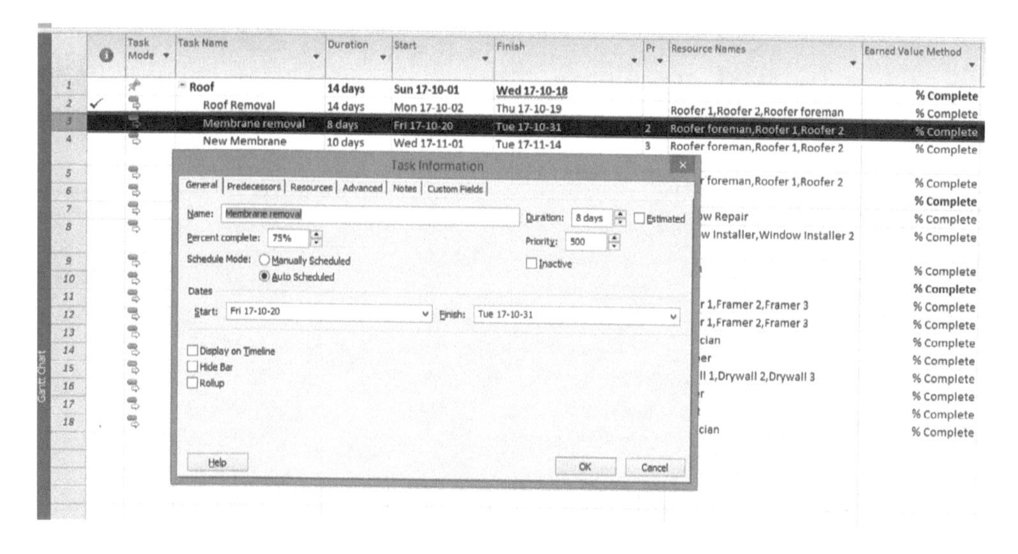


*Figure 21.9* Renovation of a historic commercial building: updating the work completed on the old roof membrane removal (75%).

Source: Microsoft Project

specific tasks. Budget, time, and deliverables all need to be tracked to guarantee that the project remains financially solvent.

## *Earned value management (EVM)*

In project management, earned value management (EVM) is an integrated approach to scheduling, budgeting, resource management, and performance (Figure 20.9). In tracking the progress of any project, it is important to know if you are spending more than you have actually budgeted. Also, if you are over budget, it is important to know if a single task is the culprit or if this is part of a trend that will impact the success of the project. Being acquainted with a few project management terms will be useful if you are engaged in project management.

**Planned value (PV)** The cost estimate of the budget resources scheduled over the lifetime of the project.

**Earned value (EV)** The cost or value of the work that has actually been completed to date.

**Actual cost (AC)** The total costs accrued for accomplishing a specific piece of work (work packages).

**Scheduled performance index (SPI)** Earned value divided by planned value of work that will be completed, ie EV/PV; can be useful in looking at the schedule for work yet to be completed.

**Cost performance index (CPI)** Earned value divided by actual cost, ie EV/AC; reveals if you are on budget for the project.

**Budgeted cost at completion (BAC)** Budget for the project.

**Scheduled variance (SV)** Earned value less planned value, ie EV – PV; a measure of performance over time.

| | | Resource Name | Work | Actual Cost |
|---|---|---|---|---|
| 1 | | Roofer foreman | 298.67 hrs | $13,693.33 |
| | | *Roof Removal* | *112 hrs* | *$7,280.00* |
| | | *Membrane remc* | *64 hrs* | *$3,120.00* |
| | | *New Membrane* | *80 hrs* | *$2,600.00* |
| | | *Tar and gravel* | *42.67 hrs* | *$693.33* |
| 2 | | Roofer 1 | 298.67 hrs | $13,693.33 |
| | | *Roof Removal* | *112 hrs* | *$7,280.00* |
| | | *Membrane remc* | *64 hrs* | *$3,120.00* |
| | | *New Membrane* | *80 hrs* | *$2,600.00* |
| | | *Tar and gravel* | *42.67 hrs* | *$693.33* |
| 3 | | Roofer 2 | 298.67 hrs | $13,693.33 |
| | | *Roof Removal* | *112 hrs* | *$7,280.00* |
| | | *Membrane remc* | *64 hrs* | *$3,120.00* |
| | | *New Membrane* | *80 hrs* | *$2,600.00* |
| | | *Tar and gravel* | *42.67 hrs* | *$693.33* |
| 4 | | Plumber | 48 hrs | $0.00 |

*Figure 22.9* Renovation of a historic commercial building: work completed to date. NB Work on the plumbing has yet to begin.

Source: Microsoft Project

Applying these concepts to our project can illustrate how a project management approach can assist in providing some oversight over the budget. As always, it is critical to have tasks clearly defined and tied to a specific budget and deliverables. When delays occur, it is important to know if they stem from getting started late or spending too much time on a task. Figure 20.9 illustrates this concept.

In this example, the planned value (PV) or estimated cost has shifted into the future. This slippage results in a delay in the completion of the task, along with an actual cost (AC) that is greater than the original estimated cost. We can use the following formula to calculate the scheduled variance (SV):

$$\text{Scheduled Variance} = \text{Earned Value} - \text{Planned Value}$$
$$SV = EV - PV$$

If we were on schedule, the difference would be equal to 0 and the SPI is equal to 1. If our AC is higher than the EV, as shown in Figure 20.9, then we are paying more than we should given what we are actually receiving. If we are on a budget, our AC would equal the EV and the difference again is 0. If we are lucky enough to have AC lower than the EV, our budget would be slightly positive and we would have extra cash in the bank account for the moment.

When applying this concept to an actual project, it is important to monitor costs and performance (deliverables) to catch overruns before they become excessive. Reviewing the schedule and budget could occur daily or weekly, but clearly waiting too long between assessments could be devastating on large projects. Project management software will use a percentage of each task completed to track progress. Many applications will default to

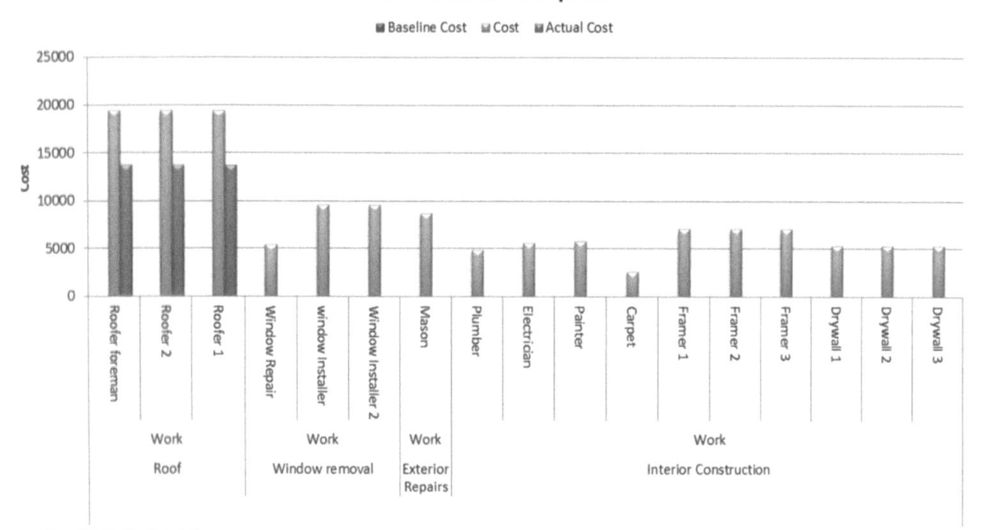


*Figure 23.9* Renovation of a historic commercial building: actual costs vs budget by each worker.

a percentage of 0, 25%, 50%, 75% or 100% as a simple way of tracking progress. Judgment should be used to establish an appropriate percentage for the EV. On small jobs, having a breakdown of 0, 50%, and 100% might be appropriate, while on bigger projects involving larger budgets, a finer subdivision of completion may be required.

Using our renovation project of a small commercial building as an example, in tracking our progress we can update the task information on the removal of the existing tar-and-gravel roof ("Membrane removal") as 75% completed. When the project is 100% completed, a checkmark will appear in the first column just to the right of the task. In this case, the old tar and gravel has been completely removed (Figure 21.9). In our resource table, we will now find the hours and expenditure for each updated task (Figure 22.9).

If we prefer to view this data as a bar chart, we see the progress that has been made for each task (Figure 23.9). However, in entering this data, the assumption was that, when we are billed at the end of a job, 100% of the task will have been successfully completed. Most contracts call for final payment only after the work is signed off. If this is not the case, then we will need to track these differences in our application.

If we are concerned about the overall progress of the project, we can use our CPI as a measure.

$$\text{CPI} = \text{Cost Performance Index} = \text{EV}/\text{AC}$$

In this case, our CPI will be equal to 1. All of the work was completed for the amount specified in our budget. If this index turns out to be less than 1, we are paying more for the work than originally budgeted, while an index greater than 1 would mean we are getting more value than originally budgeted.

Clearly, all of this monitoring of the job site and accounting takes time and money. Having every task updated on a daily basis may be difficult and expensive. The reliability of our contractors and the scale of the enterprise should determine how much attention needs to focus on the tracking progress. What is clear is that if the data entered into the system has been falsified or is inaccurate, our project management software will produce nothing more than pretty graphs and tables. Given this caveat, earned value management techniques can help in forecasting progress and completion of the project, but can also be useful in monitoring progress and informing management when the project is off-track. When the project goes awry, having a project management system in place should help get it back on track. If nothing else, focusing on solutions may make the negotiation process less contentious when things go wrong.

### *Project termination*

All projects have an end date. At that point, if you are successful, the final product of your efforts becomes the property of your client. For example, if you had purchased a building that was renovated for a small group of local investors, you would transfer title only after all the terms of the contract were met and payments received. If there were bonds to guarantee performance over the next few years, they too would need to be in place before the clients take possession. If a final audit was required by a third party, then this report will need to be submitted and reviewed by all parties before the final sign-off. In addition, if the building was a project you planned to hold and manage, you will have now entered into a new phase of the project in which tenants move in and the day-to-day operation takes over.

Sometimes, things go awry and there is an early termination. There are several reasons for not completing the project. Some may involve external forces. A sudden downturn in the economy may now make it impossible to rent the building upon completion and the only choice may be to abandon the project. Overbuilding in your area could have the same effect. Damage from catastrophic natural events may also make it difficult to complete the project. In these cases, the obvious question is "how do we get out from under this project and limit the financial loss?" Though, often, it is only the lawyers who do well financially in these situations, resolution in bankruptcy court may be the only solution. Of course, nobody likes going down that road, so if at all possible negotiating the closure of a project is the preferable choice.

## Notes

1 See: www.enr.com/
2 "Design–Build Firms: Design–Build Continues to Grow Despite Wariness and Price Concerns," *Engineering News Record* (June 12, 2006).
3 Gantt, 1903. In this article, Gantt states that the chart allows the user to see "at a glance each day just what has been done and what remains to be done, in order to enable us to lay out the work for the next day in the most economical manner".
4 Ibid.
5 Ibid.

## Bibliography

Atkinson, Roger. "Project management: cost time and quality, two best guesses and a phenomena, it's time to accept other success criteria." *International Journal of Project Management* 17, no. 6 (1999): 242–337.

Cooke-Davies, Terry. "The "real" success factors on projects." *International Journal of Project Management* 20 (2002): 185–190.
Dai, Jiukun, Paul M. Goodrum, and William F. Maloney. "Analysis of craft workers' and foremen's perceptions of the factors affecting construction labour productivity." *Construction Management and Economics* 25, no. 11 (2007): 1139–1152.
Gantt, Henry L. "A graphical daily balance in manufacture." *Transactions of the American Society of Mechanical Engineers* 24 (1903): 1322–1336.
Hegazy, Tarek. "Simplified Project Management for Construction Practitioners." *Cost Engineering* 48, no. 11 (Nov 2006): 20–28.
Magaba, Monde, Richard Cowden, and Anis Mahomed Karodia. "The Impact of Technological Changes on Project Management at a Company Operating in the Construction Industry." *Kuwait Chapter of the Arabian Journal of Business and Management Review* 5, no. 1 (September 2015): 8–37.
Mantel, S., J. Meredith, S. Shafer, and M. Sutton. *Project Management in Practice*. Hoboken, NJ: John Wiley & Sons, Inc, 2011.
Maylor, Harvey. "Beyond the Gantt Chart: Project management Moving on." *European Management Journal* 19, no. 1 (February 2001): 92–100.
Mulcahy, Rita. *PM Crash Course*. Minnetonka, MN: RMC Publications, Inc, 2006.
Munns, A. K. and B. F. Bjeirmi. "The role of project management in achieving project success." *International Journal of Project Management* 14, no. 2 (1996): 81–87.
Pinto, Jeffrey K. *Project Management, Achieving Competitive Advantage*. Boston, MA: Pearson, 2007.
Sohmen, Victor. "Project termination: Why the delay?" In PMI Research Conference. 2002.
White, Mark. "Information Technology: A Mandatory Role in Construction Project Management." *Cost Engineering* 49, no. 11 (November 2007): 18–19.
Wilson, J. M. "Gantt charts: A centenary appreciation." *European Journal of Operations Research* 149 (2003): 430–437.
Yates, J. K. and Leslie C. Battersby. "Master Builder Project Delivery System and Designer Construction Knowledge." *Journal of Construction Engineering and Management, ASCE* 129, no. 6 (October 2003): 635–644.

## Web resources

American Management Association
www.amanet.org/oe
International Project Management Association
www.ipma.world/
Project Management Association of Canada
www.pmac-agpc.ca/
Project Management Institute
www.pmi.org/

# Index